MODERN ASPECTS OF ELECTROCHEMISTRY

No. 36

Modern Aspects of Electrochemistry

Modern Aspects of Electrochemistry, No. 35:

- Impedance spectroscopy with specific applications to electrode processes involving hydrogen.
- Fundamentals and contemporary applications of electroless metal deposition.
- The development of computational electrochemistry and its application to electrochemical kinetics.
- Transition of properties of molten salts to those of aqueous solutions.
- Limitations of the Born Theory in applications to solvent polarization by ions and its extensions to treatment of kinetics of ionic reactions.

Modern Aspects of Electrochemistry, No. 34:

- Additivity principles in behavior of redox couples and corrosion processes.
- Foundation of voltaic measurements at liquid interfaces.
- Direct methanol fuel cells: current progress and emerging technology.
- Dynamics of processes in molten salts.
- Electrochemical techniques and microbially influenced corrosion (MIC).

Modern Aspects of Electrochemistry, No. 33:

- A review of the literature on the potential-of-zero charge.
- Review and discussion on nonequilibrium fluctuations in corrosion processes.
- Conducting polymers, their electrochemistry and biomimicking processes.
- Microwave photoelectrochemistry; from its origins to current research problems.
- New fluorine cell design from model development through pilot plant tests.
- The rapidly developing field of electrochemistry of electronically conducting polymers and their applications.

LIST OF CONTRIBUTORS

PANAKKATTU K. BABU
Department of Chemistry
University of Illinois at Urbana-Champaign
Urbana, Illinois 61801, USA

IVAN BOLZONELLA
Unit of Electrochemical Engineering
Institute of Chemical and Biological Process Science
School of Basic Sciences
Swiss Federal Institute of Technology
CH-1015 Lausanne, Switzerland

CHRISTOS COMNINELLIS
Unit of Electrochemical Engineering
Institute of Chemical and Biological Process Science
School of Basic Sciences
Swiss Federal Institute of Technology
CH-1015 Lausanne, Switzerland

GYÖRGY FÓTI
Unit of Electrochemical Engineering
Institute of Chemical and Biological Process Science
School of Basic Sciences
Swiss Federal Institute of Technology
CH-1015 Lausanne, Switzerland

MARC T. M. KOPER
Schuit Institute of Catalysis
Laboratory for Inorganic Chemistry and Catalysis
Eindhoven University of Technology
5600 MG Eindhoven, The Netherlands

PANOS NIKITAS
Laboratory of Physical Chemistry
Department of Chemistry
Aristotle University of Thessaloniki
54124 Thessaloniki, Greece

ERIC OLDFIELD
Department of Chemistry
University of Illinois at Urbana-Champaign
Urbana, Illinois 61801, USA

SU-IL PYUN
Department of Materials Science and Engineering
Korea Advanced Institute of Science and Technology
373-1 Guseong-dong, Yuseong-gu, Daejeon
305-701 Republic of Korea

HEON-CHEOL SHIN
Department of Materials Science and Engineering
Korea Advanced Institute of Science and Technology
373-1 Guseong-dong, Yuseong-gu, Daejeon
305-701 Republic of Korea

ANDRZEJ WIECKOWSKI
Department of Chemistry
University of Illinois at Urbana-Champaign
Urbana, Illinois 61801, USA

MODERN ASPECTS OF ELECTROCHEMISTRY

No. 36

Edited by

C. G. VAYENAS

University of Patras
Patras, Greece

B. E. CONWAY

University of Ottawa
Ottawa, Ontario, Canada

RALPH E. WHITE

University of South Carolina
Columbia, South Carolina

and

MARIA E. GAMBOA-ADELCO

Managing Editor
Superior, Colorado

Springer Science+Business Media, LLC

The Library of Congress cataloged the first volume of this title as follows:

Modern aspects of electrochemistry, no. [1]
Washington Butterworths, 1954–
v. illus., 23 cm.
No. 1–2 issued as Modern aspects series of chemistry.
Editors: No 1– J.O'M. Bockris (with B. E. Conway, No. 3–)
Imprint varies: no. 1, New York, Academic Press.—No. 2, London, Butterworths.
1. Electrochemisty—Collected works. I. Bockris, John O'M. ed. II. Conway, B. E. ed.
(Series: Modern aspects series of chemistry)

QD552.M6 54-12732 re

ISBN 978-1-4419-3396-6 ISBN 978-0-306-47927-4 (eBook)
DOI 10.1007/978-0-306-47927-4

Originally published by Kluwer Academic / Plenum Publishers, New York in 2003
Softcover reprint of the hardcover 1st edition 2003

http://www.wkap.nl/

10 9 8 7 6 5 4 3 2 1

A C.I.P. record for this book is available from the Library of Congress

Preface

This volume of *Modern Aspects* contains a remarkable spread of topics covered in an authoritative manner by some internationally renowned specialists. In a seminal chapter Drs. Babu, Oldfield and Wieckowski demonstrate eloquently the strength of electrochemical nuclear magnetic resonance (EC-NMR) to study *in situ* both sides of the electrochemical interface via the simultaneous use of ^{13}C and ^{193}Pt NMR. This powerful non-invasive technique brings new insights to both fundamental and practical key aspects of electrocatalysis, including the design of better anodes for PEM fuel cells.

The recent impressive advances in the use of rigorous ab initio quantum chemical calculations in electrochemistry are described in a remarkable chapter by Marc Koper, one of the leading protagonists in this fascinating area. This lucid chapter is addressed to all electrochemists, including those with very little prior exposure to quantum chemistry, and demonstrates the usefulness of ab initio calculations, including density functional theory (DFT) methods, to understand several key aspects of fuel cell electrocatalysis at the molecular level.

The most important macroscopic and statistical thermodynamic models developed to describe adsorption phenomena on electrodes are presented critically in a concise and authoritative chapter by Panos Nikitas. The reader is guided through the seminal contributions of Frumkin, Butler, Bockris, Guidelli and others, to the current state of the art adsorption isotherms, which are both rigorous, and in good agreement with experiment.

Electrochemical promotion (i.e., the use of electrochemistry to activate and *in situ* tune the catalytic activity and selectivity of metal and metal oxide catalysts interfaced with solid electrolytes) is discussed in a remarkable chapter by Drs. György Fóti, Ivan Bolzonella and Christos

Comninellis, who discuss both the fundamentals of this new exciting application of electrochemistry as well as recent advances including novel monolithic designs, for its practical utilization.

Lithium transport through transition metal oxides and carbonaceous materials is of paramount importance in rechargeable lithium batteries. The chapter by Drs. H. -C. Shin and Su-Il Pyun from KAIST, Korea, examines critically the "diffusion control" models, used routinely for current transients (CT) analysis, and demonstrates that, quite frequently, the cell current is controlled by the total cell impedance and not by lithium diffusion alone. This interesting chapter, rich in new experimental data, also provides a new method for CT analysis and an explanation for the existing discrepancy in Li diffusivity values obtained by the diffusion control CT analysis and other methods.

B. E. Conway

University of Ottawa
Ottawa, Ontario, Canada

C. G. Vayenas

University of Patras
Patras, Greece

R. E. White

University of South Carolina
Columbia, South Carolina

Contents

Chapter 1

NANOPARTICLE SURFACES STUDIED BY ELECTROCHEMICAL NMR

P. K. Babu, E. Oldfield, and A. Wieckowski

Chapter 2

AB INITIO QUANTUM-CHEMICAL CALCULATIONS IN ELECTROCHEMISTRY

Marc T. M. Koper

Chapter 3

MACROSCOPIC AND MOLECULAR MODELS OF ADSORPTION PHENOMENA ON ELECTRODE SURFACES

P. Nikitas

Chapter 4

ELECTROCHEMICAL PROMOTION OF CATALYSIS

György Fóti, Ivan Bolzonella, and Christos Comninellis

Chapter 5

MECHANISMS OF LITHIUM TRANSPORT THROUGH TRANSITION METAL OXIDES AND CARBONACEOUS MATERIALS

Heon-Cheol Shin and Su-Il Pyun

1

Nanoparticle Surfaces Studied by Electrochemical NMR

P. K. Babu, E. Oldfield, and A. Wieckowski

Department of Chemistry, University of Illinois at Urbana-Champaign, Urbana, IL 61801, USA.

I. INTRODUCTION

Exploring various phenomena at metal/solution interfaces relates directly to heterogeneous catalysis and its applications to fuel cell catalysis. By the late 1980s, electrochemical nuclear magnetic resonance spectroscopy (EC-NMR) was introduced[1,2] as a new technique for electrochemical surface science.[3] (See also recent reviews[4,5] and some representative references covering NMR efforts in gas phase surface science.[6-8]) It has been demonstrated that electrochemical nuclear magnetic resonance (EC-NMR) is a local surface and bulk nanoparticle probe that combines solid-state, or frequently metal NMR with electrochemistry. Experiments can be performed either under direct *in situ* potentiostatic control, or with samples prepared in a separate electrochemical cell, where the potential is both known and constant. Among several virtues, EC-NMR provides an electron-density level description of electrochemical interfaces based on the Fermi level local densities of states (Ef-LDOS). Work to date has been predominantly conducted with ^{13}C and ^{195}Pt NMR, since these nuclei

Modern Aspects of Electrochemistry, Number 36, edited by C. G. Vayenas, B. E. Conway, and R. E. White, Kluwer Academic / Plenum Publishers, New York, 2003

possess reasonable gyromagnetic ratios, ensuring adequate sensitivity. They also have a nuclear spin I = 1/2, which generally simplifies data interpretation, due to the absence of quadrupolar effects. It was demonstrated, for instance, that the ^{13}C NMR of chemisorbed CO on Pt nanoparticles yields information based on the Ef-LDOS for both 5σ and 2π* orbitals. On the other hand, ^{195}Pt NMR offers an electronic-level description of the metal surface, including information on 6*s* and 5*d* Ef-LDOS of Pt, and the variations of these orbitals with surface modifications (particle size effects, adsorption/deposition, electrode potential change, *etc.*).[4,5] Several years of active research by different research groups[1,2,9-26] have contributed to a detailed understanding of many critical issues involved in electrochemical surface science of nanoparticle electrodes, with and without adsorbates. Here, we review the results of our efforts in the EC-NMR field. A theoretical section is added as an Appendix, and includes a short description of the electron-nucleus interactions in metals, together with phenomenological models used for the analysis of EC-NMR data.

Because of difficulties associated with the low sensitivity of NMR spectroscopy, a typical high-field NMR instrument needs ~10^{18} - 10^{19} NMR active atoms, for example ^{13}C spins, to make a signal detectable within a reasonable time period. Since 1 cm^2 of a single crystal metal surface contains about 10^{15} atoms, at least 1 m^2 surface area of NMR active atoms is needed to meet the NMR sensitivity requirement. This can be met by working with large surface area samples, such as nanoparticle catalysts, including platinum black and practically all fuel cell supported catalysts. Usually, a few hundreds of milligrams of the sample is sufficient for a typical EC-NMR experiment. With such high surface area electrodes, the use of EC-NMR yields unique information on the Fermi level local density of states, on surface bonding and/or the molecular structure of adsorbates, as well as surface dynamics at electrode surfaces.

In NMR (see Appendix), a specific resonance position (a frequency) corresponds to a specific chemical environment (a structure). This frequency is the chemical shift (δ_{CS}) in a molecule or the Knight shift (K) in a metal.[27] The familiar chemical or orbital shift arises from the shielding of the external magnetic field at the nucleus by the otherwise quenched orbital angular momentum of the surrounding electrons, while the Knight shift is produced by hyperfine interactions between nuclei and electrons. This effect is specific to metals since only electron spins at the Fermi energy can be polarized and therefore create a non-zero spin density in an external magnetic field (the so-called Pauli paramagnetism).[27] Among metals, ^{195}Pt has the largest known Knight shift, -33700 ppm vs. H_2PtI_6. The Pt metal Knight shift is negative because of the major contribution from Pt $5d$ electrons to the electronic density of states at the Fermi level. Since the Knight shift (and the Korringa relationship[27]), are determined by the finite and quasi-continuous nature of the Fermi level local density of states (E_f-LDOS), measuring the Knight shift represents one of the primary probes of electronic structure in metals, and of surface molecular bonding.

Besides the chemical and/or Knight shift, there are two other major observables: the spin-lattice relaxation time in the laboratory frame (T_1), and the spin-spin relaxation time (T_2). While the chemical and/or Knight shift contains essentially static structural information, the temperature and/or magnetic field dependence of the relaxation times, both T_1 and T_2, are related to the dynamics of the observed nucleus. T_1 measures the rate at which the spin system returns back to thermal equilibrium with its environment (the lattice) after a perturbation, while T_2 measures the rate of achieving a common spin temperature within the spin system. Both T_1 and T_2 provide exceptionally important information on motions, varying typically from the 10^{-9} sec to 10^{2} sec timescales.

Below, we provide EC-NMR results for both ^{195}Pt and ^{13}C. We show the ^{195}Pt NMR lineshape for Pt nanoparticles supported on conducting carbon and a layer-model deconvolution that describes the surface, subsurface, etc., structure of these materials. We also present data obtained with unsupported Pt particle electrodes, discuss the spatially-resolved oscillation of the E_f-LDOS in a Pt catalyst and demonstrate correlations between the Pt Knight shift and the electro-negativity of the adsorbates. In ^{13}C-NMR, the observed temperature dependence of the relaxation rates obeys the Korringa equation, implying metallic behavior, just as is found in alumina supported Pt catalysts.[28] But in EC-NMR, we also show that spectra can be obtained under potential control, and that the observed spectra are sensitive to applied potentials. We describe results that indicate a linear relationship between the platinum clean surface Fermi level local density of states and the Knight shift of a ^{13}CO adsorbate. Variations of the E_f-LDOS reflects changes in the Pt-CO chemisorption bond due to both surface modifications (such as particle size change, alloying, deposition and/or chemisorption) and electrode potential change.

II. EXPERIMENTAL

Pt surfaces are routinely characterized using cyclic voltammetry (CV), usually at a sweep rate of 20 mV/min, in a preparative electrochemical cell containing a platinum gauze counter electrode, a 1 M NaCl Ag/AgCl reference electrode, and a working electrode. A typical working electrode consists of supported or unsupported platinum catalyst particles (50 to 500 mg) contained within a platinum boat.[4] Measurements are generally carried out in 0.5 M H_2SO_4 solutions under a blanket of ultra pure argon. The cell potential is controlled by a PGP201 Potentiostat/Galvanostat manufactured by Radiometer. Such Pt nanoparticle electrodes (and also a Pd electrode[16]) yield cyclic voltammograms that display features in the hydrogen region, as well as in the surface oxidation and double layer regions, that are characteristic of polycrystalline Pt electrodes. This shows that such an assembly of Pt nanoparticles can be considered as a polycrystalline electrode under well-defined, uniform potential conditions.

In a typical experiment, the originally oxidized nanoparticle Pt surfaces are first reduced by holding the cell potential at 250 mV for 1 to 2 hours, until the current decreases to an insignificant level. Then, an adlayer of ^{13}C- or ^{15}N-[4,11] adsorbate is produced on the electrode surface. In the case of surfaces covered by CO, the CO adsorbate can be produced either from electrochemical dissociative chemisorption of (99%) ^{13}C labeled methanol, or from direct ^{13}CO adsorption from CO-saturated media. The platinum sample, clean or covered by an adsorbate, together with a small portion (2-3 ml) of supporting electrolyte, is then transferred into a 10-mm diameter x 25-mm length tube and flame-sealed under reduced pressure, in a nitrogen atmosphere. In electrode-potential-controlled experiments, the sample is transferred to an NMR-electrochemical cell. Samples prepared in this way do not show any deterioration with time, and do not display any significant electrode potential change, the drift being ~10 mV in a few days. The experiments performed on these samples have yielded reproducible values for the chemical shifts and relaxation times at any time scale of interest to electrochemical studies, indicating that the electrochemical interfaces remain intact in the NMR sample tubes or EC-NMR cells.[2,9,11,16]

All NMR measurements have been carried out by using *home-built* NMR spectrometers equipped with either 3.5-inch bore 8.47 or 14.1 Tesla superconducting magnets (Oxford Instruments, Concord,

Massachusetts), Aries data acquisition systems (Tecmag, Houston, Texas), and using an Oxford Instruments CF-1200 cryostat. A Hahn echo pulse sequence ($\pi/2$—τ—π—τ—acquisition) with 16-step phase cycling to eliminate ringdown, is used for data acquisition. Typical $\pi/2$-pulse length are between 5 and 16 μs, depending on the particular sample. Chemical shifts for ^{13}C are given in ppm from tetramethylsilane (TMS) and for ^{195}Pt, the Knight shifts are reported with respect to H_2PtCl_6. Spin-lattice relaxation times are measured by using an inversion-recovery method for ^{13}C and a saturation-recovery for ^{195}Pt, followed by a Hahn-echo acquisition sequence.

III. RESULTS AND DISCUSSION

1. Selected Topics in ^{195}Pt-NMR

The ^{195}Pt NMR lineshape of a collection of Pt nanoparticles is extremely broad,[4,29] extending downfield some 4 kilogauss from the position of bulk platinum, 1.138 G/kHz, (Figure 1). This is because Pt atoms on an individual nanoparticle shell are in a different chemical environment than are those in other shells, and they resonate at a different NMR frequency. The spectra typically contain a feature on the low-field side (1.100 G/kHz), which arises from surface atoms (the frequency difference is magnetic field dependent and is about 2.5 MHz in a field of 8.5 Tesla). *Ab initio* theoretical calculations on a five-layer Pt (100) cluster have demonstrated that this surface shift must be due to a gradual drop in the *d*-like Fermi level local density of states on moving from the inside of the particle to the surface.[30] It is this distinguishable surface signal, which makes ^{195}Pt NMR of particular use in investigating the surface physics and chemistry of nanoscale platinum particles.

The ^{195}Pt NMR spectra in Figure 1 were obtained for a platinum sample having an average Pt particle diameter of 2.5 nm, supported on conducting carbon (Vulcan XC-72). Figure 1A is a spectrum of an as-received sample. A surface peak appears at 1.085 G/kHz, the same position as platinum oxide, indicating that the Pt particle surfaces are covered by an oxide species.[29] Clearly, the ability to detect Pt surface oxide species is one of the virtues of NMR spectroscopy in nanoparticle

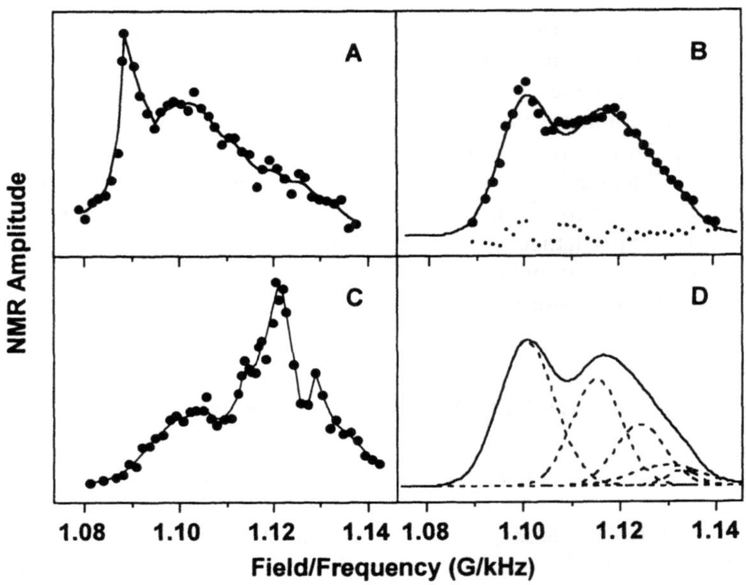

Figure 1. Point-by-point ^{195}Pt NMR spectra of 2.5 nm sample at T = 80K: A, As-received catalyst; B, electrochemically cleaned; C, sintered to nominally larger Pt size by extensive potential cycling and D, layer-model deconvolution of spectrum B. (Reprinted with permission from [15] Copyright 1999 Am. Chem. Soc.)

electrode surface characterization. (All Pt-type commercial nanoparticle catalysts come with their surfaces oxidized, which protect them from environmental contamination.) After electrochemical reduction of the same sample, done simply by holding the electrode at a potential within the double layer region,[9] one obtains the spectrum shown in Figure 1B (filled circles). Here, it is clear that the surface oxide peak disappeared giving rise to a new spectrum with the intensity near 1.100 G/kHz, attributable to the clean Pt surface,[29] and other spectral activity attributable to subsurface layers, or shells of platinum atoms (see below). The value of 1.100 G/kHz for the surface peak is to be compared to that of 1.138 G/kHz, which is assigned to bulk or core platinum atoms. Figure 1C also shows results obtained for 2.5 nm Pt particles after extensive voltammetric treatment, demonstrating pronounced sintering.[29]

The observations presented in Figure 1 are of interest since they show that, even when deposited on a conductive support and immersed in a liquid electrolyte, the small Pt particles show NMR properties typical of the same size particles deposited on non-conducting substrates,[31] or isolated in zeolites.[32] The fact that the Pt particles still retain a distinguishable surface peak also contrasts with the situation seen previously with self-supported Pt particles, where no surface peak was observed (Ref. [33], and see below). These surface properties were investigated in detail by using ^{195}Pt NMR showing,[15,18,19,25] for example, a major enhancement of the surface E_f -LDOS in the electrochemical environment with respect to its gas phase counterpart.[34,35]

Figure 1 D presents the ^{195}Pt spectrum obtained using the layer model deconvolution technique (Appendix C).[15] From the deconvolution shown in Figure 1D, the fraction of Pt atoms in surface, subsurface and interior layers can be determined.[15] The average particle size (dispersion) can also be deduced from the lineshape since the surface signal is enhanced for smaller particles. Electronic structural information can be deduced from these lineshapes, using the *healing length* concept.[7] The healing length, defined as m times the distance between two consecutive layers (0.229 nm for Pt), is equal to 0.46 nm (≅ 2 Pt layers) for the sample shown in Figure 1D. This is to be compared with 0.31 nm (≅ 1.35 Pt layers) in the gas phase case,[36] indicating a significant change in the electronic structure of the Pt sample due to the presence of the double layer (as already mentioned above).[15] As reported previously, for larger nanoparticles, the contribution of the surface peak becomes reduced and, ultimately, may give way to the bulk Pt peak.[15] An example of this effect is shown in Figure 2 for carbon supported, electrochemically cleaned 3.2 nm nanoparticles. Although the surface peak is smaller than that on the 2.5 nm sample (Figure 1B), it is clearly distinguishable from other Pt atoms in the bulk of the nanoparticle.

2. Other Pt Nanoparticles (Unsupported and Supported)

Figure 3 shows the ^{195}Pt NMR spectrum of *unsupported* Pt-black samples, reduced electrochemically as reported above. The average particle size diameters are *ca.* 3 (Figure 3A) and 6 nm (Figure 3B), as

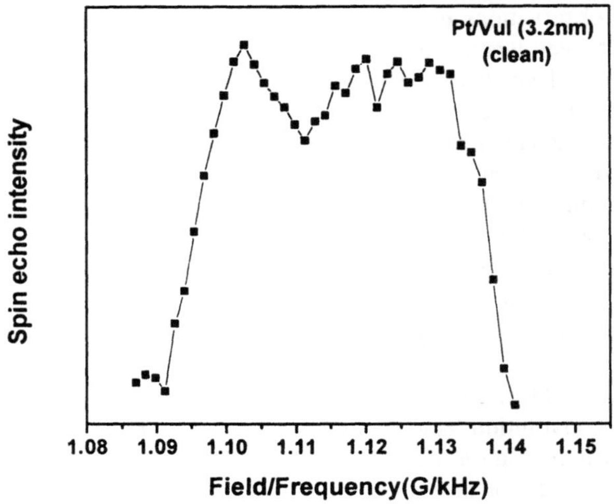

Figure 2. Point-by-point ^{195}Pt NMR spectra of 3.2 nm carbon-supported Pt nanoparticle after electrochemical cleaning.

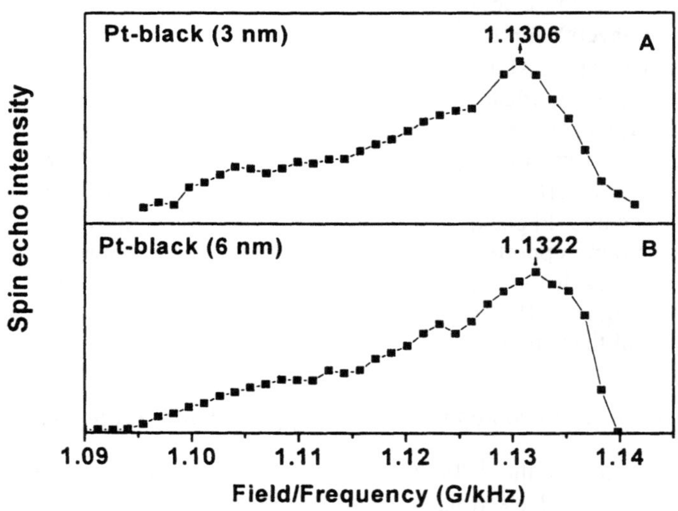

Figure 3. ^{195}Pt NMR spectra of unsupported Pt-black samples obtained point-by-point at $T = 80$ K in 8.45 T field (A) 3 nm (B) 6 nm.

obtained from TEM micrographs. Surface peaks (1.100 G/kHz) are absent, confirming previous results.[15] Nevertheless, a layer-like structure is certainly suggested since the spectra are spread over the range from 1.095 to 1.14 G/kHz, consistent with Pt atoms on the surface, sub-surface and interior layers, as described above. It is apparent, however, that the actual fraction of atoms that have a surface-like environment is quite low. An interesting observation can then be made concerning the peaks at 1.131 and 1.132 G/kHz for the smaller and larger particle sizes, respectively. Specifically, the frequencies of these peaks are much smaller than that reported for bulk Pt (1.138 G/kHz), which indicates that the smaller the particle size, the smaller the frequency of the *bulk* peak. Further modeling of such samples would be very useful since unsupported Pt nanoparticle samples are very active catalysts in fuel cells.[37]

3. Correlation Between the ^{195}Pt NMR Shift and Adsorbates Electronegativity

The influence of a series of different ligands or adsorbates (H, O, S, CN^-, CO, and Ru), adsorbed onto carbon-supported nanoscale Pt particles from the same starting batch has been investigated by ^{195}Pt NMR in an electrochemical environment.[19] The frequency difference between the resonance positions of the surface and bulk platinum atoms at 8.5 T is about 2.5 MHz, large enough to provide a convenient spectral visualization of how deep the influence of the adsorbate can go, as illustrated in the point-by-point ^{195}Pt NMR spectra (recorded at 80 K) presented in Figure 4. These spectra are normalized by equalizing their amplitudes at 1.131 G/kHz. Notice that only the Ru ligand yields a high frequency peak, which we will discuss in more detail elsewhere.[24] Another interesting observation is that the position of the high-field signal intensity (above 1.131 G/kHz), due to Pt atoms in the innermost layers, is independent of the type of the adsorbate. This adsorbate invariance provides direct experimental evidence for the validity of the Friedel-Heine invariance theorem,[38,39] which basically states that electronic properties at an atom, such as its LDOS, are determined primarily by the surrounding medium (to within a few electronic wavelengths). Although the chemical nature of the adsorbates is quite different, their influence on the LDOS is experienced by only those atoms on or very near the surface. That is, the high-field signals are invariant towards changes in surface electronic structure.

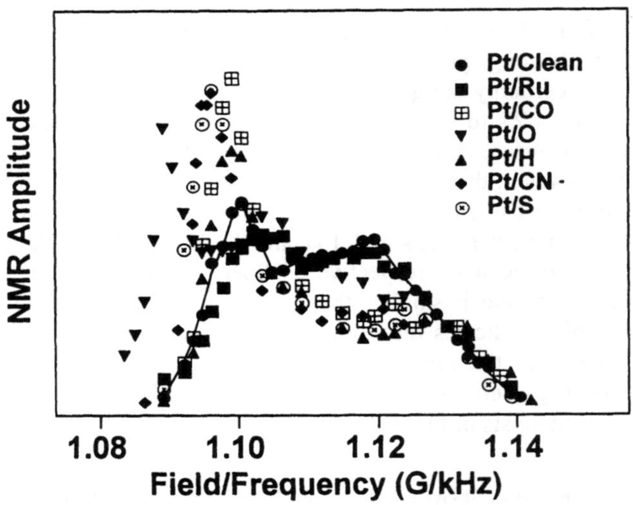

Figure 4. ^{195}Pt NMR spectra of 2.5 nm, carbon-supported Pt electrocatalyst with different adsorbates. The spectra are normalized by equalizing the amplitude at 1.131 G/kHz (indicated by the arrow). (Reprinted with permission from [19] Copyright 2000 *Am. Chem. Soc.*)

The changes induced by various adsorbates on the electronic properties of the electrode surface can be seen from the major frequency shift of the surface and subsurface NMR signals (Figure 5). The NMR layer model spectral deconvolution technique was applied to these spectra in order to obtain the variations in the surface and subsurface Knight shifts with surface adsorbate structure. The NMR parameters (such as peak position, width, relative intensity for the three central layers) were fixed to those of the clean-surface electrodes, while the surface and subsurface peak positions were allowed to vary. The results of these simulations are shown in Figure 5 and the surface and subsurface peak shift values obtained are plotted in Figure 6 as a function of the Allred-Rochow electronegativity. Allred-Rochow electronegativity is defined as the electrostatic force exerted by the nucleus on the valence electrons, expressed as:[40]

$$\chi_{AR} = \frac{3590Z^*}{r_c^2} + 0.744 \qquad (1)$$

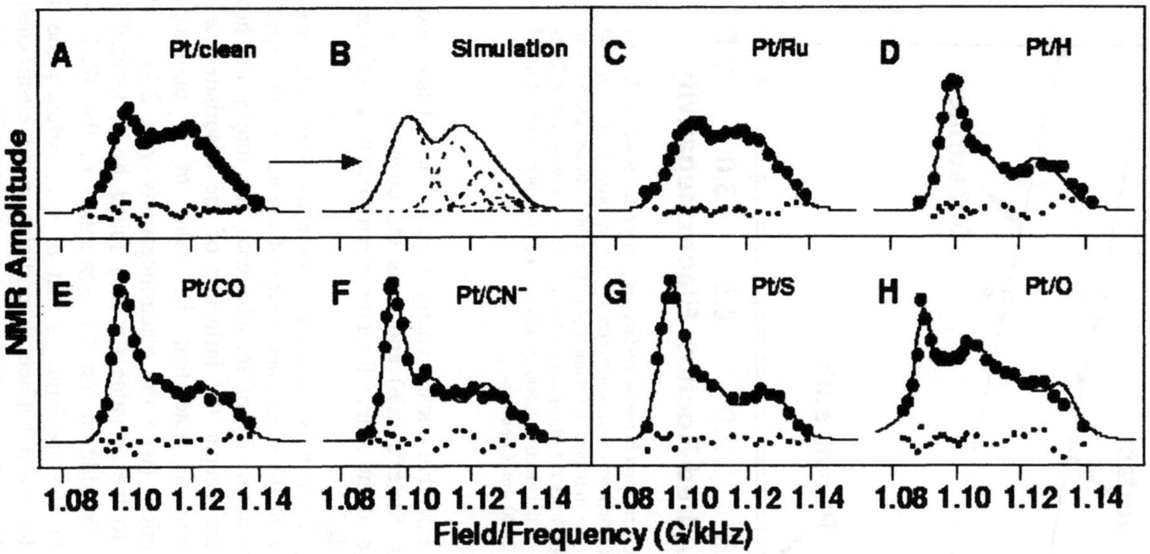

Figure 5. Layer-model simulations of the ^{195}Pt NMR spectra for different adsorbates. (Reprinted with permision from [19] Copyright 2000 *Am. Chem. Soc.*)

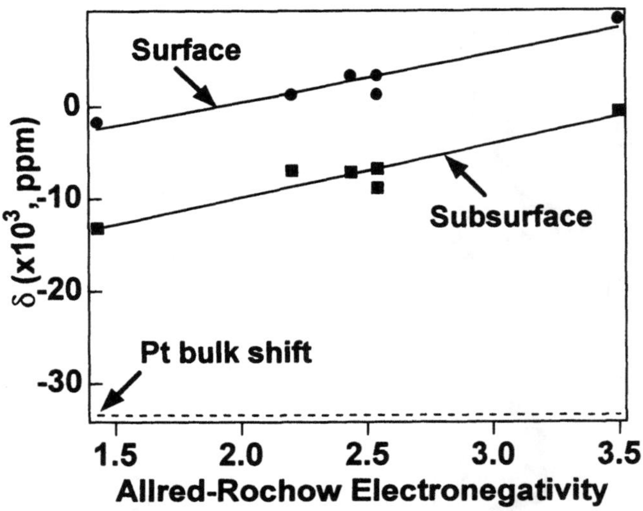

Figure 6. Correlation between surface/subsurface frequency shifts and the Allred-Rochow electronegativity of the adsorbates. The dashed horizontal line indicates the Knight shift of bulk platinum atoms. The solid straight lines are linear fits to the surface and subsurface shifts as a function of the electronegativity. (Reprinted with permission from [19] Copyright 2000 *Am. Chem. Soc.*)

where Z* is the effective nuclear charge seen by the valence electrons (obtainable from Slater's rule) and r_c is the covalent radius (pm). Tabulated values of the Allred-Rochow electronegativity of elements can be found in Ref [41].

Remarkably, both the surface and subsurface peak positions map almost linearly with the electronegativity of the adsorbate. Also, the larger the electronegativity of the adsorbate, the longer the healing length; that is the deeper the influence of the adsorbate goes. It is also interesting to note that the value of *m* obtained for hydrogen adsorption (with the electronegativity χ = 2.2) is about 2.9, very close to the *m* value (= 2.6) found by a detailed layer model analysis. In addition, for the adsorption of the alkali metal elements ($\chi \sim 1$), the prediction is that $m \sim 2$, virtually the same m as obtained for clean surface Pt particles in an electrochemical environment. One would, therefore, expect that alkali adsorption does not significantly change the ^{195}Pt NMR spectrum of

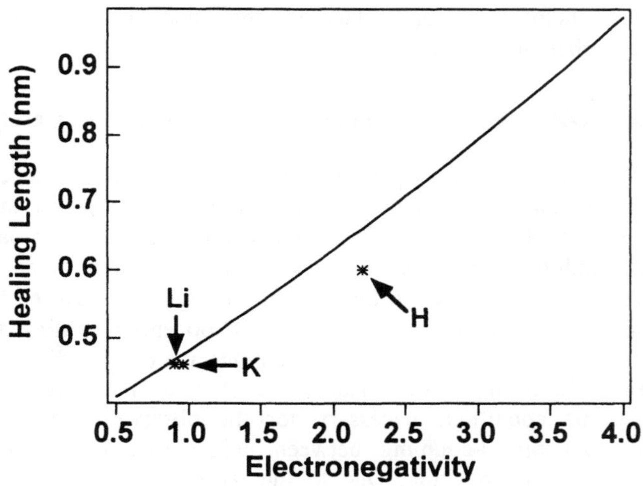

Figure 7. Healing lengths, estimated from the layer model analysis, plotted against the electronegativity. (Reprinted with permission from [19] Copyright 2000 Am. Chem. Soc.)

small Pt particles after adsorption. This is exactly what has been observed experimentally (in adsorption onto platinum from gas phase[42]). For comparison, these three experimentally estimated points are also plotted in Figure 7. (Note that the healing length values, obtained by multiplying 0.229 nm by *m*, are plotted in Figure 7 as a function of electronegativity, where 0.229 nm is the distance between two consecutive layers for Pt, see above.)

It is important to point out that the invariance observed here is in quite a different context to that of originally proposed in the Friedel-Heine invariance theorem, which refers to the invariance of electronic properties in bulk environments of dilute alloys. The situation is quite different for nanoscale metal particles from those of bulk Pt, due to the presence of the surface and the significant reductions in the particle volume. However, the E_f-LDOSs still show a remarkable invariance towards changes in surface chemical environment, even though they vary from those of bulk Pt. These results may also be expected to lead to useful general correlations between electronic properties and more conventional chemical descriptions such as ligand electronegativity, which is helpful in understanding the electronic structure of metal/adsorbate interfaces.

4. Spatially-Resolved Oscillation of the E_f-LDOS in a Pt Catalyst

^{195}Pt NMR of nanoscale Pt particles can, in principle, permit a detailed study of the spatial variation of the E_f-LDOS across a given particle. This is possible because the ^{195}Pt NMR spectra are extremely broad and nuclei in individual platinum layers resonate at different frequencies, due to significant changes in local electronic properties caused by the presence of the particle's surface. The exponential healing model (Appendix C) relates the Knight shift of atoms at a given radial position to their distance from the surface, and such a spatial dependence provides the conditions necessary for the observation of indirect nuclear spin-spin *J*-coupling between neighboring Pt atoms as a function of radial position. Due to the large *J*-coupling that exists between neighboring Pt atoms, the spin echo decay curves exhibit a characteristic *slow-beat* pattern.[31] These spin echo decay measurements can be analyzed using a 5-parameter fit[31] to extract the *J*-coupling values.

We have carried out ^{195}Pt NMR investigations of (conducting) carbon-supported Pt nanoparticles having an average particle size of 8.8 nm, together with corresponding *J*-coupling measurements.[23] Figure 8 shows the Pt NMR spectrum of the clean sample. As above, Pt atoms in the interior have their resonances centered around 1.138 G/kHz and surface atoms give rise to a peak near 1.100 G/kHz. *J* values are determined from the spin echo envelopes at a series of spectral positions (Knight shifts). Figure 9 shows typical spin echo envelopes obtained across the spectrum, which can then be analyzed using the 5-parameter fit[31] to deduce the *J*-coupling values. The distance of each layer from the particle's surface can then be obtained by employing the exponential healing model. In a metal, the conduction electron mediated *J*-coupling is almost entirely determined by the *s*-like E_f-LDOS and it is proportional to $K_s(x)$, the Knight shift caused by *s*-like electrons at a position x away from the surface.[43] Assuming that local variations in $K_s(x)$ arise solely from variations in the *s*-like E_f-LDOS, one can readily deduce the following relationship[15] between the *s*-like E_f-LDOS, $D_s(E_f, x)$, and the *J*-coupling, $J(x)$, which enables a direct measurement of $D_s(E_f, x)$:

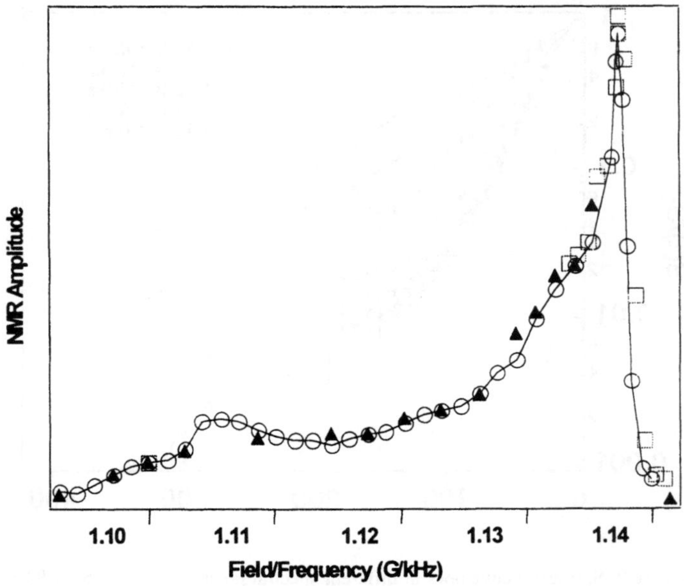

Figure 8. Point-by-point ^{195}Pt NMR spectrum of an 8.8 nm, carbon-supported clean surface Pt electrocatalyst. (After Tong *et al* [23] Copyright 2002, Elsevier Science Publishers)

$$D_s(E_f, x) = \left[0.098 + \frac{0.62}{J(x)} \right]^{-1} \tag{2}$$

The $D_s(E_f, x)$ values obtained from the *J*-coupling measurements are plotted as a function of x in Figure 10. It can be seen there that there is a clear oscillation in $D_s(E_f)$ across the spectrum. This represents the first experimental observation of a spatially-resolved oscillatory variation in $D_s(E_f)$ caused by the presence of a metal surface. Moreover, the experimental results can be fit to a simple relationship, which has the essence of a Bardeen-Friedel oscillation,[44]

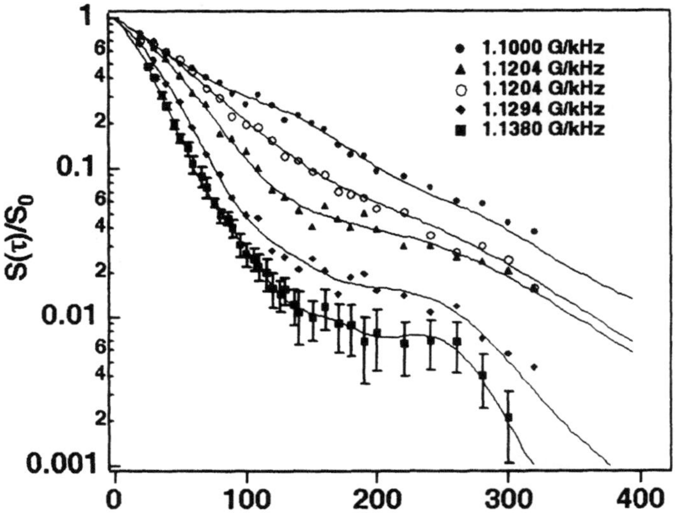

Figure 9. Spin-echo envelopes at different spectral positions, measured at 80 K for 8.8 nm, carbon-supported clean surface Pt electrocatalyst. Open symbols are from data measured from a reduced r.f. field strength. (After Tong *et al* [23] Copyright 2002, Elsevier Science Publishers)

$$D_s(E_f,x) = D_s(E_f,\infty) + b\frac{\sin(2k_F x)}{2k_F x} \tag{3}$$

The solid line in Figure 10 represents the fit to Eq. (3) which gives the following results: $D_s(E_f, \infty) = 3.5$ atom^{-1}·Ry^{-1}, $k_F = 6$ nm^{-1}, and $b = 1.5$ atom^{-1}·Ry^{-1}. Notice that $\sin(2k_F x)/2k_F x$ is an asymptotic form of the oscillation but for simplicity, it is used here to approximate the behavior over the whole range. The value of $D_s(E_f, \infty)$, 3.5 atom^{-1}·Ry^{-1}, is very close to the accepted value for bulk Pt,[45] 4 atom^{-1}·Ry^{-1}. Assuming each Pt atom contributes one *s*-like electron, the *s*-like electron density for bulk Pt would be 6.1 x 10^{22}/cm^3. By following the simplest model for free electrons, one obtains $k_F = 12$ nm^{-1}, about twice the value we find. In other words, the value of 6 nm^{-1} for k_F indicates a much longer decay length than that would be expected from the Jellium model.[23] The oscillatory variation in $D_s(E_f)$ beyond 0.4 nm (Figure 10)

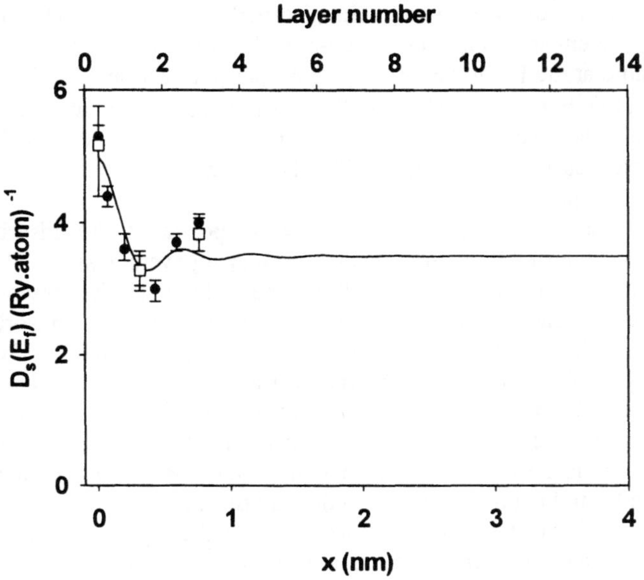

Figure 10. The s-like E_f -LDOS as a function of distance from the particle surface, showing oscillatory behavior. The solid line is a fit to Equation (3). (After Tong *et al* [23] Copyright 2002, Elsevier Science Publishers

indicates that the influence of the metal surface goes at least three layers inside the particles, in contrast to predictions based on the Jellium model. This conclusion is consistent, however, with the NMR results obtained for other nanoparticles and suggests that much of the observed broadening of the bulk-like peak in ^{195}Pt NMR spectra of such systems can be attributed to the spatial variation of the $D_s(E_f)$.

IV. ^{13}C NMR AT THE ELECTROCHEMICAL INTERFACE

1. ^{13}C NMR Knight Shift

An *ab initio* theoretical calculation shows that the rather localized 5σ HOMO and 2π* LUMO orbitals of CO before chemisorption become extended bands after being chemisorbed onto a Pt(111) surface.[46] Before chemisorption, CO is non-metallic since there are no orbitals available at the Fermi energy, E_f. But upon interaction with the *d*-band of Pt(111), both 5σ and $2\pi^*$ of CO on Pt(111) have orbitals available at E_f, *i.e.*, chemisorbed CO becomes metallic. This implies that for chemisorbed CO the Knight shift arises from conduction electrons in 5σ and/or $2\pi^*$ orbitals.

In Figure 11, we show the ^{13}C NMR spectra of CO adsorbed on carbon-supported Pt nanoparticles with varying average sizes. The peak at ~160 ppm downfield from TMS can be assigned to ^{13}C present in the nonmetallic carbon support. The frequency shift of ^{13}CO adsorbed on platinum is at ~330 ppm from TMS, an approximately 170 ppm downfield shift from the 160 ppm values seen in many terminal or on top bonded diamagnetic metal carbonyls. This indicates the presence of a large Knight shift, *i.e.*, that the carbon atoms of CO adsorbed on the platinum surface are in the metallic state, as noted previously for CO on Pt in the gas phase (*i.e.*, in dry samples) by the Slichter group.[28]

As will be described in more detail in the Appendix, the Knight shift and the chemical shift are always measured together as a combined frequency shift. Fortunately, however, for CO adsorbed onto a metal surface, it is possible to use the Korringa relationship to detect the presence of conduction electrons and hence a Knight shift at the observed nucleus. The Korringa relationship indicates that $1/T_1 \propto T$, where T is the absolute temperature of the sample. In Figure 12, we show the relationship between the spin-lattice relaxation rates ($1/T_1$) of chemisorbed ^{13}CO (squares) and ^{13}CN (circles) on platinum.[4] The linearity of the ($1/T_1$) vs. T plot in chemisorbed CO and CN, respectively, and the Korringa products of $T_1T = 82$ s·K and 135 s·K, prove unambiguously that the carbon atom on the platinum surface in these two adsorbates is in a metallic state.

2. EC-NMR Under Potential Control

We found that the ^{13}C spectra of adsorbed CO and CN on Pt nanoparticle electrodes depend on the value of the electrode potential.[11,16] For EC-NMR under direct potential control, again, a few hundred milligrams of powdered Pt (fuel cell grade platinum black) was used as the working electrode in a three-electrode NMR electrochemical cell. The cell was positioned inside the NMR coil with the leads connected to an external potentiostat. The NMR electrochemical cell also incorporates

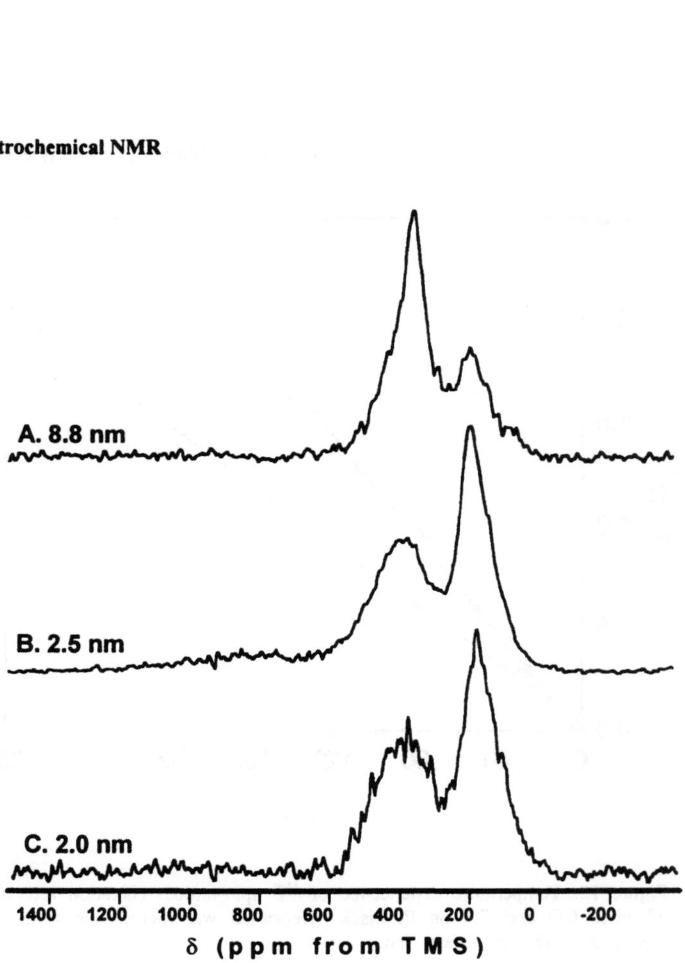

Figure 11. ^{13}C NMR spectra of chemisorbed CO on (A) 8.8 nm (B) 2.5 nm and (C) 2.0 nm particle size Pt-carbon samples (Reprinted with permission from [18]. Copyright 2000, Am. Chem. Soc.)

the ability to exchange the electrolyte without removing the probe from the magnet. CO was adsorbed onto the electrode surface by catalytic decomposition of ^{13}C enriched methanol. For CN^- adsorption studies, the 0.5 M H_2SO_4 was replaced by 0.5 M Na_2SO_4 adjusted to pH 11.2 with NaOH. A concentrated solution of ^{13}C-enriched sodium cyanide was then added to the basic solution to give a final concentration of 0.2 M CN^-. The adsorption of CN^- was carried out at -300 mV vs. a 3 M NaCl Ag/AgCl reference electrode. The data in Figure 13 demonstrate

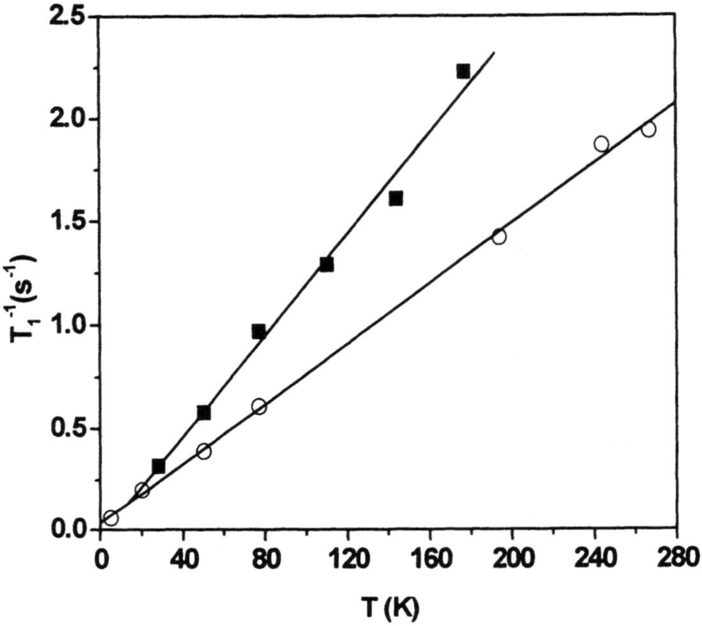

Figure 12. Temperature dependence of ^{13}C spin-lattice relaxation rate of adsorbed CO and CN on Pt black. (Reprinted with permission from [4]. Copyright 1998, Am. Chem. Soc.)

that ^{13}C resonances for both adsorbates became more shielded when the applied potential was increased in the positive direction, with the slope -71 ppm/V for ^{13}CO and -50 ppm/V for ^{13}CN. For CO on Pd, the slope was found to be even larger: about -136 ppm/V, as discussed below.[11]

3. Correlation of NMR to FTIR Data

In the field of FTIR of electrode surfaces, it is well known that the C–O vibrational frequency, ν_{CO}, decreases when the electrode potential becomes more negative. This variation of the vibrational frequency with electrode potential is referred to as the vibrational Stark effect.[47] The Stark effect can now be considered in the context of our NMR

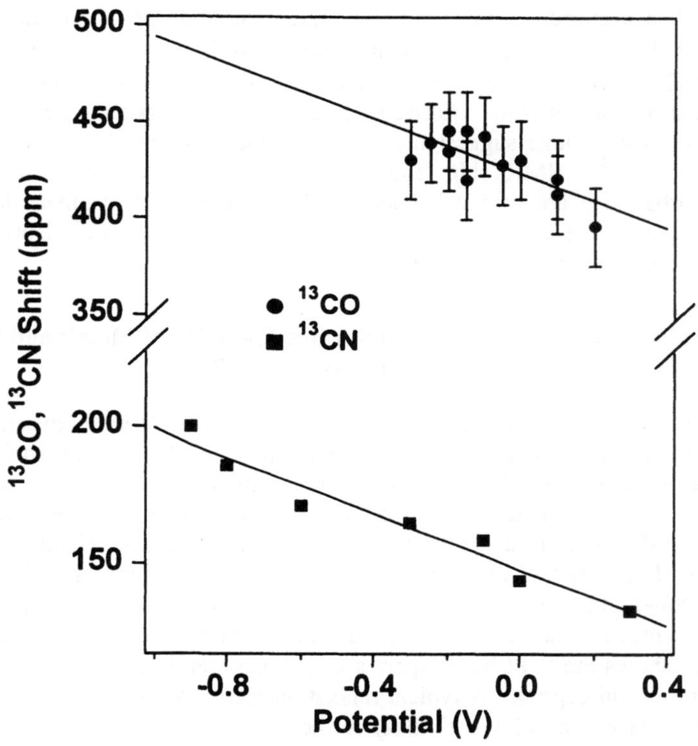

Figure 13. Potential dependence of the chemical shift of ^{13}CO and ^{13}CN adsorbed on Pt nanoparticles. (After Wu *et al* [11], Copyright 1997 Royal Society of Chemistry).

data. Here, we note that, first, the response of ^{13}C NMR to the electrode potential is due essentially to the Knight shift.[4,16] Second, the magnitude of the clean-surface E_f-LDOS of the Pt substrate correlates very well with the extent of the Knight shift as well as the vibrational stretch frequency of chemisorbed CO: the higher the clean surface E_f-LDOS, the larger is the ^{13}C Knight shift, or C–O stretch frequency of the CO. Both relationships are linear, with slopes of about 12 ppm·Ry·atom and -4 cm^{-1}Ry·atom respectively, giving a ratio of -3 ppm/cm^{-1}.[12,36] This is to be compared with the ratio of -2.8 ppm/cm^{-1}, obtained from the slopes (-71 ppm/V and 25 cm^{-1}/V respectively) of δ_{cs}

(^{13}C) in electrochemistry, and ν_{CO} versus electrode potential relationships for CO on Pt black. Third, variations in ν_{CO} can be directly correlated with changes in $2\pi^*$ back-donation: the higher the back-donation, the lower the CO stretch frequency. When taken together, all these observations suggest that the electrode potential dependence of both the ^{13}C NMR shift and the vibrational stretch frequency of adsorbed CO are primarily electronic in nature, and originate from changes in the E_f-LDOS at the metal surface and at the adsorbate, induced by electrode polarization.

4. Correlation Between Clean Surface E_f-LDOS of Metals and the Adsorbate Knight Shift

It is of interest to see if there exists any type of correlation between the clean surface E_f-LDOS of the metal and the ^{13}C Knight shift of chemisorbed CO. The former is an important surface attribute defining the ability of a metal surface to donate as well as to accept electrons, while the latter measures the degree of metallization of chemisorbed CO. Experimental data[15] for ^{13}C NMR of CO adsorbed onto 2.5 nm carbon-supported Pt particles and ^{195}Pt NMR of the corresponding clean metal surfaces were analyzed using the two-band models. Figure 14A shows the ^{195}Pt NMR spectra of a clean metal surface before and after CO adsorption. A typical relaxation recovery curve, measured at the surface peak (1.100 G/kHz) is shown in Figure 14B. A double-exponential fit (solid line) to the data gives the same relaxation times for the slow component before and after CO adsorption, indicating that the slow component arises from Pt atoms that are not bonded to CO. Therefore, relaxation data for the fast recovering component, from Pt atoms that are bonded to CO, can be analyzed with the two-band model.

A ^{13}C NMR spectrum of CO adsorbed onto 2.5 nm Pt particles is shown in Figure 15A.[15] The peak at 170 ppm again originates from the graphite support while the one at 380 ppm arises from CO adsorbed on platinum. For the adsorbed CO, the spin-lattice relaxation is found to be single exponential (inset of Figure 15A). The purely orbital contributions (chemical shifts) for atop CO on Pt and on Pd were calculated by using density functional theory (DFT) on model $COPt_7$ and $COPd_7$ clusters in which CO sits atop the central metal atom, which is coordinated by the remaining six metal atoms (inset in Figure 15B). For $COPt_7$, a value of 160 ± 20 ppm from TMS was obtained, while for

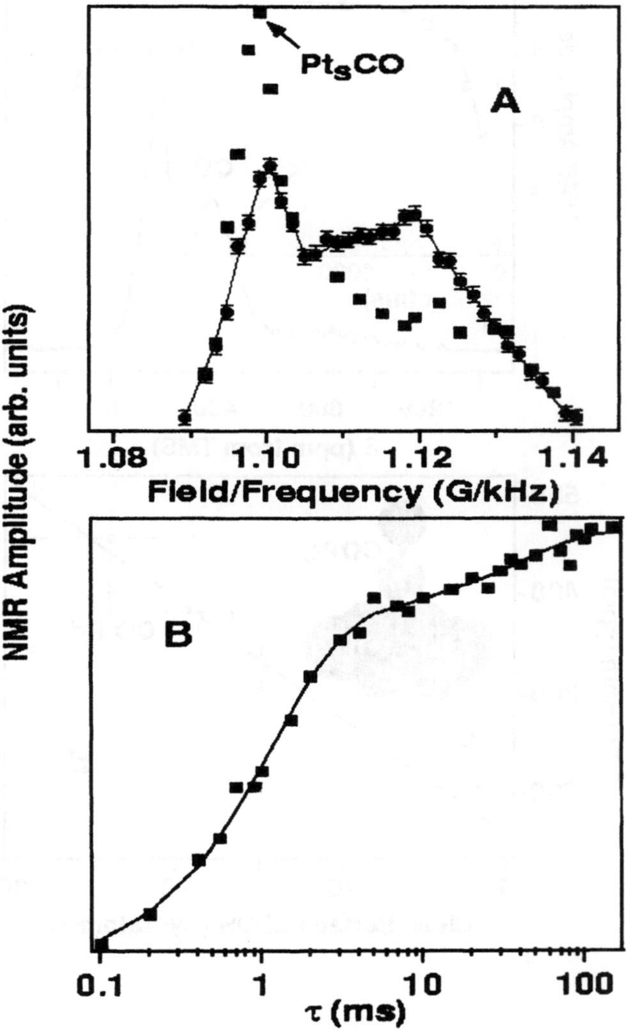

Figure 14. (A) Point-by-point ^{195}Pt NMR spectrum of clean Pt particles before and after CO adsorption, (B) Typical nuclear magnetization recovery profile for ^{195}Pt for the surface peak. The solid line is a two-exponential fit. (Reprinted with permission from [15] Copyright 1999 Am. Chem. Soc.)

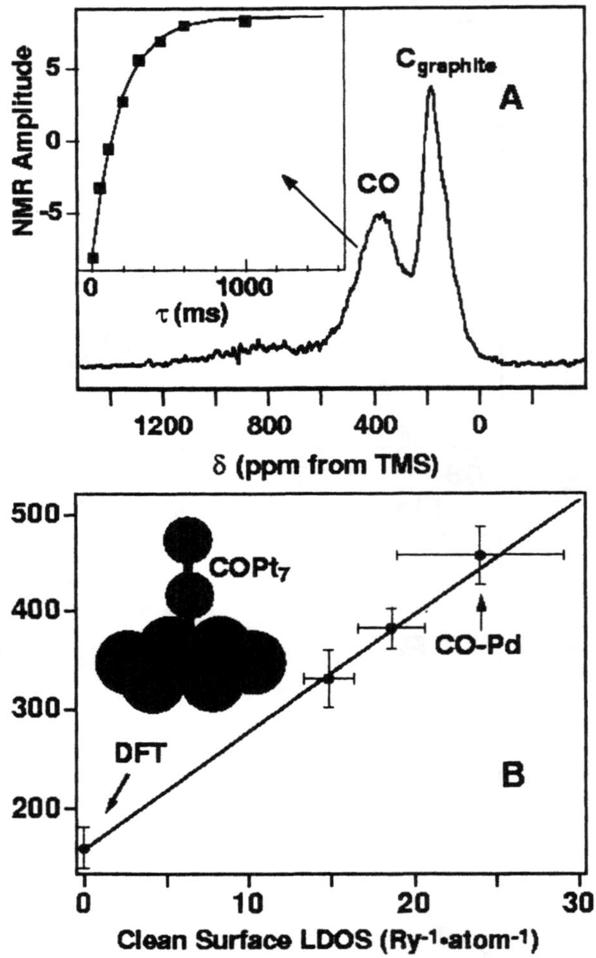

Figure 15. (A) ^{13}C NMR spectra of chemisorbed CO on Pt nanoparticles of size 2.5nm . Inset shows the magnetization recovery for ^{13}C of chemisorbed CO. (B) ^{13}C shift plotted as function of clean metal surface E_f-LDOS. Inset shows the 7-atom Pt cluster with adsorbed CO, used for the DFT calculations. (Reprinted with permission from [15] Copyright 1999 Am. Chem. Soc.)

$COPd_7$, the value was 203 ± 20 ppm. These calculated values are in surprisingly good agreement with available experimental data. The standard deviations account for uncertainties due to use of different functionals and basis sets. The clean surface E_f-LDOS for Pd was estimated based on the changes in magnetic susceptibility of small Pd particles with respect to their bulk value.

As may be seen in Figure 15B, the ^{13}C NMR shift of CO scales linearly with the clean surface E_f-LDOS before adsorption. A slope of 12 ppm·Ry·atom is obtained from the linear fit to the data. This linear relationship can be readily related to the frontier orbital interpretation of the Blyholder model.[48] A higher clean surface E_f -LDOS means more metal electrons and holes are available to engage in 5σ-forward and 2π*-back donation. As the E_f -LDOS increases, CO becomes more metallic, which results in a larger Knight shift. A very similar correlation to this has also been found to exist in the gas phase, where the C–O vibrational stretch frequency (after chemisorption) was shown to correlate linearly with the Pt clean surface E_f-LDOS before chemisorption.[12] The higher the surface E_f-LDOS, the lower the C–O stretching frequency. The linear relationship shown in Figure 15B is important because it demonstrates the validity of the frontier orbital-interaction picture of metal surface chemistry, where the importance of the clean surface E_f-LDOS is highlighted, in our electrochemical context. In addition, it puts ^{13}C NMR spectroscopic studies of chemisorbed CO on a firmer footing, by probing the electronic properties of transition metal surfaces before CO chemisorption.

5. NMR Comparison of CO Adsorbed on Pt-Black from Different Sources

^{13}C NMR of adsorbed CO on a metal nanoparticle electrode surface has been the main system investigated so far from the perspective of probing the properties of the Pt surface and Pt-CO interactions. In most of these NMR experiments, CO was chemisorbed onto metal nanoparticle catalyst surfaces through the catalytic decomposition of methanol. However, an ad-layer of CO can also be produced by immersing clean nanoparticles into electrolyte solutions saturated with CO. In order to understand whether the electronic properties of chemisorbed CO depended on the source from which it was chemisorbed, we have carried out ^{13}C NMR experiments on CO chemisorbed onto Pt-black from methanol or from CO-saturated solutions.[49] CO was chemisorbed onto about 500 mg of catalyst from ^{13}C (99%) enriched CO solutions at open circuit. After 1 hour of adsorption, excess CO was removed by purging with argon and rinsing with 0.1 M D_2SO_4.

The ^{13}C EC-NMR spectrum of CO chemisorbed from the CO-saturated solution is shown in Figure 16A.[50] A single resonance peak around 370 ppm is found, basically as seen before in experiments where CO was generated from methanol. For comparison, the ^{13}C NMR spectrum of CO adsorbed via methanol oxidation is given in Figure 16B. The slight increase in the width of the NMR spectrum for CO chemisorbed from the gas phase may arise from the increased dipolar interaction between the CO molecules as a consequence of the higher CO coverage. The relaxation rates and their temperature

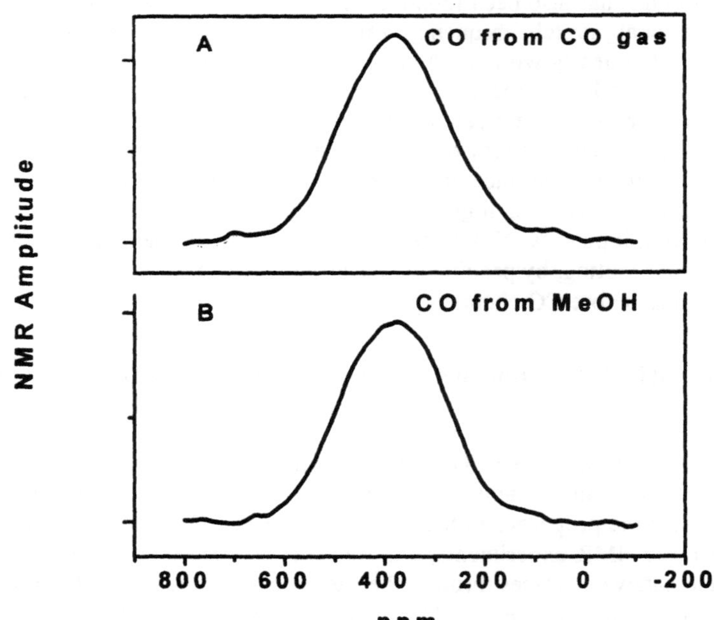

Figure 16. ^{13}C NMR spectra of CO adsorbed on clean Pt-black (A) CO generated from methanol oxidation (B) CO from CO gas (Reprinted with permission from [49] Copyright 2002 Am. Chem. Soc.)

Table 1
Comparison of the ^{13}C NMR Parameters of CO Produced from CO Gas and Methanol Oxidation on a Nanocrystalline Platinum Catalyst

	CO (From MeOH)	CO (From CO gas)
Peak (ppm)	370	375
T_1T (s.K)	59	57
$D_{5\sigma}(E_f)$ (Ry.atom)$^{-1}$	0.6	0.6
$D_{2\pi^*}(E_f)$ (Ry.atom)$^{-1}$	7.0	7.3

variation (see Table 1) are found to be essentially identical in both systems, implying that CO adsorbed on Pt has the same electronic properties irrespective of its source of generation. That is, the chemisorbed product from CO solutions and from methanol decomposition, are the same.

6. Effect of Surface Charge on the Chemisorption Bond: CO Chemisorption on Pd

The electrode potential dependence of the ^{13}C NMR spectra of CO adsorbed on Pd powder provides essential additional evidence for alterations to the metal/solution interface due to the application of an electric field.[4,16] Samples consisted of Pd powder with an average particle size of about 8 nm, produced by a gas condensation technique. Pd powder was cleaned in 0.1 M NaOH through multiple cycles of potential controlled reduction and oxidation. An adlayer of CO was produced from ^{13}CO saturating the supporting electrolyte. The cleaned powder and electrolyte were introduced into a sealed vessel and stirred under ^{13}CO gas.

The ^{13}C NMR spectrum of CO adsorbed on Pd is shown in Figure 17. CO adsorbed on Pd has a much larger Knight shift than CO on Pt (*c.f.*, for instance, Figure 16). This is attributed to a molecular orbital very near the Fermi level of Pd.[51] The peak position of the ^{13}C NMR spectrum shows a change as the electrode potential is varied and has a slope of -136 ppm/V. An estimate of the surface electric field was made from the IR stretching frequency of adsorbed CO to be about 10^9 V/m. A similar value is obtained from simple considerations of the

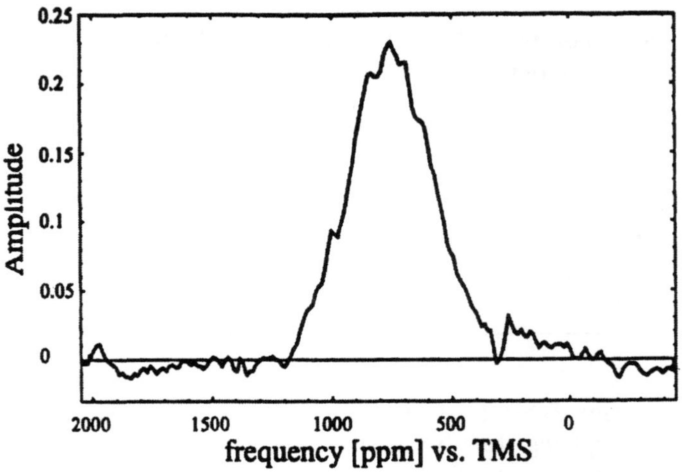

Figure 17. ^{13}C NMR spectrum of CO on Pd for a freshly prepared sample. (After Vuizzov *et al* [16] Copyright 1999, *Am. Phys. Soc.)*

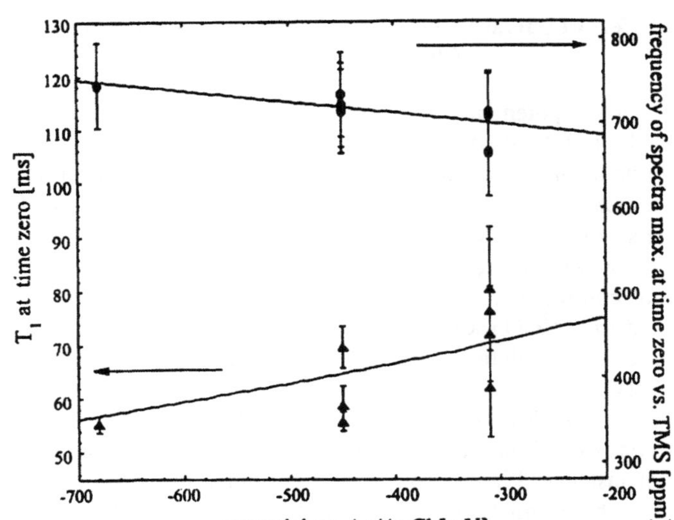

Figure 18. ^{13}C NMR line position (top) and spin-lattice relaxation time T_1 (bottom) of CO on Pd as a function of the electrode potential. (After Vuizzov *et al* [16] Copyright 1999, *Am. Phys. Soc.*)

measured capacitance. From the experimentally measured shift *vs.* electrode potential it was found that such electric fields could produce only about a –15 ppm/V change in the chemical shift. Therefore, it was concluded that the potential dependent NMR peak shift arises from changes in the Knight shift due to application of the electric field (see the above discussion for the Pt electrode). In addition, the potential dependence of the ^{13}C T_1 confirms this idea. Figure 18 shows both T_1 and the peak shift as a function of the applied potential. The straight line is a fit to the data with the T_1 values calculated using the Korringa relation (see above and Appendix). A good fit to both data sets is only obtained when the Knight shift is allowed to vary by the rate obtained above. Since the Knight shift is proportional to the E_f -LDOS, the changes in the Knight shift reflect the alteration to the E_f-LDOS by the electrode potential (or more specifically the electric field). Thus, EC-NMR under potential control serves as a prime tool to elucidate electronic changes in the electrochemical interface due to the presence of electric fields.

7. Pt Electrodes Modified by Ruthenium: A Study in Tolerance

Addition of Ru to Pt electrode surfaces is known to produce catalysts with enhanced CO tolerance. (For example, the CO stripping peak on Pt/Ru electrodes occurs at less positive potentials than on clean Pt electrodes.[52]) We have produced a series of Pt/Ru electrode surfaces using spontaneous deposition[22] of ruthenium onto commercial nano-particle platinum supports, and monitored their catalytic activity.[52] The spontaneous deposition method[37] used here, involves immersing platinum in a ruthenium chloride solution (in 0.1 M $HClO_4$), rinsing the sample in the cell with supporting electrolyte, then galvanically breaking the chemisorbed precursors to metallic/oxide ruthenium forms. Once a Pt surface decorated by metallic ruthenium is obtained in the first deposition, the deposition can be repeated many times in order to obtain the desired Ru packing density.[52] This method has several important advantages over other approaches: It is simple, yields active surfaces decorated by nano-sized islands of (predominantly) monatomic height and provides a good way for adjusting ruthenium coverage in the submonolayer deposition regime.

Results of methanol-electrooxidation reactivity measurements on four Pt/Ru samples (with Ru coverages of 0, 0.14, 0.35 and 0.53) are shown in Figure 19. The clean nanoparticle Pt electrode produces a very small methanol oxidation current, due to severe CO poisoning. On the other hand, surfaces obtained by spontaneous deposition of ruthenium show enhanced tolerance to CO poisoning, and hence much higher steady-state currents. The extent of enhancement depends on ruthenium coverage (Figure 19), with maximum activity at a 0.4 - 0.5 fraction Ru packing density.[52] Two distinct CO stripping peaks can be seen (Figure 20) in the CV of the Pt/Ru samples, whereas for pure Pt black there is only a single peak, irrespective of CO coverage. This was the first observation of two CO stripping peaks on a bimetallic nanoparticle electrode. The total CO packing on the surface, obtained from CV, was estimated to be about 0.9 per metal surface site, indicating that CO is adsorbed on both Pt and Ru sites. Experimental results from Pt and Pt/Ru alloy electrodes[53] as well as numerical simulations[54]

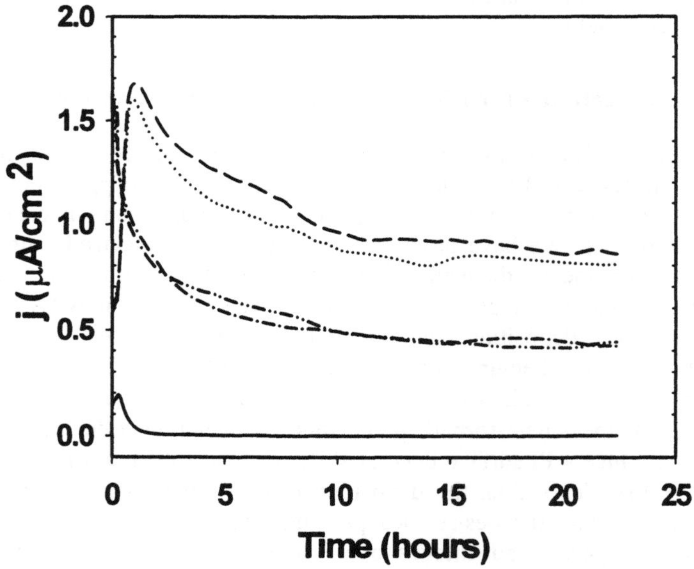

Figure 19. Methanol oxidation current density on Pt (—) and Pt/Ru: after 1st (– · –), 2nd (– ·· –), 3rd (···) and 4th (– –) deposition in 0.5 M methanol and 0.5 M H_2SO_4. Ruthenium packing density was 0.22, 0.23, 0.36, 0.40, respectively. (After Waszczuk *et al* [52] Copyright @2001 by Academic Press)

have shown that CO oxidizes at less positive potentials on Pt/Ru than it does on pure Pt electrode surfaces. Therefore, the low-potential CO stripping peak was assigned to CO adsorption on the Pt/Ru phase, and the higher potential peak, to CO adsorbed on Pt patches not occupied by ruthenium.[22,54-56] Notably, the results also show that progressively more CO's from the Pt phase are oxidized at low potential as the Ru coverage is increased.

In order to try to better understand the mechanism of enhanced CO tolerance, we undertook a ^{13}C NMR investigation of chemisorbed CO on these Ru modified electrode surfaces.[22] ^{13}C NMR spectra recorded in a 14.1 T magnetic field showed the presence of two broad peaks for

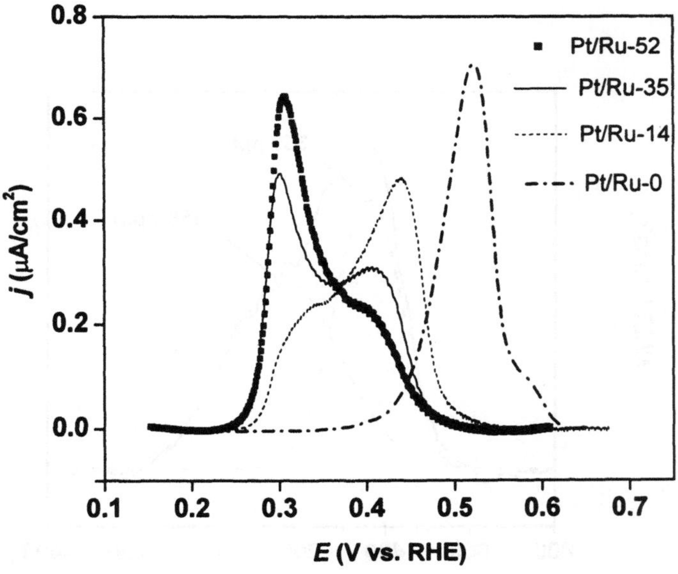

Figure 20. CO stripping from nanoparticle Pt/Ru electrode obtained at a scan rate of 10 mV s^{-1} after 12h of CO adsorption. With increasing Ru coverage, a new peak emerges to appear in the low potential region of the CV. (Reprinted with permission from [22] Copyright 2002 Am. Chem. Soc.)

each of several Pt/Ru samples (Figure 21) while only a single peak was seen for a pure Pt sample (Figure 16). Thus, EC-NMR also detects two CO populations: CO chemisorbed on pure Pt atoms and Ru atoms on the Pt/Ru electrodes, as was seen previously using infrared methods.[57-59] Since the electrochemical measurements suggest the presence of two CO domains (Pt, Ru/Pt) one can expect them to have different electronic properties. In fact, we were able to simulate the NMR spectra of a series of Pt/Ru samples using just two Gaussians. In Figure 21, we show such a double- Gaussian deconvolution of the NMR spectrum of one of these samples. Remarkably, the volume fractions of CO estimated from these NMR measurements are essentially identical to those obtained by CV, confirming the two-domain model proposed from the electrochemical measurements. The peak positions (obtained

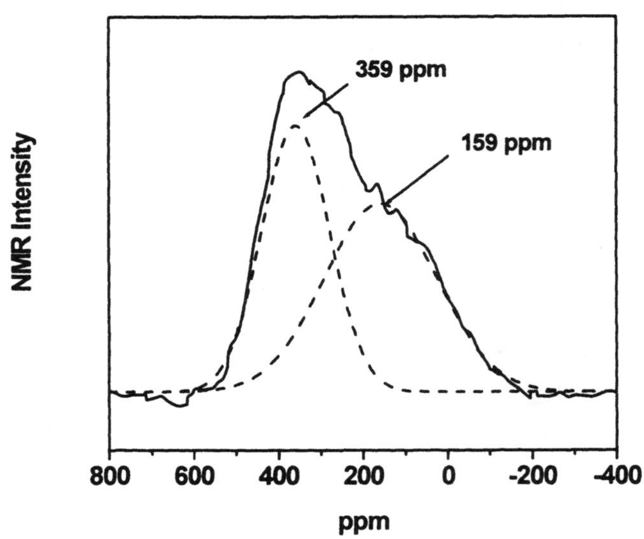

Figure 21. ^{13}C NMR spectrum of Pt/Ru-35 obtained in 14.1 T field. Dotted lines represent the double-Gaussian deconvolution that indicates the presence of two peaks, corresponding to the two different CO populations.

from deconvolution) and the spin-lattice relaxation data (measured at the two peak positions) are given in Table 2. The position of the low field peak changes by only 20 ppm as a function of Ru content. Also, the changes in the corresponding Korringa constants, T_1T, are relatively small. These results indicate that CO molecules resonating at low field,

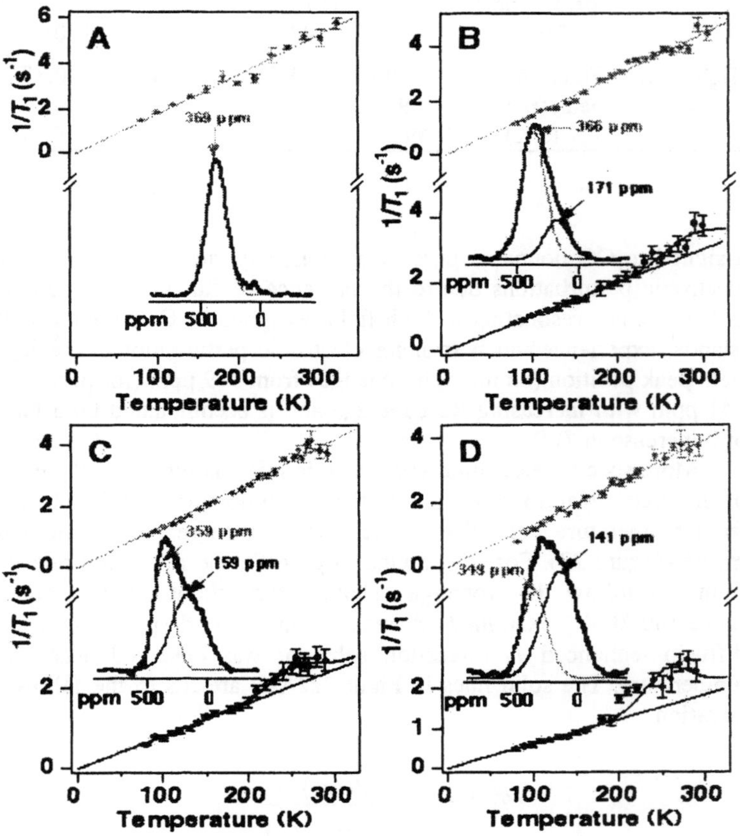

Figure 22. Temperature dependence of the ^{13}C nuclear spin-lattice relaxation rates for CO adsorbed on (A) Pt; (B) Pt/Ru-14; (C) Pt/Ru-35; (D) Pt/Ru-52. Insets show the double-Gaussian deconvolution. Relaxation rates are measured at the peak positions indicated by the arrows. (Reprinted with permission from [22] Copyright 2002 Am. Chem. Soc.)

Table 2
Double-Gaussian Deconvolution of ^{13}C NMR Data and the Corresponding T_1 Data.

		Pt/Ru-0	Pt/Ru-14	Pt/Ru-35	Pt/Ru-52
Low Freq.	Fraction	0	0.23	0.54	0.63
	peak (ppm)		171	159	141
	T_1T (s.K)		98	120	160
High Freq.	Fraction	1.00	0.77	0.46	0.37
	peak (ppm)	369	366	359	349
	T_1T (s.K)	59	67	71	75

assigned to CO bound to pure Pt regions, experience relatively weak electronic perturbations due to the presence of Ru. In sharp contrast, CO molecules resonating at high-field, assigned to COs bound to Ru islands, experience much stronger electronic perturbations. The high-field peak position (on the right) changes from 369 ppm (for pure Pt) to 141 ppm with increasing Ru coverage and is accompanied by a factor of 2 increase in T_1T.

More specific electronic and dynamic information about the nature of the metal-ligand interactions in these two domains can be obtained from temperature dependent nuclear spin-lattice relaxation measurements (Figure 22). For COs bound to pure Pt, the relaxation rate was found to follow the Korringa behavior throughout the temperature range (80-300K) examined whereas for the second species of CO, a diffusion-enhanced T_1 relaxation behavior was observed near room temperature. The solid lines in Figure 22A-D are fits to the following equation

$$T_1^{-1} = aT + \frac{2(\Delta\omega)^2\tau}{(1+\omega_0\tau^2)} \tag{4}$$

The first term is the standard Korringa contribution in which T is the absolute temperature and a^{-1} is the Korringa constant, obtained from the straight line fit at $T < 200$ K. The second term accounts for surface

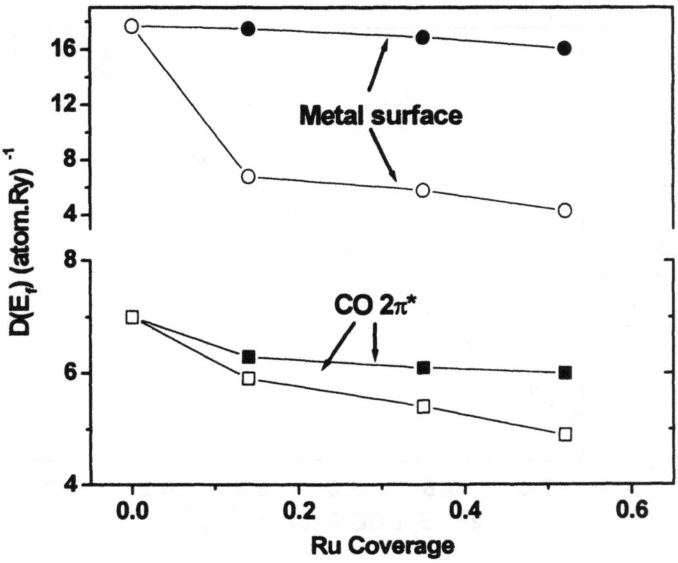

Figure 23. The metal surface E_f-LDOS before chemisorption and the $2\pi^*$ E_f-LDOS at ^{13}C after chemisorption for (remote) Pt (solid symbols) and Pt/Ru (open symbols) sites, as a function of Ru coverage. (Reprinted with permission from [22] Copyright 2002 Am. Chem. Soc.)

diffusion, in which $\Delta\omega$ is the average local field inhomogeneity seen by a moving CO molecule, ω_0 is the NMR Larmor frequency, and τ is the correlation time. The correlation time is related to the activation energy for diffusion (E_{diff}) through the Arrhenius relation $\tau = \tau_0 \exp(E_{diff}/RT)$, where τ_0 is the preexponential factor, taken to be 10^{-13} s. The fit gives two parameters: the local field inhomogeneity $\Delta\omega$ and the activation energy for diffusion (E_{diff}). The values obtained for $\Delta\omega$ and E_{diff} are 3.3, 2.7, and 3.6 kHz and 5.7, 4.9, and 5.1 kcal/mol, for Pt/Ru-14, Pt/Ru-35, and Pt/Ru-52, respectively. For comparison, the activation energy for CO diffusion on pure Pt under the same electrochemical conditions has been previously determined to be 7.8 kcal/mol.[4,9,11] Clearly, the presence of Ru substantially enhances surface diffusion for the CO molecules on Pt/Ru sites, but for the Pt sites, the effects are very small. Thus, COs bound to Ru are more mobile than those adsorbed on Pt and

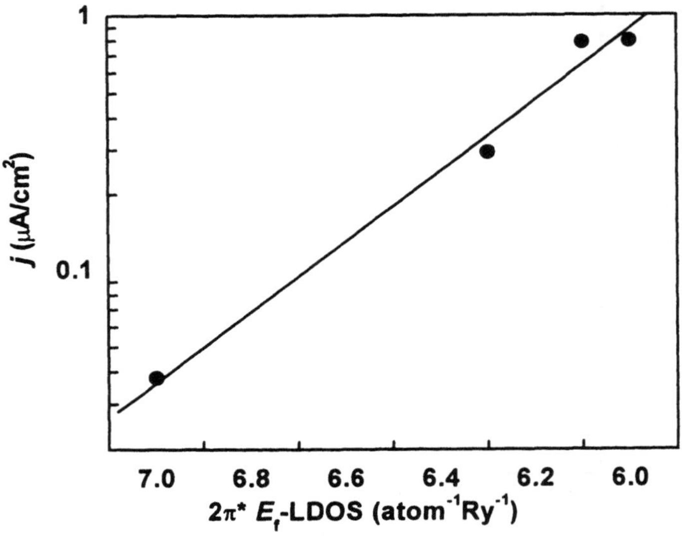

Figure 24. Graph showing the correlation between methanol oxidation current density (after reaching the steady-state) and 2π* E_f-LDOS. (Reprinted with permission from [22] Copyright 2002 Am. Chem. Soc.)

there is no exchange between the two CO species, at least on this NMR time scale. Analyzing the spin-lattice relaxation data with the help of the two-band model, we then obtained the average Fermi level local density of states (E_f-LDOS) at the clean metal surfaces before chemisorption, and the $2\pi^*$- E_f-LDOS at ^{13}C after chemisorption (Figure 23). As Ru coverage increases, the clean metal E_f-LDOS decreases, causing a clearly discernable reduction in $2\pi^*$ back-donation, which weakens CO-metal bonding. Another interesting correlation can be seen when the methanol oxidation current density is plotted against the $2\pi^*$- E_f-LDOS, as shown in Figure 24. The oxidation current density shows a marked increase with the reduction in the $2\pi^*$- E_f -LDOS, clearly indicating the importance of electronic alterations in enhancing the CO tolerance of such Ru deposited catalysts.

From the EC-NMR data,[22] we conclude that diffusion of CO on Pt/Ru islands is fast and that the exchange between these two different

CO populations on the NMR and voltammetric timescale[22] is very slow. In summary, these observations demonstrate that there are two major populations of CO on Pt/Ru surfaces prepared by the spontaneous deposition method: COs located on Ru islands undergoing fast, thermally activated diffusion, and COs on Pt sites further away from Ru, undergoing slower diffusion.[4,11] With increasing Ru coverage, the E_f -LDOS for the clean metal surface and the $2\pi^*$ orbital of chemisorbed CO decrease, indicating there are strong electronic perturbations caused by the Ru addition. The increase in the catalytic activity of the Ru modified Pt nanoparticles appears to have a direct correlation with these electronic modifications.

8. EC-NMR of Pt/Ru Alloy Nanoparticles

As discussed above, Ru is an excellent promoter of Pt catalysts for methanol oxidation and Pt nanoparticles decorated with Ru show better performance in the oxidation process than do commercially available Pt/Ru (1:1) alloy nanoparticles.[22,52] In order to compare the electronic properties of these two systems, we undertook further ^{13}C NMR investigations of the Pt/Ru alloy nanoparticles covered by CO,[24] and have obtained new insights into the electronic alterations caused by Ru addition to Pt surfaces.

Cyclic voltammograms for CO stripping for the Pt/Ru alloy and commercial Pt black electrodes, obtained at a scanning rate of 10 mV/min, are shown in Figure 25. In the Pt/Ru alloy, the CO stripping peak is clearly shifted (as is well known) towards low potential, similar to the behavior seen previously with Pt black samples having Ru deposited through spontaneous deposition.[22] However, the two well resolved CO stripping peaks observed with the Ru-deposited samples (Figure 20) are now absent, Figure 25. Another interesting feature is that there is an increase in peak width for the Pt/Ru sample, suggesting that the Ru induces a broad range of local electronic structures in the Pt substrate, although unlike the situation with spontaneous deposition, the distribution is continuous rather than bimodal.

We show in Figure 26, the ^{13}C NMR spectra of CO adsorbed on Pt/Ru and Pt-black. For the alloy nanoparticles, Figure 26B, the whole spectrum has moved to a much lower overall shift (239 ppm) from TMS than that observed for CO on the Pt-black sample (375 ppm). In addition, there is a noticeable increase in the full-width at half-

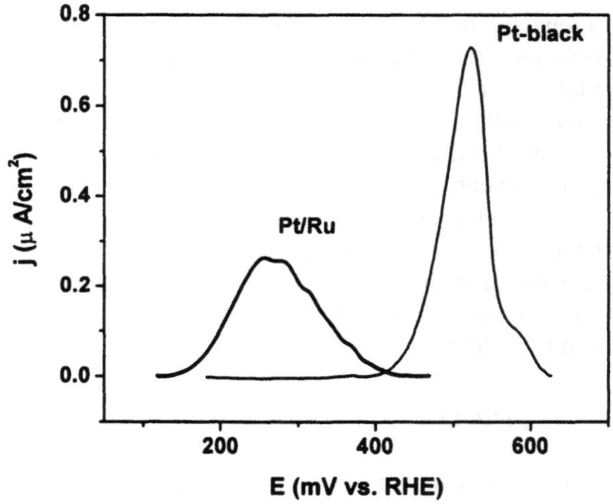

Figure 25. CO stripping voltammograms obtained after 12 hrs of methanol adsorption onto commercial Pt-black and Pt/Ru alloy nanoparticles. Scan rate was 10 mv/min. The surface areas are normalized.[24]

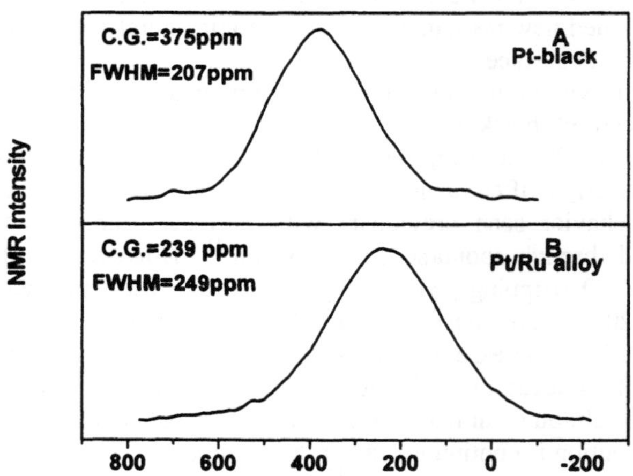

Figure 26. ^{13}C NMR spectra of chemisorbed CO on (A) clean Pt black and (B) clean Pt/Ru (1:1) commercial alloy nanoparticles. The FWHM changes from 207 to 249 ppm while the center of gravity shifts from 375 to 239 ppm as a result of Ru addition.[24]

maximum height (from 207 to 249 ppm). These results are consistent with the CV results. That is, in Pt black, the CV for CO stripping is narrow while that for the Pt/Ru alloy is broad, indicating a broad distribution of CO binding sites in the alloy, corresponding to different local Pt/Ru bonding environments (*e.g.* Pt/Ru compositional ratios, numbers of coordinating Ru atoms, Pt-Ru bond lengths). The actual shifts in the CV trace implies a weaker Pt-CO bond in the Pt/Ru alloy, consistent with the shift in the ^{13}CO NMR resonance (a low E_f-LDOS), due to a change in $2\pi^*$-back bonding. Moreover, the addition of Ru to Pt to form an alloy results in a downfield shift from the bulk Pt resonance position (1.138 to 1.106 G/kHz, see below), due to a decrease in the Pt E_f-LDOS. The decrease in overall ^{13}C shift of the adsorbed CO on the Pt/Ru alloy also implies that the E_f-LDOS of Pt atoms decreases due to the presence of Ru, consistent with the ^{195}Pt NMR results.

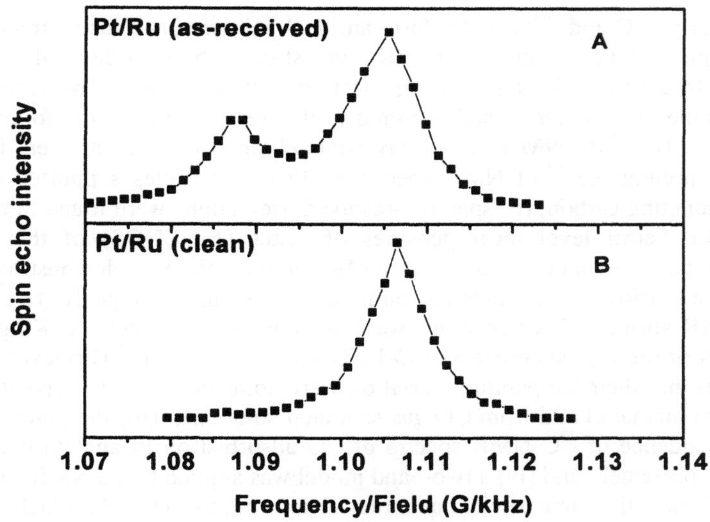

Figure 27. ^{195}Pt NMR spectra of commercial Pt-Ru (1:1) alloy, $T = 80$ K in 8.45 T field (A) As-received (B) Electrochemically cleaned.[24]

In Figure 27, we show the ^{195}Pt NMR spectra of an *unsupported*, commercial alloy (1:1) Pt/Ru catalyst obtained from Johnson-Matthey and having an average Pt particle diameter of 2.5 nm. The as-received material has a strong surface oxide peak (Figure 27A), which disappears after electrochemical reduction. The Pt NMR spectrum of the oxide-free Pt/Ru is completely different to that of the oxide-free and ruthenium-free Pt-samples discussed above, see Figure 1B. It consists of a symmetric, single Gaussian peak centered at 1.106 G/kHz, close to the pure Pt surface atom frequency. There are no Pt atoms resonating in the vicinity of 1.138 G/kHz, (the ^{195}Pt NMR bulk Pt frequency) and there are no subsurface peak characteristics of pure Pt particles either. Possibly, most Pt atoms are deposited on top of the nanoparticle, which has a dominant ruthenium (and/or ruthenium oxide[60]) core. More research is planned to test these ideas.

V. SUMMARY AND CONCLUSIONS

Clearly, ^{13}C and ^{195}Pt NMR form an ideal pair of non-invasive microscopic probes with which to investigate both sides of the electrochemical interface using electrochemical NMR. The results obtained by EC-NMR and reported in this review include the following. **In ^{195}Pt NMR**: (i) a layer-model analysis was applied for interpreting the ^{195}Pt NMR spectra of Pt nanoparticles supported on conducting carbon, (ii) spatially-resolved oscillations were found in the *s*-like Fermi level local densities of states (E_f -LDOS) of the Pt nanoparticles, and (iii) the ^{195}Pt NMR surface peak was demonstrated to be sensitive to the chemical nature of an adsorbate (a ligand). **In ^{13}C NMR studies**: (i) correlations were demonstrated between ^{13}C Knight shift of the CO adsorbate and E_f-LDOS of platinum, (ii) ^{13}C relaxation rates and their temperature variation were compared for CO deposited from methanol and from CO gas saturated solutions, (iii) the potential dependence of ^{13}C NMR spectra of CO adsorbed on Pt and Pd black was presented, and (iv) a two-band model was applied to the NMR shift and relaxation time to yield quantitative information about E_f-LDOS of Pt-CO surface chemical bonds, including data on the E_f-LDOS of both 5σ and $2\pi^*$ orbitals.

We believe we have given a fairly comprehensive account of the use of EC-NMR in electrochemical surface science and documented that EC-NMR provides information which cannot be obtained by any other experimental method at a comparable level of detail, although in some cases the conclusions reached support previous work by XANES using synchrotron radiation.[61] Our results clearly demonstrate that due to a quantum mechanical electron density spillover from platinum, the interface is metallized, as evidenced by Korringa relaxation and Knight shift behavior. Thus, the adsorbate on a platinum electrode belongs to the metal part of the platinum-solution interface, and most likely other *d*-metal interfaces, and should be considered as such in any realistic models of the structure of the electrical double layer of interest to electrocatalysis.

We find that a layer model analysis can adequately describe the ^{195}Pt NMR spectrum of nanoscale electrode materials. The shifts of the surface and sub-surface peaks of ^{195}Pt NMR spectra correlate well with the electronegativity of various adsorbates, while the Knight shift of the adsorbate varies linearly with the E_f-LDOS of the clean metal surface. The ^{195}Pt NMR response of Pt atoms from the innermost layers of the nanoparticles does not show any influence of the adsorbate present on the surface. This provides experimental evidence, which extends the applicability of the Friedel-Heine invariance theorem to the case of metal nanoparticles. Further, a spatially-resolved oscillation in the *s*-like E_f-LDOS was observed via ^{195}Pt NMR of a carbon-supported Pt catalyst sample. The data indicate that much of the observed broadening of the bulk-like peak in ^{195}Pt NMR spectra of such systems can be attributed to spatial variations of the $D_s(E_f)$. The oscillatory variation in $D_s(E_f)$ beyond 0.4 nm indicates that the influence of the metal surface goes at least three layers inside the particles, in contrast to the predictions based on the Jellium model.

Comparing the results of EC-NMR and IR investigations, we find that the potential dependence of ^{13}C NMR shift and the vibrational frequency of adsorbed CO are primarily electronic in nature, and originate from changes in the E_f-LDOS. ^{13}C NMR results show that CO adsorbed on Pt, either directly from CO gas or from methanol oxidation, have the same electronic properties. That is, the chemisorbed product (surface CO) from CO solutions and from methanol decomposition is the same. The electrode potential dependence of the ^{13}C NMR spectra of CO adsorbed on Pt and Pd nanoparticles provide direct evidence for electric field induced alterations in the E_f-LDOS. In relation to fuel cell catalysis, EC-NMR investigations of Pt nanoparticles decorated with Ru show that there exist two different kinds of CO populations having markedly different electronic properties. COs adsorbed on the Ru phase of Pt/Ru electrodes undergo faster diffusion and have a reduced E_f-LDOS, indicating that the CO-metal interaction is weakened due to the presence of Ru. The enhanced catalytic activity of Pt/Ru electrodes shows a direct correlation with the electronic alterations revealed by EC-NMR studies. Taken together, these results suggest that prospects for further studies are bright and indeed the method may have broader applications to improve our understanding of the electronic structure of electrochemical interfaces. Therefore, further utilization of this technique —with particular focus on electron-density level insights in to electrode adsorption and on studies of surface diffusion at electrochemical interfaces— may be expected.

ACKNOWLEDGEMENTS

This work was supported by the Department of Energy Grants DEFG02-96ER45439 and DEGF-02-99ER14993, as well as by the National Science Foundation under Grant NSF CTS 97-26419.

APPENDIX

A. Electron-Nucleus Interactions in Metal NMR

In NMR experiments, one observes the resonant absorption of electromagnetic radiation between nuclear spin energy levels are split due to the Zeeman interaction between the nuclear magnetic moments and the external magnetic field. However, this resonance frequency can be shifted from what one might expect due solely to the Zeeman interaction. In diamagnetic compounds shifts arises due to the interaction between the orbital magnetic moment of the electron with the nuclear spin. This is known as the chemical shift. In metals, in addition to this orbital interaction, it is possible to have a large perturbation of the electron spin magnetism due to the presence of conduction electrons. That is, since *s*-like conduction electron wavefunctions have finite overlap with the nuclei, there exists a strong interaction between the conduction electron and the nuclear spins, which leads to the Knight shift (K) in metals.

At thermal equilibrium, the nuclear spin energy levels are populated according to the Boltzmann distribution. Application of a radio frequency (RF) pulse that satisfies the resonance condition causes absorption of radiation and therefore, alteration in the spin population distribution. Thermal equilibrium populations are re-established when nuclear spins exchange their energies with the surroundings. This process is known as the nuclear spin-lattice relaxation and it is characterized by a time constant called T_1. T_1 is an important parameter whose numerical value and temperature dependence are characteristics of a material. In diamagnetic materials, molecular motions can produce the necessary fluctuating magnetic fields for the nuclear spin-lattice relaxation. In metals, electron-nuclear spin interactions that produce the Knight shift are often primarily responsible for the spin-lattice relaxation.

The Hamiltonian for the electron-nucleus interaction can be written as:

$$H = H_{FC} + H_D + H_{ORB} \tag{A1}$$

The first term is the Fermi-contact interaction given by

$$H_{FC} = \frac{8\pi}{3} \cdot \gamma_e \gamma_n \hbar^2 \mathbf{I} \cdot \mathbf{S} \delta(\mathbf{r}) \tag{A2}$$

where γ_e and γ_n are the gyromagnetic ratios of the electron and nuclear spins. **I** is the nuclear spin operator, **S** is the electron spin operator and $\delta(\mathbf{r})$ is the Dirac delta function, where **r** corresponds to the radius vector of the electron with the nucleus taken as the origin. Fermi-contact interaction causes a large shift in the NMR frequency of a nucleus compared to a diamagnetic reference. This interaction can scatter the conduction electrons near the Fermi level to vacant states by exchanging energy with nuclear spins, providing a very effective channel for nuclear spin-lattice relaxation. The energy associated with nuclear spin quanta is less than a micro electron-volt whereas a typical Fermi energy in metals is of the order of a few electron-volts. Since nuclear spin-lattice relaxation involves the conservation of energy, electrons can be scattered only to the vacant states that are lying very close to the Fermi level. In simple *s*-band metals, K and $1/T_1$ are related by the well-known Korringa equation:

$$T_1TK^2 = B\left(\frac{\gamma_e}{\gamma_n}\right)^2 \frac{\hbar}{4\pi k_B} \tag{A3}$$

where T is the absolute temperature and B is a constant, equal to unity when the many-body effects are negligible. This unique temperature dependence of $1/T_1$ is the NMR fingerprint of a metallic state. It results from the fact that only conduction electrons around the Fermi level can satisfy energy conservation for the electron-nuclear spin *flip-flop* relaxation process, and the fraction of these electrons is proportional to k_BT.

The second term of Eq. (A1) is the electron-nucleus dipole-dipole coupling and is given by

$$H_D = \gamma_e\gamma_n\hbar^2\mathbf{I}.\left(\frac{\mathbf{S}}{r^3} - \frac{3\mathbf{r}(\mathbf{S}.\mathbf{r})}{r^5}\right) \tag{A4}$$

This dipolar interaction does not cause any shift in the NMR spectrum (or more precisely, the center of gravity). However, its time-dependent part can contribute to nuclear spin-lattice relaxation $[(1/T_1)_{\text{dip}}]$. The third term of Equation (1), which represents the coupling of the orbital magnetic moment of the electron to the nuclear spin, is given by

$$H_{ORB} = \gamma_e\gamma_n\hbar^2\left(\frac{\mathbf{I}.\mathbf{L}}{r^3}\right) \tag{A5}$$

where **L** is the orbital angular momentum of the electron. The isotropic part of H_{ORB} produces the chemical shift. In metals, the fluctuations in the orbital interaction could also contribute to the spin-lattice relaxation $[(1/T_1)_{\text{orb}}]$. The Knight shift and the chemical shift are always measured together as a combined frequency shift, and there is practically no experimental distinction to be made between them. When δ_{CS} is of the same order as K, as with CO adsorbed onto a metal surface, one needs to use the Korringa relationship to confirm the presence of conduction electrons (and therefore of a Knight shift) at the observed nucleus.

B. Local Density of States and the Two-Band Model

In a typical EC-NMR experiment, the shift in the NMR frequency (K) and the nuclear spin-lattice relaxation time (T_1) are measured as function of temperature. The total shift in the NMR frequency consists of both the orbital part (K_{orb}) and the part due to the Fermi-contact interaction. For simple metals with only one type of electrons at the Fermi energy (E_f), the Knight shift is a function of the local density of states (LDOS) at E_f:

$$K \propto H_{hf} D(E_f) \tag{B1}$$

$$T_1 T K^2 = S \tag{B2}$$

where H_{hf} is the hyperfine field, $D(E_f)$ is the LDOS at the Fermi level and S is the Korringa constant appearing on the right hand side of Eq. (A3). The LDOS is defined by the following equation:

$$D(E_f) = 2\sum_i |\psi_i(\mathbf{x})|^2 \delta(E_i - E) \tag{B3}$$

where $|\psi_i(\mathbf{x})|^2$ is the probability of finding the ith electron with energy E_i at location **x**. The energy integral of the LDOS gives the total electron density at site **x**. In complex systems like the metal-adsorbate pair in EC-NMR, the LDOS at E_f can have contributions from more than one type of electron. For ^{195}Pt, there are contributions from 6s and 5d electrons, while for ^{13}C (in the chemisorbed CO) the LDOS at E_f consists of contributions from $2\pi^*$ and 5σ orbitals. In such cases, phenomenological two-band models[45,62] can be used to obtain the LDOS at E_f from the NMR parameters. For ^{195}Pt NMR, the total Knight shift can be written as

$$\begin{aligned} K &= K_s + K_d + K_{orb} \\ &= (1-\alpha_s)^{-1}\mu_B D_s(E_f) H_{hf,s} \\ &\quad + (1-\alpha_d)^{-1}\mu_B D_d(E_f) H_{hf,d} + \frac{\chi_{orb}}{\mu_B} H_{hf,orb} \end{aligned} \tag{B4}$$

and the Korringa relation becomes,

$$S(T_1T)^{-1} = k(\alpha_s)K_s^2 + k(\alpha_d)K_d^2 R_d + (\mu_B D_d(E_f)H_{hf,orb})^2 R_{orb} \tag{B5}$$

where

$$k(\alpha) = (1-\alpha)\left(1+\frac{\alpha}{4}\right) \tag{B6}$$

For ^{13}C NMR of CO, the corresponding equations become,

$$\begin{aligned} K &= K_\sigma + K_\pi + K_{orb} \\ &= (1-\alpha_\sigma)^{-1}\mu_B D_\sigma(E_f)H_{hf,\sigma} \\ &\quad + (1-\alpha_\pi)^{-1}\mu_B D_\pi(E_f)H_{hf,\pi} + K_{orb} \end{aligned} \tag{B7}$$

and

$$S(T_1T)^{-1} = k(\alpha_\sigma)K_\sigma^2 + \left(\frac{1}{2}\right)k(\alpha_\pi)K_\pi^2 + \left(\frac{13}{5}\right)(\mu_B D_\pi(E_f)H_{hf,orb})^2 \tag{B8}$$

In this formalism, the many-body effects are incorporated through the Stoner enhancement factor

$$\alpha_i = I_i D_i(E_f), \qquad i = s, d, \sigma, \pi \tag{B9}$$

and the Shaw-Warren de-enhancement factor $k(\alpha_i)$, where I_i is the exchange integral.[43] Using the parameters given in Table 3, it is possible to solve the set of simultaneous equations (for K and $S(T_1T)^{-1}$) to obtain $D_s(E_f)$, $D_d(E_f)$, $D_\sigma(E_f)$, and $D_\pi(E_f)$.

C. Layer Model Deconvolution and Exponential Healing

^{195}Pt NMR spectra of platinum nanoparticles are usually very broad, extending from 1.07 G/kHz to 1.138 G/kHz which corresponds to a range of almost 4 MHz at an 8.45 T field, which makes it possible to distinguish NMR responses of surface Pt atoms from those of bulk and

Table 3
Parameters Used fr E_f–LDOS Analysis

	K_{orb} (ppm)	$I_{s/\sigma}$ (Ry)	$I_{d/\pi}$ (Ry)	$H_{hf,s/\sigma}$ (kG)	$H_{hf,d/\pi}$ (kG)	$H_{hf,orb}$ (kG)
^{195}Pt	2100	0.098	0.037	2700	-1180	1180
^{13}C	160	0.086	0.078	363	17	50

other intermediate positions. The layer model deconvolution[36] is a very convenient method for analyzing ^{195}Pt NMR spectra of catalysts and takes into account the actual size distribution of particles. In this model, atoms in a nanoparticle sample are divided into groups belonging to various atomic layers. Particle size histograms obtained from transmission electron microscopy (TEM) are interpreted in terms of FCC cubooctahedra to estimate the fraction of atoms in each layer. Atoms from a given layer are assumed to produce a Gaussian NMR peak having a width of the order of one MHz. The observed NMR spectrum is then reconstructed by convoluting a set of Gaussians, each representing a given layer of the cubooctahedron. The peak position is taken as a function of the layer number[7] while their integrals are constrained to be proportional to the fraction of atoms in each layer. The widths of these Gaussians are taken as fitting parameters.

Particle size distribution histograms are constructed from the TEM micrographs by converting the image diameters (d_i) into the total number of atoms (N_T) in a particle using the following equation:

$$N_T = \pi\frac{\sqrt{2}}{6}\left(\frac{d_i}{2r_s}\right)^3 \tag{C1}$$

For platinum, $2r_s$ = 0.277 nm where r_s is the hard sphere radius. For each size distribution histogram obtained from the TEM, it is possible to find the number of cubooctahedral layers (l) by solving the equation

$$N_T = \left(\frac{10}{3}\right)l^3 - 5l^2 + \left(\frac{11}{3}\right)l - 1 \tag{C2}$$

The number of surface atoms in a given layer is calculated from,

$$N_S = 10l^2 + 20l - 12 \tag{C3}$$

The number of atoms in the subsurface layers is obtained by replacing l in Equation (C3) by $(l - 1)$, $(l - 2)$ and so on.

According to the *exponential healing theorem*,[7] the Knight shift of the n-th layer (K_n) heals back to the bulk value (K_∞), exponentially as a function of the layer number (n). Therefore, the Knight shift of the n-th layer can be written as

$$K_n - K_\infty = (K_0 - K_\infty)\exp(-n/m) \tag{C4}$$

where the dimensionless constant m is known as the *healing length* for the Knight shift.

REFERENCES

[1]K.W.H. Chan and A. Wieckowski, *J. Electrochem. Soc.* **137** (1990) 367.

[2]P.J. Slezak and A. Wieckowski, *J. Magnetic Resonance, Series A* **102** (1993) 166.

[3]A. *Interfacial Electrochemistry: Theory, Experiment, and Applications*, Ed. by A. Wieckowski, Marcel Dekker, New York, 1999.

[4]Y.Y. Tong, E. Oldfield, and A. Wieckowski, *Analytical Chemistry* **70** (1998) A 518.

[5]Y.Y. Tong, E. Oldfield, and A. Wieckowski, in: *Encyclopedia of Electrochemistry: Interfacial Kinetics and Mass Transport*, Vol. 2, Ed. E.J. Calvo, Wiley-VCH, 2002.

[6]C.P. Slichter, *Ann. Rev. Phys. Chem.* **37** (1986) 25.

[7]J.J. Van der Klink, *Advances in Catalysis* **44** (2000) 1.

[8]Y.Y. Tong and J.J. van der Klink in: *Catalysis and Electrocatalysis at Nanoparticle Surfaces*, Ed. by A. Wieckowski, E. Savinova, and C. Vayenas, Marcel Dekker Inc., New York, 2002.

[9]J.B. Day, P.A. Vuissoz, E. Oldfield, A. Wieckowski, and J.P. Ansermet, *J. Am. Chem. Soc.* **118** (1996) 13046.

[10]M.S. Yahnke, B.M. Rush, J.A. Reimer, and E.J. Cairns, *J. Am. Chem. Soc.* **118** (1996) 12250.

[11]J.J. Wu, J.B. Day, K. Franaszczuk, B. Montez, E. Oldfield, A. Wieckowski, P.A. Vuissoz, and J.-P. Ansermet, *J. Chem. Soc. Faraday Trans.* **93** (1997) 1017.

[12]Y.Y. Tong, P. Meriaudeau, A.J. Renouprez, and J.J. van der Klink, *J. Phys. Chem. B* **101** (1997) 10155.

[13]C. Belrose, Y.Y. Tong, E. Oldfield, and A. Wieckowski, *Electrochem. Acta* **43** (1998) 2825.

[14]P.A. Vuissoz, J.-P. Ansermet, and A. Wieckowski, *Electrochem. Acta* **44** (1998) 1397.

[15]Y.Y. Tong, C. Rice, N. Godbout, A. Wieckowski, and E. Oldfield, *J. Am. Chem. Soc.* **121** (1999) 2996.

[16]P.A. Vuissoz, J.-P. Ansermet, and A. Wieckowski, *Phys. Rev. Letters* **83** (1999) 2457.

[17]C. Rice, Y. Tong, E. Oldfield, A. Wieckowski, F. Hahn, F. Gloaguen, J.M. Leger, and C. Lamy, *J. Phys. Chem. B* **104** (2000) 5803.

[18]Y. Y. Tong, C. Rice, A. Wieckowski, and E. Oldfield, *J. Am. Chem. Soc.* **122** (2000) 1123.
[19]Y. Y. Tong, C. Rice, A. Wieckowski, and E. Oldfield, *J. Am.Chem. Soc.* **122** (2000) 11921.
[20]B. M. Rush, J. A. Reimer, and E. J. Cairns, *J. Electrochem. Soc.* **148** (2001) A137.
[21]K. S. Han and O. H. Han, *Electrochimica Acta* **47** (2001) 519.
[22]Y. Y. Tong, H.-S. Kim, P. K. Babu, P. Waszczuk, A. Wieckowski, and E. Oldfield, *J. Am. Chem. Soc.* **124** (2002) 468.
[23]Y. Y. Tong, P.K. Babu, A. Wieckowski, and E. Oldfield, *Chem. Phys. Lett.* **361** (2002) 181.
[24]P. K. Babu, H. S. Kim, A. Wieckowski, and E. Oldfield, (2002) In Preparation.
[25]Y. Y. Tong, A. Wieckowski, and E. Oldfield, *J. Phys. Chem. B* **106** (2002) 2434.
[26]Y. Y. Tong, E. Oldfield, and A. Wieckowski, in: *Encyclopedia of Surface & Colloid Science*, Ed. by A. Hubbard, Marcel Dekker, Inc., 2002, p. 3713.
[27]C. P. Slichter, *Principles of Magnetic Resonance*, 3rd ed., Springer-Verlag, 1992.
[28]S. L. Rudaz, J.-P. Ansermet, P. K. Wang, and C. P. Slichter, *Phys. Rev. Lett.* **54** (1985) 71.
[29]Y. Y. Tong, C. Belrose, A. Wieckowski, and E. Oldfield, *J. Am. Chem. Soc.* **119** (1997) 11709.
[30]M. Weinert and A. J. Freeman, *Phys. Rev. B* **26** (1983) 6262.
[31]H. T. Stokes, H. E. Rhodes, P. K. Wang, C. P. Slichter, and J. H. Sinfelt, *Phys. Rev. B* **26** (1982) 3575.
[32]Y.Y. Tong, D. Laub, G. Schultz-Ekloff, A. J. Renouprez, and J. J. van der Klink, *Phys. Rev. B* **52** (1994) 8407.
[33]I. Yu, A. V. Gibson, E. R. Hunt, and W. P. Halperin, *Phys. Rev. Lett.* **44** (1980) 348.
[34]H. E. Rhodes, P. K. Wang, H. T. Stokes, C. P. Slichter, and J. H. Sinfelt, *Phys. Rev. B* **26** (1982) 3559.
[35]J.J. van der Klink, J. Buttet, and M. Graetzel, *Phys. Rev. B* **29** (1984) 11038.
[36]J. P. Bucher, J. Buttet, J. J. van der Klink, and M. Graetzel, *Surf. Sci.* **214** (1989) 347.
[37]P. Waszczuk, A. Crown, S. Mitrovski, and A. Wieckowski, in: *Fuel Cell Handbook: Fuel Cell Electrocatalysis*, Ed. by A. Lamm, H. Gasteiger, W. Vielstich, Wiley-VCH, in press.
[38]J. Friedel, *Adv. Phys.* **3** (1954) 446.
[39]V. Heine, *Solid State Physics*, Academic Press, New York, 1980; Vol. 35.
[40]A. L. Allred, E. G. Rochow, *J. Inorg. Nucl. Chem.* **5** (1958) 264.
[41]J. E. Huheey, *Inorganic Chemistry: Principle of Structure and Reactivity*, Harper Collins College Publishers, New York, 1993.
[42]Y. Y. Tong, G. A. Martin, and J. J. van der Klink, *J. Phys. Condens. Matter* **6** (1994) L553.
[43]J. J. van der Klink, and H. B. Brom, *Progress in Nuclear Magnetic Resonance Spectroscopy* **36** (2000) 89.
[44]R. L. Kautz and B. B. Schwartz, *Phys. Rev. B* **26** (1976) 2017.
[45]J. P. Bucher and J. J. van der Klink, *Phys. Rev. B* **38** (1988) 11038.
[46]B. Hammer, Y. Morikawa, and J.K. Nørskov, *Phys. Rev. Lett.* **76** (1996) 2141–2144.
[47]D. K. Lambert, *Electrochim. Acta* **41** (1996) 623.
[48]G. Blyholder, *J. Phys. Chem.* **68** (1964) 2772.
[49]C. Lu, C. Rice, R. I. Masel, P. K. Babu, P. Waszczuk, H. S. Kim, E. Oldfield, and A. Wieckowski, *J. Phys. Chem. B.* (2002) in press.
[50]P. K. Babu, Y. Y. Tong, H. S. Kim, and A. Wieckowski, *J. Electroanal. Chem.* **524-525** (2002) 157.

[51]J. Andzelm and D. R. Salahub, *J. Quantum Chem.* **29** (1986) 1091.
[52]P. Waszczuk, J. Solla-Gullon, H. S. Kim, Y. Y. Tong, V. Montiel, A. Aldaz, and A. Wieckowski, *J. Catalysis* **203** (2001) 1.
[53]H. A. Gasteiger, N. Markovic, P. N. Ross, Jr., and E.J. Cairns, *J. Electrochem. Soc.* **141** (1994) 1795.
[54]M. T. M. Koper, J. J. Lukkien, A. P. J. Jansen, and R. A. van Santen, *J. Phys. Chem. B.* **103** (1999) 5522.
[55]H. Massong, H. S. Wang, G. Samjeske, and H. Baltruschat, *Electrochem. Acta* **46** (2000) 701.
[56]G.-Q. Lu, P. Waszczuk, and A. Wieckowski, *J. Electroanal. Chem.* (2002) In Press.
[57]K. A. Friedrich, K. P. Geyzers, U. Linke, U. Stimming, and J. Stumper, *J. Electroanal. Chem.* **402** (1996) 123.
[58]W. F. Lin, M. S. Zei, M. Eiswirth, G. Ertl, T. Iwasita, and W. Vielstich, *J. Phys. Chem. B* **103** (1999) 6968.
[59]T. Iwasita, F. C. Nart, and W. Vielstich, *Ber. Bunsenegs. Phys. Chem.* **94** (1990) 1030.
[60]A. Crown, H. Kim, G. Q. Lu, I. R. de Moraes, C. Rice, and A. Wieckowski, *J. New Mat. Electrochem. Systems* **3** (2000) 275.
[61]J. McBreen and S. Mukerjee, in: *Interfacial Electrochemistry: Theory, Experiment, and Applications*, Ed. by A. Wieckowski, Marcel Dekker, New York, 1999, p. 895.
[62]J.J. van der Klink, *J. Phys.- Cond. Matt.* **8** (1996) 1845.

2

Ab Initio Quantum-Chemical Calculations in Electrochemistry

Marc T. M. Koper

Schuit Institute of Catalysis, Laboratory for Inorganic Chemistry and Catalysis, Eindhoven University of Technology, 5600 MB Eindhoven, The Netherlands

I. INTRODUCTION

According to a famous statement by Paul Dirac,

> "...the underlying physics necessary for the mathematical theory of (...) the whole of chemistry are known, and the difficulty is only that the exact application of these (quantum-mechanical, MK) laws leads to equations much too difficult to be soluble."[1]

In spite of Dirac's pessimistic viewpoint on the applicability of laws of quantum mechanics to chemistry, the quantum-mechanical description of chemical bonds and reactions has been one of the most prominent and active areas of theoretical chemistry since the early days of quantum mechanics. As anticipated by Dirac, applying the laws of quantum mechanics to systems of chemical interest was frustrated by great computational difficulties for many years, with the exception perhaps of the simplest molecules. However, with recent developments both in conceptual quantum chemistry, *i.e.* the application of density

Modern Aspects of Electrochemistry, Number 36, edited by C. G. Vayenas, B. E. Conway, and Ralph E. White, Kluwer Academic / Plenum Publishers, New York, 2003

functional theory (DFT), and the computational power of modern-day computers, the tools of computational ab initio quantum chemistry have become indispensable in many of the traditional areas of chemistry. This is most clearly illustrated by the award of the 1998 Nobel Prize to two pioneers in computational quantum chemistry, John Pople[2] and Walter Kohn.[3]

Applying the methods of ab initio quantum chemistry to electrochemistry has a more recent history than their application to such fields as gas-phase chemistry or organic chemistry. This is undoubtedly related to the inherent complexity of the electrochemical interface. One of the main reasons for the recent upsurge in using ab initio quantum chemistry in modeling electrochemical interfaces is the degree of success that has been achieved in applying ab initio quantum-chemical methods to processes and reactions at metal-gas interfaces (for recent reviews in this area, see Refs.4-7). This has motivated many theoretically inclined electrochemists to use similar methods and ideas to model adsorption and reactions at electrified metal-liquid interfaces, and has also attracted theoreticians from the field of surface science to electrochemistry.

The purpose of this review is to present to the reader, typically an electrochemist with relatively little prior exposure to quantum chemistry, an overview of the different ab initio quantum-chemical methods currently available to the computational chemist. The largest part of the Chapter is devoted to the discussion of some selected applications of these different methods to electrochemical problems, focusing in particular on the type of physical insight that these methods may give that cannot be obtained by any other (experimental) method. It is exactly in the latter area that the power and added value of these methods lie. I will discuss only ab initio methods, and no other so-called semi-empirical methods of quantum chemistry. The initial applications in 1970s of computational quantum chemistry to electrochemistry made use of semi-empirical methods (for a relatively recent review, see the paper of Nazmutdinov and Shapnik[8]), but these methods have become somewhat outdated and are currently only used for very specific problems involving a very high number of atoms or molecules. In principle, this situation would apply also to electrochemistry, but there are still few examples of these large-scale semi-empirical calculations to electrochemistry. In any case, the main interest in applied quantum chemistry lies in the use of ab initio methods, in which the only parameters that enter the calculation are the atomic numbers of the constituting particles.

In the next section, the different aspects and methods of ab initio quantum chemistry will be discussed. This will include a brief and incomplete discussion of the basics of solving a quantum-chemical problem, *i. e.* an electronic structure calculation. The practical calculation may proceed using either wave-function-based methods or density functional theory-based methods, and my focus will be on the latter as at the present time these methods are by far the most popular, giving the best accuracy for large systems in a limited amount of computer time. Different methods of analyzing the outcome of an ab initio calculation will also be discussed briefly. Finally, a development will be discussed which has become very important for modeling processes in condensed phases, namely the combination of electronic structure calculations (which are usually static and apply to a temperature of 0 K) with molecular dynamics (which are dynamic and involve a finite temperature). These methods come under the name of ab initio molecular dynamics or Carr-Parrinello molecular dynamics, and in principle give access to all the finite-temperature information one might be interested in by combining the methods of quantum chemistry and statistical mechanics. These methods are *state-of-the-art* and are still restricted to very small system sizes, but hold great promise for the future.

After the brief introduction to the modern methods of ab initio quantum chemistry, we will discuss specific applications. First of all, we will discuss some general aspects of the adsorption of atoms and molecules on electrochemical surfaces, including a discussion of the two different types of geometrical models that may be used to study surfaces, *i. e.* clusters and slabs, and how to model the effect of the electrode potential in an ab initio calculation. As a first application, the adsorption of halogens and halides on metal surfaces, a problem very central to interfacial electrochemistry, will be dealt with, followed by a section on the ab initio quantum chemical description of the adsorption of a paradigmatic probe molecule in both interfacial electrochemistry and surface science, namely carbon monoxide. Next we will discuss in detail an issue uniquely specific to electrochemistry, namely the effect of the electric field, *i. e.* the variable electrode potential, on the adsorption energy and vibrational properties of chemisorbed atoms and molecules. The potential-dependent adsorption of carbon monoxide will be discussed in a separate section, as this is a much studied system both in experimental electrochemistry and ab initio quantum electrochemistry. The interaction of water and water dissociation products with metal surfaces will be the next topic of interest. Finally, as a last application we will discuss some recent results on the application of ab initio quantum chemistry and ab initio molecular dynamics to the modeling of electrochemical reactions.

As mentioned, the Chapter will deal uniquely with applications of ab initio quantum chemistry to electrochemistry. There are, of course, many other theoretical and computational methods available to the study of electrochemical problems, such as classical molecular dynamics, Monte Carlo methods, and the more traditional coarse-grained or continuum-type theoretical or computational approaches. Several recent reviews cover these techniques and the advances made in their application in the field of interfacial electrochemistry.[8-13]

II. AB INITIO QUANTUM CHEMISTRY

1. General Aspects of Quantum Chemistry and Electronic Structure Calculations

The aim of this section is to discuss briefly (and incompletely) the main ideas and concepts involved in carrying out ab initio quantum-chemical electronic structure calculations. Good introductions into the conceptual and practical aspects of quantum chemistry can be found in many textbooks. The author's favorite books are those by Lowe[14] and Jensen[15] for their clear and practical approach.

Electronic structure calculations as we will consider them here concentrate on solving the Schrödinger equation. This equation takes into account one of the main postulates of quantum mechanics, namely that of the wave-particle duality. All particles, in particular the very light ones such as the electrons, must be described in terms of a wave function, Ψ. The square of this wave function gives the probability of finding the particle at a given position. In the time-independent Schrödinger equation, which we will consider here, Ψ depends only on the location of all particles $\boldsymbol{r}$, and is solved from:

$$H(\boldsymbol{r})\Psi(\boldsymbol{r}) = E\Psi(\boldsymbol{r}) \tag{1}$$

where $H(\boldsymbol{r})$ is the Hamilton operator and E the total energy of the general N-particle system. The Hamilton operator consists of a kinetic (T) and a potential (V) energy part for all particles:

$$H = T + V \tag{2a}$$

$$T = \sum_{i=1}^{N} T_i = -\sum_{i=1}^{N} \frac{\hbar}{2m_i} \nabla_i^2 \tag{2b}$$

$$\nabla_i^2 = \left(\frac{\partial^2}{\partial x_i^2} + \frac{\partial^2}{\partial y_i^2} + \frac{\partial^2}{\partial z_i^2} \right) \tag{2c}$$

$$V = \sum_{i=1}^{N} \sum_{j>1}^{N} V_{ij} \tag{2d}$$

$$V_{ij} = \frac{q_i q_j}{r_{ij}} \tag{2e}$$

where i and j are indexing all the particles in the system, m_i is the mass of particle i, q_i the charge of particle i, and $\hbar$ is Planck's constant divided by 2π. Since the nuclei in the system are much heavier than the electrons, the Schrödinger equation can to a good approximation be replaced by an electronic Schrödinger equation describing the electronic wave function for a fixed nuclear geometry $\boldsymbol{R}$. This gives the electronic energy of the system. The classical potential energy for the nuclear geometry is added to obtain the total energy, as for most practical purposes the wave-like nature of particles other than electrons can be neglected, certainly when it concerns total energy calculations. This procedure to decouple electronic motion from nuclear motion is known as the Born-Oppenheimer approximation. The nuclei move on the potential energy surface obtained from solving the electronic Schrödinger equation for different nuclear geometries. This approximation is usually very good, and is used in all the results described in this Chapter.

Solving the electronic Schrödinger equation for a particular nuclear geometry is usually referred to as an electronic structure calculation. Calculating an accurate electronic energy is, however, a demanding task. Broadly, there are two different methods of computing the electronic energy: a method based on the accurate calculation of the electronic wave function, and a second method which is based on the accurate calculation of the electronic density rather than the electronic wave function. The former methods, based on more classical quantum-chemical concepts, will be referred to as wave-function-based methods, whereas the latter methods, which have been borrowed from solid-state physics, are known as density-functional theory methods. We will briefly touch upon the main principles of wave-function-based methods in the next subsection. We will spend more time explaining the density-functional-theory (DFT) methods as most contemporary quantum-chemical calculations involving large molecules and systems, including the vast majority of the ones to be described in Section III, are based on these methods.

2. Wave-Function-Based Methods

The key idea behind solving the electronic Schrödinger equation is to start with an initial guess for the total electronic wave function with some freely adjustable parameters, and vary these parameters until the lowest energy is found. The variational principle then states that this energy is the ground-state energy.

The total electronic wave function is approximated from single electron wave functions, known as orbitals. The simplest approximation would be to make the total electronic wave function a product of the orbitals, but since another postulate of quantum mechanics requires the electronic wave function to be antisymmetric, the next simplest approximation is a so-called Slater Determinant of orbitals. If a single Slater Determinant is used to calculate the trial wave function and the energy, this is known as the Hartree-Fock approximation. In essence, the Hartree-Fock approximation, sometimes also referred to as the Mean-Field approximation, implies that electron correlation is neglected, or equivalently, the Coulomb repulsion that an electron feels from the other electrons is treated on a smeared-out, averaged level. Since electron correlation is very important for calculating accurate energies, especially those of chemical bonds, several more accurate treatments have been developed. These are either based on perturbation methods (such as Møller-Plesset MP theory) or on approximating the electronic wave function by a sum of Slater Determinants representing different excited states, the so-called Configuration Interaction (CI) methods. However, the computational time involved with these methods scales poorly with the number of electrons in the system, N, making them unattractive for dealing with very large systems.

3. Density Functional Theory Methods

Density functional theory (DFT) methods have become a very popular alternative for wave-function-based methods, especially since DFT allows an approximate but reasonably accurate inclusion of electron correlation at relatively low computational cost. Hohenberg and Kohn laid the foundation for DFT in 1964,[16] showing that the ground-state electronic energy is completely and uniquely determined by the electron density $\rho(r)$. In other words, there is one-to-one correspondence between the electron density of the system and its energy. This is of potential great significance computationally, as the electron density does not depend on the number of electrons in the system, whereas with wave-function methods the computational demands will always scale with the number of electrons. However, the "recipe" on how to calculate the electronic energy from the electron density is not known. Fortunately, some very significant advances have made in the last 10-20 years in designing "functionals" which connect electron density to electronic energy.

It turns out to be useful to separate the energy functional into three different parts, which can be recognized from the Schrödinger equation: a kinetic energy part $T[\rho]$, the attraction between nuclei and electrons $E_{ne}[\rho]$, and the electron-electron repulsion $E_{ee}[\rho]$ (the nuclear-nuclear repulsion adds a constant energy term to the total energy in the Born-Oppenheimer approximation). The electron-electron repulsion energy is further divided into two parts: the classical electron-electron repulsion term $J[\rho]$, *i. e.*

$$J[\rho]=\frac{1}{2}\int\int\frac{\rho(r)\rho(r')}{|r-r'|}dr\,dr' \tag{3}$$

and a "quantum-mechanical" interaction energy known as the exchange-correlation energy $E_{xc}[\rho]$. We will return to the more exact definition of this energy term below.

In order to make DFT a practical and accurate approach, Kohn and Sham[17] reintroduced the use of orbitals in computational DFT. The reason for this is twofold: most importantly, the kinetic energy term can be calculated exactly from the orbital approach (but will clearly only give the right number if the orbitals themselves are exact), and the electron density can be simply obtained by summing the square of the orbital wave functions:

$$\rho(\boldsymbol{r}) = \sum_{i=1}^{N} |\phi_i(\boldsymbol{r})|^2 \tag{4}$$

To this end, Kohn and Sham assumed that the electrons in these so-called Kohn-Sham orbitals are non-interacting, such that the total electronic wave function can be written as a Slater Determinant. This allows the kinetic energy functional to be split into two parts, one of which, T_S, can be evaluated exactly, in a fashion very similar to the way it is done in Hartree-Fock theory, and a small correction term, which is formally absorbed in the exchange-correlation energy term. Thus, a general DFT energy expression can be written as

$$E_{DTF}[\rho] = T_S[\rho] + E_{ne}[\rho] + J[\rho] + E_{xc}[\rho] \tag{5}$$

If E_{DFT} is the exact energy, this equation may be regarded as the definition of the exchange-correlation energy.

The exact meaning of the exchange-correlation energy is a difficult one, partly because the DFT definitions of exchange and correlation are not exactly the same as those used in wave-function methods. As mentioned in the previous section, electron correlation arises from the correlated behavior between electrons that is not accounted for in the mean-field Hartree-Fock approach. The exchange energy is the total electron-electron repulsion minus the Coulomb repulsion, and is basically a consequence of the Pauli principle, which states that no two electrons can have the same quantum numbers, *i. e.* two electrons in the same orbital must have opposite spin.

The great advantage of DFT over wave-function-based methods is that, if the exact exchange-correlation functional is known, or some good approximation to it, it has the potential of including the computationally difficult part in wave-function methods, the correlation energy, at a computational cost similar to that of a Hartree-Fock calculation, which does not include any electronic correlation at all. Hence, the "holy grail" of DFT is to derive or find a suitable functional for the exchange-correlation energy. Unfortunately, the exact exchange-correlation functional is not known, nor is it possible to derive it systematically. However, over the years some very good approximations have been developed, which have proved to be very useful and quite accurate in predicting binding energies. The most popular family of functionals is derived in the so-called Generalized Gradient Approximation (GGA). These functionals include not only the local electronic density as most of older DFT calculations (referred to as DFT-LDA, LDAstanding for "Local Density Approximation"), but also its derivatives. In this way, not only short-range electronic correlation is included in the functional, but also some longer-range electronic correlation effects, which is of importance in accurately describing chemical bonds. However, very long-range electronic correlation effects, such Van der Waals forces, are not treated well by the DFT-GGA methods. There are several DFT-GGA functionals, carrying acronyms such BP86, BLYP, B3LYP, PW91, or PBE. Their exact nature is not interest here, and is discussed extensively in the literature.[15,18-20] Normally there are no great differences in the results obtained with these different functionals, although certain functionals give better results or are more popular for certain applications. B3LYP is very popular in computational organic chemistry,[15] whereas PW91 or PBE are normally preferred when a metal-adsorbate system is studied.[20] For liquid water, BLYP is the preferred functional,[21] although PW91 gives very similar results.[22] This makes PW91 suitable for metal-water interfaces. Currently, the accuracy of DFT-GGA methods in predicting bond energies may be roughly estimated as *ca.* 10 %, or 10-20 kJ/mol, depending somewhat on the total strength of the bond. This is quite acceptable from the quantitative point-of-view, and very good for qualitative purposes, such as comparing systems or studying trends as specific features of the system are varied.

Since there is an element of arbitrariness in carrying out a DFT calculation, namely in the choice of functional, one may argue that DFT is not a truly *ab initio* method. However, DFT is based on an exact formalism, and once a functional has been chosen, all that enters a calculation are the atomic numbers of the atoms in the system (apart from the choice of basis set and effective potentials, see next section). Therefore, in this respect DFT methods are certainly *ab initio*. Nevertheless, in spite of the impressive results that can be obtained with DFT-GGA methods, a major drawback of DFT is that there is no systematic way to improve the calculation such that the exact result is approached, something which is formally (though usually not practically) possible in wave-function-based methods.

4. Basis Sets and Effective Potentials

Apart from the choice of method or DFT functional, a practical electronic structure calculation also involves choosing a set of functions in which the electronic orbitals are expanded, the so-called basis set. Moreover, certain atoms in the system under consideration may contain many electrons, of which the ones with the lowest energy do not make any significant contribution to the bond energy. These atoms and their core electrons are usually replaced by effective potentials that describe the interaction of valence electrons with the nuclear core and the low-energy electrons. The low-energy core electrons themselves are then no longer included in the calculation, thereby saving computation time.

The choice of basis set is a rather technical issue that we do not want to pursue here. However, there is a fundamental difference in the type of basis set for two different types of geometric models used in the literature. The so-called cluster model uses a cluster of metal atoms (typically 5-50 atoms) as a clip-out of the metal surface, and studies the interaction of a certain adsorbate with that cluster. These calculations are usually carried out with standard quantum-chemical packages using localized atomic orbitals as a basis set. These localized atomic orbitals are functions with their origin on the atoms in the cluster, and the quality of the basis set depends on the number of functions used for expanding the wave function. By contrast, the so-called slab model uses a periodic box (or "supercell") in which layers of atoms in part of the box represent the surface, whereas the rest of the box is left empty ("the vacuum"). The box has periodic boundary conditions, such that an (semi-infinite) system is created which serves as a model for an extended surface. The basis set must also satisfy these periodic boundary conditions, and the simplest type of function for this purpose are (sinusoidal) plane waves.[23] This basis set is delocalized, although electrons can be localized by expanding the wave functions in a large number of plane waves. The quality of the plane-wave basis set is determined by the highest frequency plane wave, usually referred to as the energy cut-off.

Effective potentials also depend on the type of basis set used. In atomic orbital calculations, they are sometimes referred to as frozen-core potentials. In most cases, only the highest-energy *s*, *p* and *d* electrons are included in the calculation. In plane-wave calculations, effective potentials are known as pseudopotentials[23] They come in different varieties; *soft* or *ultrasoft* pseudopotentials need only a relatively low energy cut-off as they involve a larger atomic core.[23,24]

5. Structure, Energetics, and Vibrational Frequencies

The basic output of an electronic structure calculation is the electronic and the total energy of the system for a certain nuclear geometry. When the energy is minimized as a function of a number of internuclear distances or bond angles, this is referred to as a (constrained) geometry optimization. This will normally give the equilibrium structure under the restraints of the geometrical model.

The bond energy between two fragments, for instance between a hydrogen atom and a metal surface, is obtained by subtracting the energy of the individual fragments, *i. e.* the H atom in vacuum and the clean metal surface, from the energy of the chemisorbed H on the metal surface after the appropriate geometry optimization. By calculating the total energy for a number of geometries close to the equilibrium geometry, a potential energy surface (PES) can be obtained. By fitting this PES to polynomials, one can compute the vibrational characteristics of the system. These latter quantities are of great interest in comparison to experiment, as binding energies are difficult to obtain in electrochemical experiments.

6. Methods of Analysis

Apart from calculating the (expectation value of the) energy, there are other physically interesting quantities that can be obtained from an electronic structure calculation. Sometimes these quantities give more insight into the particular problem at hand, or help in understanding trends or tracing the origin of a certain effect.

One such quantity is the charge on a certain atom or fragment. Sometimes it is useful to know where charge has flown in the formation of a bond, in order to characterize if a bond could be termed ionic or covalent. However, there is a certain amount of arbitrariness in how to divide electronic charge between atoms. A well-known method by Mulliken suggests to sum up the contributions from all atomic orbitals located on the atom, where the contributions from orbitals involving basis functions on different atoms are equally divided. However, quantum chemists currently consider this method to suffer from several drawbacks. For instance, the Mulliken charge is known to be rather sensitive to the choice of basis set.[14] An alternative method suggested by Hirshfeld estimates the charge as a weighted spatial integral of the self-consistent field charge density associated with the atom.[25] This method has been advocated by some quantum chemists as being more robust and realistic.[26] Another related analysis makes use of projecting the entire wave function onto a certain orbital, such that an estimate is obtained of the amount of charge in that orbital. These projection techniques are obviously useful when multiple orbitals are involved in the formation of a bond.

In analyzing the charge distribution or charge flow in a molecule, the often preferred, since less arbitrary, method is to calculate the molecule's dipole characteristics. As we will illustrate in some detail below, analyzing a system's static and dynamic dipole holds a great deal of useful information on the polarity of a bond that is less ambiguous than the information obtained from a charge analysis. Evidently, the most complete picture is often obtained by combining both methods of analysis.

Another insightful method of analysis is energy decomposition. Every (bond) energy formally consists of multiple contributions ascribable to different physical or chemical mechanisms. In many quantum chemistry texts bonds are usually analyzed in terms of the steric Pauli repulsion between two fragments, and the bonding, non-bonding, or anti-bonding contributions due to the interaction between the different orbitals on the fragments. Bagus *et al.*[27] introduced an energy decomposition scheme applicable to SCF-HF calculations called the Constrained Space Orbital Variation (CSOV) method. By allowing different subsets of orbital space to vary whereas others are kept fixed, the CSOV analysis makes it possible to distinguish the consequences of intraunit charge polarization and interunit covalent and ionic bonding. A conceptually similar energy decomposition analysis, which is incorporated in a DFT code called ADF, subdivides the total energy into two main contributions, *i. e.*, the steric and orbital bonding energies.[28] The steric part of the bonding energy is commonly defined as the energy difference between the separate fragments and the composite system, where the latter is described by the determinantal wave function, which is the antisymmetrized product of overlapping fragment orbitals. No electron relaxation due to the formation of bonds is accounted for, only the rise in kinetic energy due to the orthogonalization of the fragment orbitals. The steric repulsion consists of an attractive electrostatic part and a repulsive Pauli contribution, and since the latter dominates, the steric repulsion energy is a positive quantity. The difference between the total bonding energy and the steric repulsion is due to orbital interactions, and these can be subdivided into the contributions of orbitals with different symmetry. Significantly, the energy decomposition analysis is not limited to decomposing energies, but can also be used to decompose potential energy surfaces. In this way, changes in vibrational frequencies can be analyzed in a very similar fashion, and be decomposed into the various contributions related to steric and orbital effects.

A final method of analysis to be mentioned is the so-called density of states, or projected density of states. This is basically the solid-state-equivalent to the energy level diagram. Since in a solid-state system there are infinite energy levels, energy states are characterized by their density per unit of energy, and a plot of the density as a function of energy is a so-called density-of-states diagram. A projected density-of-states projects a particular state or orbital that one is interested in onto all the other states in the system, to study the interaction of the orbital with the metal states in terms of orbital broadening, orbital shift, or orbital splitting. A nice book by Hoffmann gives a full account of the usefulness of this method in studying bonding at surfaces.[29]

7. Ab Initio Molecular Dynamics

A very prominent development in DFT has been the coupling of electronic structure calculations (which, when the ground state is concerned, apply to zero temperature) with finite-temperature molecular dynamics simulations. Carr and Parrinello published the founding paper in this field in 1985.[30] Carr and Parrinello formulated effective equations of motion for the electrons to be solved simultaneously with the classical equations of motion for the ions. The forces on the ions are calculated from first principles by use of the Hellman-Feynman theorem. An alternative to the Carr-Parrinello method is to solve the electronic structure self-consistently at every ionic time step. Both methods are referred to as *ab initio* Molecular Dynamics (AIMD).

AIMD is still a very time-consuming simulation method, and has so far mainly been used to study the structure and dynamics of bulk water,[31] as well as proton transfer[32] and simple S_N2 reactions in bulk water.[33] AIMD simulations are as yet limited to small system sizes and "real" simulation times of not more than a few picoseconds. However, some first applications of this technique to interfacial systems of interest to electrochemistry have appeared. There is no doubt that AIMD simulations of electrochemical interfaces will become increasingly important in the future, and some interesting first results will be described in Sections III.7 and III.8.

III. SELECTED APPLICATIONS

1. Clusters and Slabs

Since extended surfaces are large systems from the atomistic point of view, the question arises what kind of molecular ensemble would constitute a reasonable model for the adsorbate-metal interaction. Here, essentially two different types of approaches have been taken in the literature. One approach is to model the surface as a small cluster of atoms, a typical cluster size being 10-50 atoms. It is important to note that this size is still smaller than that of a typical catalytic nanoparticle, which consists of at least several hundreds of atoms (and usually more).[34,35] A second approach that has become popular in the last ten years is to model the surface as a periodic slab, in which a certain ensemble is periodically repeated in three dimensions by applying periodic boundary conditions in the computational setup (see Figure 1). A surface, and a molecule-surface ensemble, is then modeled as a number of layers (typically 3-5 layers) of atoms in one part of the periodic cell, and a vacuum in the other part of the cell. The vacuum region is chosen large enough such that the different periodic images of the surface have a negligible interaction with each other.

The advantages and disadvantages of using clusters or slabs are well established.[4-7] The main advantages of using clusters lie in the relatively low computational cost and the fact that "classical" quantum-chemical techniques can be used in analyzing the chemisorption bond because of the use of localized atomic orbitals. The main disadvantages of cluster models are edge and finite-size effects, leading to an electronic charge distribution quite different from extended metal surfaces. Also, the metal's band structure is not represented well and charge-screening effects such as image forces of ions and dipoles near conducting surfaces are not treated adequately. The use of large clusters may overcome some of these problems, at the cost of increased computation time. Slab models are then a better alternative, provided the slab is thick enough (usually at least 3 or 4 layers for well-converged binding energies, but much thicker for bulk quantities like the work function or a fully converged band structure) and the vacuum region chosen large enough so that the interaction between slabs can be neglected. It is now generally agreed upon that slab calculations give more reliable quantitative results when the binding energetics of the molecule-surface interaction is concerned. More local information,

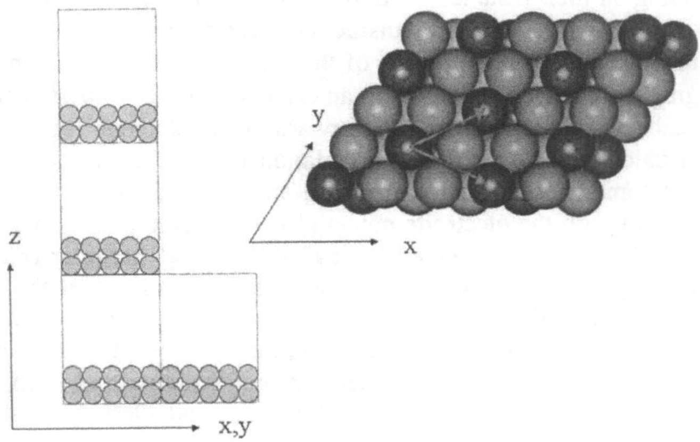

Figure 1. Geometrical arrangement of the slab model. The periodic supercell is replicated into three directions; one side of the contains the metal slab, the other side the vacuum or water phase. The top view of the surface shows a typical arrangement for a √3x√3 unit cell, in this case representing an alloy A_2B.

such binding distances and vibrational properties, may still be calculated with good accuracy using cluster models.

2. How to Model the Electrode Potential

A critical issue in the ab initio description of electrochemical interfaces is the correct incorporation, or the realistic modeling, of the effect of the electrode potential. Essentially two different approaches have been employed in the literature, which, in statistical-mechanical terms, may be referred to as "canonical" and "grand canonical".[36]

In the canonical or "constant number of electrons" mode, an isolated cluster or slab is placed in an external electric field. At the level of the Schrödinger equation, the electric field is simply added as an extra term to the Hamiltonian. As a result of the external field, the cluster/slab becomes polarized, with a positive surface charge on one ter/slab, the high electronic polarizability of the metal will screen the electric field, leading to a constant (but ill-defined) electrochemical potential. Outside the cluster/slab, electronic energy levels will shift depending on their distance from the cluster/slab, and depending on the polarizability of the medium outside the cluster/slab (which is zero for vacuum). Changing the strength of the electric field is then similar to changing the electrode potential in an electrochemistry experiment. The approach is simple and straightforward to implement (certainly in cluster calculations), but the direct relation with the electrode potential remains somewhat elusive. A practical and empirical point-of-view to the relation with the electrode potential is to assume that the field-free calculation corresponds to the potential of zero free charge or zero total charge, depending on whether adsorbates are present on the metal surface or not, and to connect the applied external field to changes in the electrode potential by assuming some effective thickness of the double layer (say, 3-4 Å).[37] The canonical or "external field" approach has been used in the vast majority of ab initio calculations of electrochemical surfaces.

From the conceptual point-of-view, the grand-canonical or "constant electrochemical potential" mode is preferable and in many cases closer to the actual experimental situation. However, its practical implementation is less straightforward, even though the grand canonical DFT formalism is well established. Several authors have addressed the problem of carrying out ab initio simulations at constant electrochemical potential,[36,38-40] and we will briefly discuss here the general treatment given recently by Lozovoi *et al.*[36] In order to allow for a constant chemical potential, the density functional theory has to be extended to finite temperatures in the sense that the Helmholtz free energy functional is minimized:

$$F[\rho] = \Omega[\rho] + \mu N_e + E_{ii} \tag{6}$$

where μ is the chemical potential, N_e the total number of electrons (which does not have to be an integer), E_{ii} the energy of the ion-ion interaction, and $\Omega[\rho]$ the grand canonical density functional defined at a finite temperature T. Without going into the detailed equations, the calculation at constant μ now proceeds by calculating the electron density at each self-consistent iteration as a sum of partial densities over Kohn-Sham orbitals with eigenvalues up to a given μ, thus rendering ρ and N_e dependent on μ. In this type of simulation, the total number of electrons is not zero and may even not be an integer. In order to avoid an artificial electrostatic interaction between the different charged supercells, a compensating charge correction has to be made, the energy of which is substracted afterwards. Note that this approach amounts to having a non-zero charge on the slab and placing the exact countercharge somewhere else in the cell, but the amount of charge is dictated by the fixed chemical potential, rather than vice versa. In the grand canonical mode, it is in principle possible to have a well-defined electrode potential by referring the chemical potential of the slab to a plane (or "reference electrode") at a finite distance from the slab where the effective electric field vanishes (this obviously depends on the particular system and geometry considered).[36] The grand canonical DFT formalism still has to be tested extensively for electrochemical systems, but it looks like a very promising method to incorporate the electrode potential in ab initio calculations in a realistic way.

3. Chemisorption of Halogens and Halides

The first applications of ab initio quantum-chemical calculations to systems of electrochemical interest were concerned with the adsorption of halides at metal surfaces. The adsorption of halides is of great experimental and practical importance as many electrolyte solutions contain halide anions, which tend to adsorb specifically at the metal-water interface, especially at more positive electrode potentials. Issues of interest in halide chemisorption are the nature of chemical bond with the surface (*i. e.*, covalent or ionic), the strength of the interaction of the different halides on different substrates, the preferred binding geometry on single-crystal surfaces, the effect of the electrode potential, and the importance of including the solvent (water) in correctly modeling the properties of the chemisorption bond.

Bagus *et al.*[41] were the first to give a detailed ab initio description of the halogen-silver bond using Hartree-Fock wave functions. The silver surface was modeled as a four- or ten-atom cluster exposing the (111) plane (see Figure 2), and the total system, *i. e.* metal cluster plus halogen, was charge neutral. The halogen was adsorbed in the threefold hollow site. As the most important result, Bagus *et al.* clearly established that the halogen bond to the silver surface is ionic. Evidence for the bond ionicity in the silver-halogen interaction was obtained by four different methods.

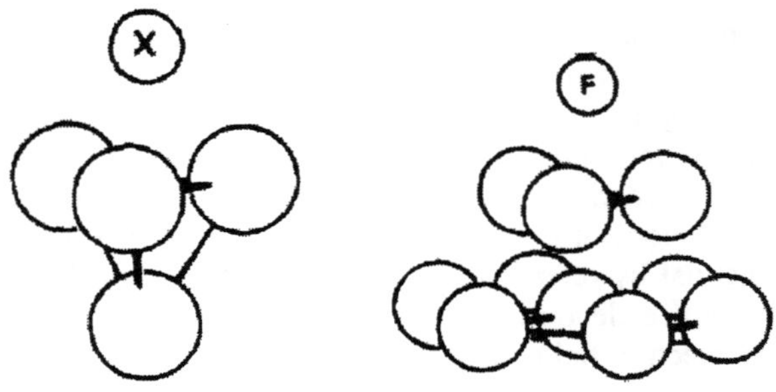

Figure 2. Schematic representation of the Ag_4-X; X = F, Cl, Br, and of the Ag_{10}-F clusters modeling the hollow site on the Ag(111) surface. The halogen atom is above the center of the triangle formed by the three Ag surface atoms. Reprinted with permission from P. S. Bagus *et al.*, *J. Chem. Phys.* **90(8)**, (1989) 7287. Copyright 1989, American Institute of Physics.

First of all, the extent to which the halogen orbitals are occupied in the many-electron wave function of the total system, was calculated by projection techniques. In all cases, *i. e.* Ag_4-F, Ag_{10}-F, Ag_4-Br, and Ag_4-Cl, the charge on the halogen was calculated to be very close to −1, suggesting that the chemisorbed halogen is essentially a halide.

In a second method of analysis, Bagus *et al.* computed so-called dipole moment curves of the silver-halogen bond. This curve plots the dipole moment of the total system $\mu(Ag_n\text{-}X;z)$ as a function of the distance z between metal surface and halogen. Such curves can be Taylor expanded about the equilibrium binding distance $z = z_e$:

$$\mu(Ag_n - X; z) = M_0 + M_1(z - z_e) + M_2(z - z_e)^2 + M_3(z - z_e)^3 + \ldots \tag{7}$$

where the first term M_0 is $\mu(Ag_n\text{-}X;z_e)$. In principle, an estimate of the charge on the halogen can be obtained from the latter quantity by

$$q_X \approx [M_1 - \mu(Ag_n)]/z_e \tag{8}$$

However, it turns out that this quantity is a rather poor indicator of the bond ionicity, as the system hardly behaves like a classical dipole

However, it turns out that this quantity is a rather poor indicator of the bond ionicity, as the system hardly behaves like a classical dipole consisting of two point charges, which is also illustrated by the fact that M_0 depends strongly on cluster size. A much better indicator of the bond polarity is contained in the quantities M_1, M_2, and M_3. For all systems studied, $M_1 \approx -1.4$ to -1.6, and M_2 and M_3 are small, and furthermore the cluster size dependence of these quantities is much less than that of M_0. (This points to a general property of cluster calculations already alluded to in Section III.1: whereas absolute quantities may depend sensitively on cluster size, trends and slopes of trend curves show a much weaker cluster size dependence.) This linear behavior is consistent with an ionic bond where the slope, M_1, is related to the extent of ionicity. The dipole moment of two point charges $+q$ and $-q$ will be $\mu = -q \times r$ and $d\mu/dr = M_1 = -q$. Hence for an ideal fully ionic halide bond, the curve is expected to be a straight line with slope $M_1 = -1$.

The fact $|M_1| > 1$ in the calculations is explained by the fact that the presence of the (halide) ion in front of the cluster leads to a redistribution of charge within the metal cluster, leading to an effective image charge inside the metal so as to screen the field of the ion in the metal. In image charge theory, $q = M_1/2$, giving a qualitative explanation for the large $|M_1|$. One way to account for this effect is to compare the computed dipole moment curves and M_1 of the halogens to the dipole moment curve and M_1 of an ideal point charge with $q = -1$ in front of the same metal cluster with charge +1 (for an example, see Figure 3). Bagus *et al.* have suggested another way, which essentially amounts to looking only at the charge in the cluster orbitals that are mostly located on the halogen X. The slopes of the corresponding dipole moment curves are indeed close to –1 for fluoride, and *ca.* –1.1 for bromide and chloride. The latter deviation may be ascribed to the larger polarizability of bromide and chloride compared to fluoride.

A third method of analysis used by Bagus *et al.* consisted in separating the different bonding contributions to the interaction energy by the CSOV method mentioned in section II.6. It was found that the electrostatic interaction and the cluster polarization alone account for about 80-90 % of the whole interaction energy. This again points towards an ionic bond between the halogen and the metal cluster. The main difference between F, Cl and Br is that, because of the larger equilibrium bond distances, the electrostatic interaction and the Ag

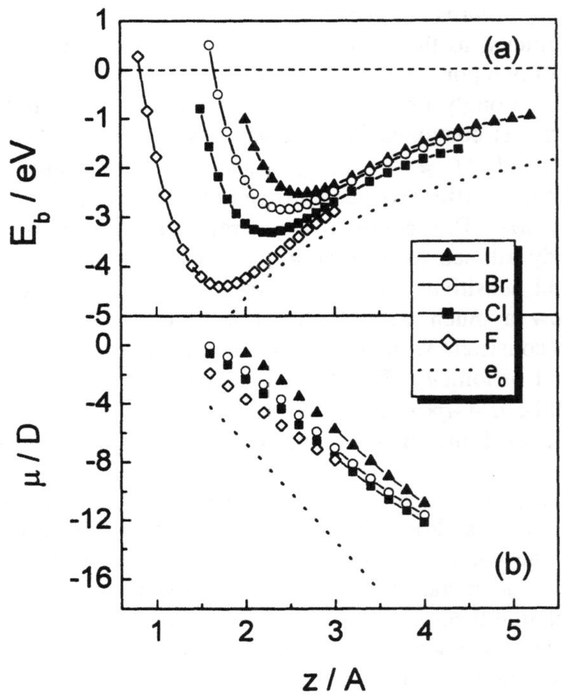

Figure 3. (a) Potential energy surface of the interaction of I, Br, Cl, and F with a $Ag_{19}(111)$ cluster; dashed line is the interaction of a negative point charge interacting with the $Ag_{19}^{+}(111)$ cluster (b) Corresponding dipole moment curves.

expected on the basis of the larger dipole polarizabilities of these anions.

Finally, as a last method of analysis, Bagus *et al.* studied the effect of the electric field on the equilibrium bond distances. With a negative applied field, the halogens are pushed away from the surface, as expected for a negatively charged adsorbate. With a positive field, the ions are pulled towards the surface but the equilibrium distances vary to a smaller extent as compared to the negative field case because of the increasing steric Pauli repulsion ("wall effect") occurring when the ion approaches the metal surface. Electric field effects on ion and atom chemisorption will be dealt with in more detail in Section III.5.

Several studies following the classical paper by Bagus *et al.* came to essentially the same conclusions. Illas *et al.*[42,43] studied halogen and halide chemisorption on mercury clusters using Hartree-Fock theory and concluded that the F, Cl, Br and I bond to mercury is mainly ionic, even for electronic states that formally separate into a neutral metal cluster and a neutral halogen. Bagus and Pacchioni[44] looked in more detail at the electric field effect on the halogen-silver bond, including field-dependent binding energies and Ag_n-X vibrational frequencies. They found that the Ag_n-X vibrational frequency has a positive Stark tuning slope, *i. e.* increases with more positive field, which they ascribe primarily to the wall effect. The vibrational frequency increases as the adsorbed atom approaches the surface (positive field) and decreases as it moves away from the surface (negative field). As mentioned, we will discuss a more general treatment of field effects in Section III.5.

The first DFT-GGA study of halide adsorption on metal clusters was published by Ignaczak and Gomes.[45] Especially the binding energies obtained from these calculations should be more reliable than those obtained using the self-consistent Hartree-Fock method. Ignaczak and Gomes consistently find that the binding to the (100) plane of 12-atom copper, silver and gold clusters gets weaker as the size of the halide increases, *i. e.* from fluoride to iodide. This trend is the opposite to that observed in electrochemical experiments, which is ascribed to the absence of water in the calculations. From the Mulliken charges they find that there is significant charge transfer from the halide to the metal cluster, the fluoride retaining most of its negative charge (q_F = -0.5 to –0.6), and the iodide retaining least (q_I = -0.3 to –0.4). No further analysis of the ionicity of the bond was carried out. The vibrational frequencies decrease with increasing mass of the halide, as expected and observed experimentally, and are higher on gold than on silver, which also agrees with experiment. This may be related to the higher adsorption energy of the halides to gold, although one should in general be careful with assuming relations between binding energy and vibrational frequency.

An extensive analysis of the halogen-metal bond ionicity using DFT-GGA calculations was performed by Koper and van Santen.[46] Figure 3 shows the binding energy and dipole moment curves for the four halogens interacting with the hollow site of the (111) plane of a Ag_{19} cluster. The dashed line in both plots gives the same curve for a negative point charge interacting with a Ag_{19}^+ cluster. The same binding energy trend as reported by Ignaczak and Gomes[45] is observed: the binding energy gets more negative going from iodine to fluorine. Interestingly, this seems to be mainly due to the decreasing size of the halogens in that order, as further away from the surface the interaction energies of all four halogens are very similar. The dipole moment curves suggest a predominantly ionic bond for all halogens, although the slope of the curve is not as steep as for the point charge. Interestingly, both a Mulliken and Hirshfeld charge analysis suggest that fluorine is the most negatively charged halogen, and iodine the least negatively charged, in agreement with earlier DFT results of Ignaczak and Gomes, but this degree in bond ionicity is not reproduced by the dipole moment analysis, where the effect is much smaller and the trend is in fact reversed.

At this point, it may be good to comment on the two different approaches that have been taken in modeling the halogen surface bond by cluster calculations. Some authors have considered a totally charge-neutral system, which may be referred to as halogen adsorption, whereas others have considered the adsorption of a X^- halide, making a cluster with a total charge of −1, which may be referred to as halide adsorption. The binding energies from these different calculations are not the same, although the trends are often very similar. Some authors have expressed a preference for doing calculations with an overall charge-neutral system, as this situation corresponds to an uncharged surface (which may be linked to the electrochemical potential scale by the potential of zero (total) charge), since it does not depend on the ability of the cluster to delocalize the (net) charge, and since the adsorbed states for halides and halogens should be identical in the limit of an infinitely large cluster, regardless of the dissociation limit. The latter effect was indeed confirmed by Sellers *et al.*[47,48] and Koper and van Santen[46] for halogen and halide adsorption on clusters of increasing size, for which the differences in equilibrium bond distances and adsorbate charge tend to vanish with larger cluster size. Sellers *et al.*[46] have also shown that it is possible to convert the single atom (halogen) binding energies for the corresponding ion (halide) by a simple thermodynamic cycle. For an anion X and a metal surface M, the cycle is

$$X^-(g) \rightarrow X(g) + e^- \qquad \Delta E = EA(X) \tag{9a}$$

$$X(g) + M \rightarrow X-M \qquad \Delta E = \Delta E_b(X) \tag{9b}$$

$$\mathrm{X-M}+e^- \rightarrow (\mathrm{X-M})^- \qquad \Delta \mathrm{E} = -\Phi_M \tag{9c}$$

so that

$$\Delta \mathrm{E_b}(\mathrm{X}^-) = \Delta \mathrm{E_b}(\mathrm{X}) + \mathrm{EA}(\mathrm{X}) - \Phi_\mathrm{M} \tag{10}$$

with EA(X) the halogen's electron affinity and Φ_M the metal work function, both defined as positive quantities. The bare metal work function appears in this formula because it is assumed that the presence of one adatom does not alter the work function.

In most of the cluster calculations of halogen adsorption, it is found that the preferred adsorption geometry is a multifold coordination site, *i. e.* threefold hollow on (111) surfaces and fourfold hollow on (100) surfaces.[45] It seems that one may interpret this as another effect of the strong ionicity of the halogen-metal surface bond, which will drive the adsorbate to the surface site where it will be closest to the metal, so as to optimize the electrostatic interaction energy. Recent slab calculations for Cl adsorption on Ag(111) also predict both fcc and hcp hollow sites to be energetically favored over bridge and atop, and have shown that the bond should be considered ionic.[49] Experimentally, an interesting deviation from this rule is observed for Au(100), where apparently the bridge site is the preferred geometry.[50] The cluster calculations by Ignaczak and Gomes[45] were not able to reproduce this effect. For such rather subtle binding energy effects, slab calculations are needed, as recently presented by Wang and Rikvold for Br adsorption on Ag(100) and Au(100).[51] These authors indeed found that on Ag(100), the hollow site is preferred, whereas on Au(100), the bridge site is the lowest in energy. From an analysis of their results, they conclude that this change in binding site is due to steric effects. The more diffuse Au core orbitals push Br out off the hollow site to the bridge site, where the steric repulsion is less.

Mitchell and Koper[52] have recently analyzed the polarity of the halogen-metal bond by DFT-GGA slab calculations of Br on Au(100) at 0.25 ML coverage. Figure 4 shows the so-called charge transfer function, defined as the z-dependent charge density of the halogen-substrate system minus that of the clean substrate and the isolated halogen, for Br at the atop, bridge, and hollow sites on Au(100). The charge transfer function gives an indication of the amount of electronic charge that has flown from the metal to the halogen after the bond has

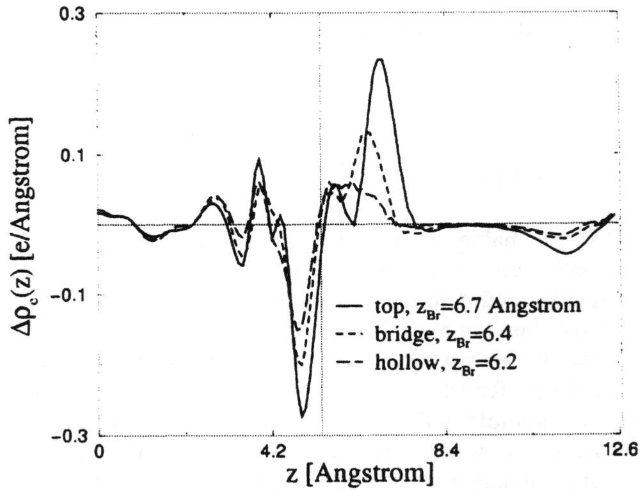

Figure 4. Charge density transfer function for Br at the atop, bridge, and hollow site on Au(100). The last layer of metal atoms is centered about the vertical line at ca. 5.4 Å.

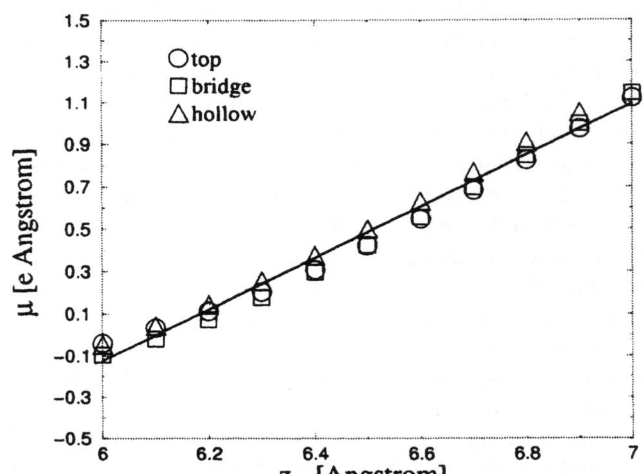

Figure 5. Dipole moment curves for Br at the atop, bridge and hollow site of the Au(100) surface. The distance z is measured from the middle of the slab. The last layer is centered about 5.4 Å.

formed. It is observed that the amount of charge transferred is mainly a function of the distance of the halogen from surface, and therefore increases in the order hollow, bridge to atop, as can be seen in Figure 4. The static dipole is also higher the further away the halogen is from the surface, but the dipole moment curves given in Figure 5 show that the dynamic dipole M_l is essentially the same in all three adsorption sites, implying that the polarity of the bond when analyzed in terms of dynamic dipole moment does not depend on its coordination. The slope of the dynamic dipole moment curve is approximately 1.2, slightly lower than values obtained earlier with cluster calculations of Br on Ag(111).[41]

In all of the above mentioned studies, the effect of the solvent on halogen adsorption was not considered. When a comparison to electrochemical interfaces is envisaged, this is obviously an enormous disadvantage. As already noted, the binding energy trend among the halides is opposite to that observed in an electrochemical cell, which is undoubtedly related to the absence of solvent molecules in the calculations. Also, some authors have used potential energy curves such as those presented in Figure 3 as model potentials for the halide-surface interaction in molecular dynamics simulations of halide adsorption at metal-water interfaces.[53-56] Even though a good alternative to this approach may not be readily available, the real interaction potential is expected to be substantially different in the presence of water, making this a somewhat questionable approach.

Kairys and Head[57] have as yet been the only ones to study the co-adsorption of water molecules and a halogen on the Al(111) surface (although there has been some similar older semi-empirical work by Lorenz and co-workers.[58,59]) For this purpose, they compared fluoride and iodide adsorption on an $Al_{19}(111)$ surface in the absence and in the presence of 1 or 3 water molecules. In agreement with earlier studies, in the absence of water fluoride has a stronger bond to the Al cluster than iodide, the difference in adsorption energy being *ca.* 290 kJ/mol in favor of fluoride. When the fluoride and iodide are co-adsorbed with 1 water molecule on the negatively-charged Al cluster, some interesting effects are observed. When co-adsorbed with one water molecule, the adsorption energy of $F^-(H_2O)$ is only 230 kJ/mol more favorable than that of $I^-(H_2O)$. Co-adsorbed with three water molecules (in a restricted geometry to keep the computation time reasonable) the adsorption energy of $F^-(H_2O)_3$ is only 66 kJ/mol more favorable than that of $I^-(H_2O)_3$. (All quoted adsorption energies correspond to HF-MP2 calculations.) These numbers illustrate that as the chemisorbed halogen becomes hydrated, there is a change in trend that seems to approach the experimental observations. In fact, it is observed that the adsorption energy of I^- is not very different from that of $I^-(H_2O)$ or $I^-(H_2O)_3$. The biggest changes in energy hence come from the change in the geometry of the hydration shell of F^- as it becomes adsorbed on the Al(111) surface. Because of the smaller radius of the fluoride, the water molecules are pulled too close to the surface, leading to a destabilization of the Al-F bond. Iodide is a much larger ion, and hence the water molecules are much better able to keep their original distances and geometries with respect to the adsorbed iodide. Figure 6 shows the optimized geometries obtained by Kairys and Head for both the $F^-(H_2O)_3$ and the $I^-(H_2O)_3$ on the Al(111) cluster surface, illustrating the effect of the water on straining the Al-F bond.

A different ab initio DFT-GGA approach to the effect of the solvent on the anion chemisorption bond was taken by Wasileski and Weaver.[60] These authors were mainly interested in the effect of the solvent on the bond polarity. To this end, the solvent was modeled as an infinite number of solvent dielectric spheres of a diameter close to that of water. The metal was modeled as a 13- or 33-atom cluster exposing the (111) plane, with the halogen adsorbed in the threefold hollow site. As metals, Pt, Au, Ag and Hg were considered, while Cl, Br, and I were the studied halogens. Wasileski and Weaver investigated the influence of the solvent on the Hirshfeld charge of the chemisorbed halogen and on the metal-adsorbate static dipole moment, evaluated as the difference in the cluster dipole perpendicular to the surface plane in the presence and absence of the adsorbate. The latter quantity is of interest to electrochemists as it determines the dependence of the binding energy on the external field (see Section III.5). Unfortunately, the static dipole moment shows a relatively poor convergence with cluster size, but some trends could be recognized among the different halides and metals. Mercury exhibits the most polar halide bond of the four metals, in agreement with the experimentally observed smaller electrosorption valencies for halides on mercury. On Pt, Au and Ag, iodide has the least polar bond of three halides, as expected from the smaller Hirshfeld charges computed for I compared to Cl and Br. Addition of the solvent dielectric spheres leads to a significant increase of the static dipole moment of metal-halide bond and the Hirshfeld charge on the chemisorbed halogen, although the effect is small for those systems for which the static dipole moment was already small without a solvent (*i. e.* I on Pt and Au). This is expected since the dielectric continuum solvent will stabilize a charge separation. The

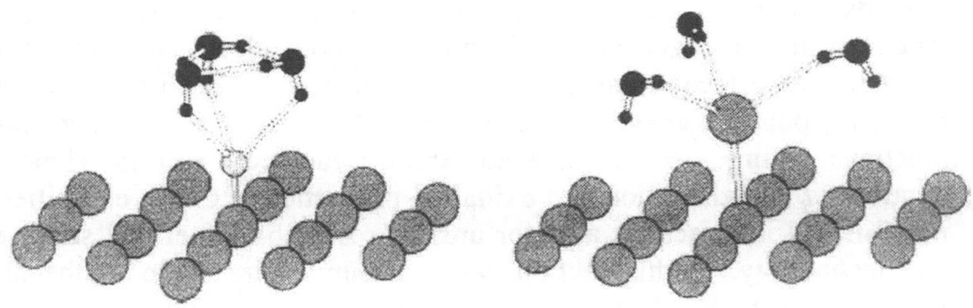

Figure 6. Optimized structures for $F^-(H_2O)_3$ (left) and $I^-(H_2O)_3$ (right) on the $Al_{19}^-(111)$ cluster with the C_3 symmetry constraint.

computed static dipole moments were in reasonable accordance with the experimental dipole moments estimated from the experimental electrosorption valencies.

Finally, it is important to note again the difference between halogen and halide adsorption. As the dipole moment of the halogen-metal surface bond is negative, this implies that the halogen binds stronger at more negative field, *i. e.* more negative electrode potential. This seems contrary to electrochemical intuition, which says that *halides* bind stronger with more positive electrode potential. This difference between atom (halogen) and corresponding ion (halide) binding simply reflects the surface potential-dependent energy of moving a charge ze (with z the charge number of the ion in solution) completely across the interface, required in order to transform between the charged and uncharged bulk-phase reference states. The mathematical relationship between the field-dependence of atom and ion binding follows by deriving Eq. (10) with respect to the field or potential:

$$\frac{d\Delta E_b(X^-)}{e_0 d\phi} = \frac{d\Delta E_b(X)}{e_0 d\phi} + z \tag{11}$$

with e_0 the electronic charge, ϕ the electrode potential, and z = -1 for a halide.

4. Chemisorption of Carbon Monoxide on Metals and Alloys

The adsorption of carbon monoxide on metal surfaces is a paradigmatic model system for molecular adsorption in both electrochemical and UHV-based surface science. It is also an important adsorbate from the practical point-of-view, as adsorbed CO poisons electrochemical reactions taking place at low-temperature fuel cell anodes. Hence, controlling the adsorption and oxidation properties of CO at electrified metal-liquid interfaces is a major area of research in fuel cell science and technology. In this section, we will summarize some of the ab initio quantum-chemical work concerning the adsorption of CO at uncharged metal and bimetallic surfaces. Bimetallic surfaces and alloys are of interest as practical fuel cell anodes consist of platinum alloyed with 1, 2 or even more (metallic) components. In subsequent sections, we will study in more detail the electric field dependence of CO binding to metals (Section III.6), and discuss the ab initio studies that have attempted to shed light on the electrochemical oxidation mechanism of chemisorbed CO (Section III.8).

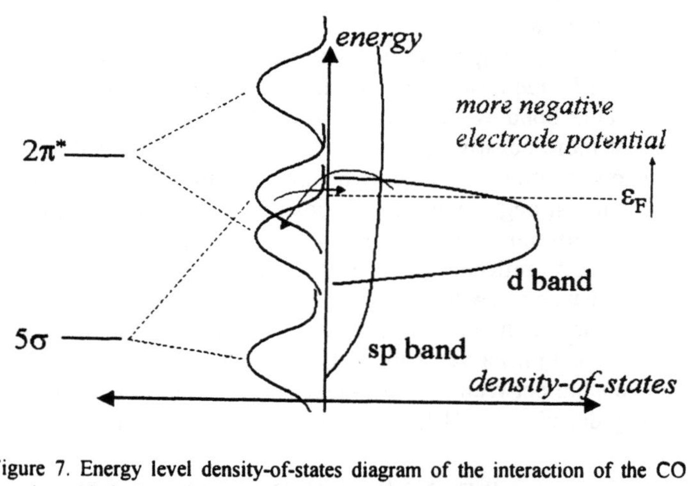

Figure 7. Energy level density-of-states diagram of the interaction of the CO frontier orbitals 5σ and $2\pi^*$ with a transition metal surface. The arrows indicate the donation of charge from the 5σ and the back donation of charge into the $2\pi^*$. The effect of the electrode potential is to shift the metal energy levels up with respect to the adsorbate and the solution as the potential becomes more negative. Reprinted with permission from M. T. Koper *et al.*, *J. Chem. Phys.*, **113**, (2000) 4392. Copyright 2000, American Institute of Physics

The adsorption of carbon monoxide on metal surfaces can be qualitatively understood using a model originally formulated by Blyholder.[61] A simplified molecular orbital picture of the interaction of CO with a transition metal surface is given in Figure 7. The CO frontier orbitals 5σ and $2\pi^*$ interact with the localized d metal states by splitting into bonding and antibonding hybridized metal-chemisorbate orbitals, which are in turn broadened by the interaction with the much more delocalized sp metal states.

Hammer, Morikawa, and Nørskov[62] have shown how this simple picture emphasizing the interaction of the CO frontier orbitals with the metal d states can explain trends in the binding energies of CO to various (modified) metal surfaces. Their model for CO chemisorption is based on an earlier, more general model for chemisorption suggested by Hammer and Nørskov.[63] The Hammer- Nørskov model singles out three surface properties contributing to the ability of the surface to make and break adsorbate bonds: (i) the center ε_d of the d band, (ii) the degree of filling f of the d band, and (iii) the coupling matrix element

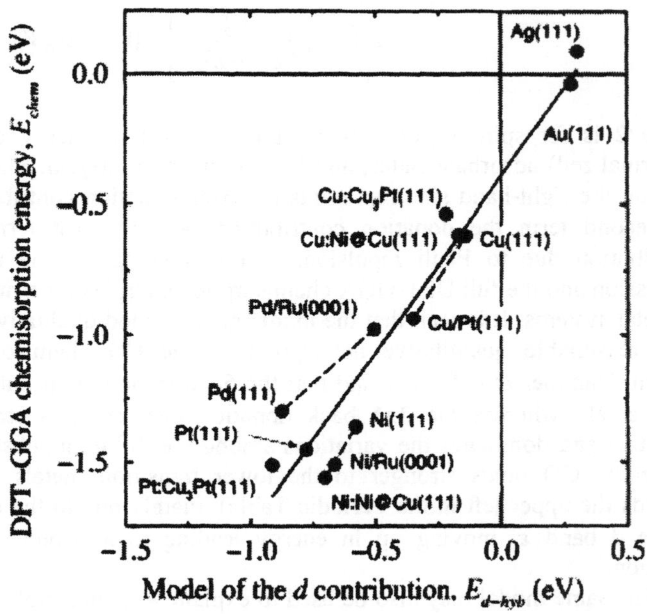

Figure 8. Comparison of the full DFT-GGA and model chemisorption energies for CO on a number of metal and metal-overlayer systems.

V_{ad} between the adsorbate states and the metal d states. These quantities have been calculated from extensive quantum-chemical calculations for a significant portion of the Periodic Table.[64] The basic idea of the Hammer-Nørskov model is that trends in the interaction and reactivity are governed by the coupling of the adsorbate states with the metal d states, since the coupling with the metal sp states is essentially the same for the transition and noble metals, and mainly acts to renormalize (i.e, shift and broaden) the energy of the adsorbate orbital.

In the Blyholder model, CO interacts with the metal states through two different states, the 5σ and $2\pi^*$ orbitals, but due to their different symmetry they interact with different metal *d* orbitals, and the two interactions can be treated independently. Hammer, Morikawa, and Nørskov[62] used the following expression to model the *d* contribution to the CO chemisorption energy:

$$E_{d-hyb} = -4\left[f\frac{V_\pi^2}{\varepsilon_{2\pi}-\varepsilon_d} + fS_\pi V_\pi\right] - 2\left[(1-f)\frac{V_\sigma^2}{\varepsilon_d - \varepsilon_{5\sigma}}\right] + (1+f)S_\sigma V_\sigma \quad (12)$$

where 2 is for spin, $\varepsilon_{2\pi}$ and $\varepsilon_{5\sigma}$ are the positions in energy of the (renormalized) adsorbate states, and S is the overlap integral. The first term on the right-hand side represents the back donation contribution, the second term the donation contribution, and the last term the contribution due to Pauli repulsion. Figure 8 compares the model expression and the full DFT-GGA chemisorption energies for a number of metal systems. It is seen that the model gives a good qualitative and even reasonable quantitative description of the CO chemisorption system. Hammer *et al.*[62] also find that the 5σ donation contribution is quite small, whereas the $2\pi^*$ back donation interaction is strongly attractive and dominates the variations among the different substrates. Generally, CO binds stronger to the lower transition metals (*i. e.*, towards the upper left in the Periodic Table), mainly due to the center of the d band ε_d moving up in energy leading to a stronger back donation.

The same model may also be used to explain why on metals in the upper-right corner of the Periodic Table (Pd, Ni) CO prefers multifold coordination, whereas towards the lower left corner (Ru, Ir) CO preferentially adsorbs atop.[37] Pd and Ni combine a high d band filling factor *f*, which is favorable for a strong back donation, with a low steric repulsion as the d states on these metals are more contracted. Since the back donation interaction is a bonding interaction, it will prefer to coordinate to as many surface atoms as possible. A low steric repulsion is also a favorable condition for multifold coordination. Ru and Ir, on the other hand, combine a relatively lower d band filling with much more diffuse d band states leading to a strong steric Pauli repulsion. Both factors will favor atop coordination on these metals. Pt and Rh are borderline cases and indeed on these metals there is no strong energetic preference for either atop, bridge or hollow coordination of CO.

One aspect of CO chemisorption at which quantum-chemical calculations have been particularly successful in explaining its origin, is the redshift observed in the C-O intramolecular vibrational frequency upon adsorption. Depending on the substrate and the type of coordination, the C-O intramolecular vibration frequency in the chemisorbed state may be 100-400 cm^{-1} lower than its uncoordinated vacuum value (2170 cm^{-1}) (for a detailed comparison of UHV and electrochemical data, see Refs.65 and 66). A simple but incomplete explanation for this observation refers to the Blyholder model. In its chemisorbed state, CO mainly interacts through its $2\pi^*$ orbital. This antibonding orbital is unoccupied for uncoordinated CO, but becomes partially occupied in its chemisorbed state. Note that this back donation interaction is *bonding* in terms of the substrate-adorbate interaction, but *antibonding* for the intramolecular C-O bond. This weakening of the C-O bond due to the metal to CO back donation is accompanied by a lowering of the C-O stretch frequency. This picture suggests a simple relationship between C-O stretching frequency and its adsorption strength: a lower C-O stretch frequency would imply a stronger back donation and hence a more favorable adsorption energy. This relationship is simple and attractive and hence very popular in the literature, but it is misleading and generally wrong.

Bagus and Pacchioni[44] were the first to suggest using ab initio SCH-HF calculations that not only the strength of the back donation determines the C-O stretching frequency, but also the fact that in its chemisorbed state, the C-O mode stretches against a "surface wall" of non-bonding metal electrons. This "wall effect" leads to a stiffening of the C-O potential energy surface and hence to a blueshift of the C-O stretching frequency. A more detailed analysis of the origin of the C-O vibrational shift on a 4-atom Pt(111) cluster was carried out by Illas and co-workers,[67,68] making use of the CSOV analysis in ab initio SCF-

HF calculations. The absolute values of the C-O vibrational frequencies calculated in their work are too high due to the lack of electron correlation, but the vibrational shift upon CO chemisorption on the 4-atom Pt(111) cluster, when calculated for the ground state, was in good agreement with experiment, and hence the origin of the shift could be studied. In the initial step of the CSOV analysis the expectation value of the energy is calculated using the initial "frozen orbital" wave function. No orbital hybridization is allowed for, so that only steric Pauli repulsion effects are taken into account. This leads to a 399 cm^{-1} blueshift of the C-O frequency in the adsorbed state, as the PES is stiffened by the presence of the above-mentioned "surface wall" of non-bonding metal electrons. Next, the substrate electrons are allowed to polarize due to their interaction with the CO charge density, leading to a redshift of 87 cm^{-1}. (These numbers actually depend on the basis set and the electronic state considered – the numbers chosen are for a representative calculation.) In a third step, there is substrate to CO donation (*i. e.*, mainly back donation into $2\pi^*$), which leads to a large additional redshift of 260 cm^{-1}. The effect of internal CO polarization is very small, but the CO to substrate donation (mainly from the 5σ) leads to another significant lowering of 108 cm^{-1}. Apart from some small remaining terms, this leads to a total redshift of the C-O stretch of 79 cm^{-1} upon adsorption onto the Pt(111) cluster. This analysis shows that the dominant contributions to the vibrational shift come from the surface wall effect (*i. e.*, increased Pauli repulsion) and the back donation contribution, but also that *all* contributions are important in accounting for the total shift. Interestingly, the redshift caused by the donation contribution is contrary to expectation, as removing an electron from the 5σ orbital in the vacuum leads to a blueshift in the C-O frequency. Illas *et al.*[68] ascribe this to the fact that the 5σ orbital is directed to the surface and is therefore to a large extent responsible for the surface wall effect. The 5σ donation therefore reduces the repulsion with the non-bonding metal electrons and hence leads to a decrease in the frequency.

A similar analysis of the origin of the vibrational shift of the CO and NO stretching frequency on a series of transition metal surfaces was carried out using DFT-GGA cluster calculations by Koper *et al.*[37] It was observed that the DFT-computed C-O and N-O stretching frequencies were in good agreement with the experimental frequencies, the DFT values being slightly lower (*ca.*20-60 cm^{-1}), as shown in Figure 9. Table I gives some representative and illustrative results of the frequency decomposition analysis for CO adsorbed in the atop and hollow site of a 13-atom Pt(111) cluster, and at the atop site of a 13-atom Ru(0001) cluster. Comparing the frequency decomposition for CO adsorbed atop and hollow on the Pt(111) cluster shows that the lower frequency in the hollow site is primarily due to the more negative back donation contribution there. Comparing CO adsorbed atop on Pt(111) and Ru(0001) shows that the C-O stretch frequency on Ru(0001) is lower due to the smaller steric contribution to the overall frequency change. This is in contrast with the common interpretation in the literature that a lower frequency is related to a stronger back donation, and hence a stronger chemisorption bond. A detailed discussion of the absence of a clear-cut relation between the C-O stretching frequency and the chemisorption energy was recently given by Wasileski and Weaver.[69]

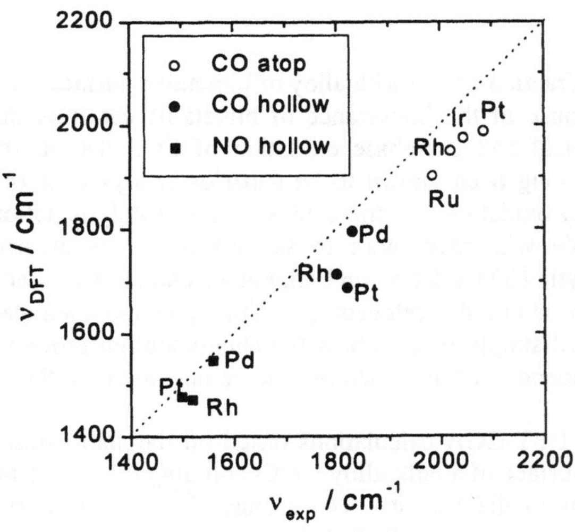

Figure 9. Comparison of the DFT-GGA calculated C-O and N-O internal stretching frequencies on a number of metals compared to the experimental values obtained at low coverage in UHV. Reprinted with permission from M. T. Koper *et al.*, *J. Chem. Phys.*, **113**, (2000) 4392. Copyright 2000, American Institute of Physics

Table 1
Decomposition of the Zero-Field Vibrational Frequency Shifts (in cm^{-1}) Compared to the Calculated Vacuum Values into Steric and Orbital Components for CO Chemisorbed in the Atop and Hollow Site of Pt(111) and CO Adsorbed in the Atop Site of Ru(0001)[a]

System	Steric	Donation	Back donation	Rest	Final
Pt-CO (atop)	+743	-328(-486)	-569(-411)	+46	1987
Pt-CO (hollow)	+645	-198(-371)	-910(-737)	+82	1714
Ru-CO (atop)	+670	-349(-471)	-562(-440)	+47	1901

[a]The far-left column in the cluster-adsorbate system. Adjacent three columns (steric/donation/back donation) give in the change in frequency due to each contribution, calculated with respect to the uncoordinated DFT frequency of 2095 cm^{-1} for CO. The main entries have been calculated in the order as given in the Table, i.e., steric-donation-back donation, whereas the entries within parentheses refer to the order steric-back donation-donation. The column "rest" refers to the residual frequency shift upon adsorption not accounted for by the sum of steric/donation/back donation. The far-right column gives the coordinated C-O vibrational frequency.

The interaction of CO with alloy or bimetallic surfaces is of special interest because of the importance of bimetallic catalysts in both the electrochemical and gas-phase oxidation of CO. Platinum-ruthenium alloys have long been known to be superior catalysts for the electrochemical CO oxidation, but the details of their catalytic action are still disputed. We will have more to say about the mechanism of the electrocatalytic CO oxidation on both metals and alloys in section III.8, in particular about the relevant ab initio quantum-chemical studies. Here, we will simply discuss how the chemisorption properties of CO on PtRu depend on the structure and composition of the bimetallic surface.

Recent DFT-GGA calculations based on the slab geometry show that on the surface of a bulk alloy the CO binding to the Pt site weakens whereas that to the Ru site gets stronger.[70,71] For example, on the surface of a homogeneous $Pt_2Ru(111)$ alloy, the CO atop binding to Pt is weakened by about 0.2-0.3 eV, and that to Ru strengthened by about 0.1 eV. However, real catalytic surfaces are not homogeneous but may show the tendency to surface segregate. Figure 10 shows the binding energy and vibrational properties of CO coordinated to an atop Pt site on a series of surfaces with a pure Pt top layer, but for which the bulk composition changes from pure Pt, to Pt:Ru 2:1, to Pt:Ru 1:2, to pure

Ru. It is observed that a higher fraction of Ru in the bulk causes a weakening of the CO bond to the Pt overlayer. This electronic alloying effect can be understood on the basis of the Hammer-Norskov d band shift model. Alloying Pt with Ru causes a flow of electrons from Ru to Pt, as evidenced from a charge analysis of the slab calculations,[55] which pushes the local d band center on Pt downward with respect to the Fermi level. Interestingly, Figure 10 also illustrates that there is no correlation between the binding energy and the C-O stretching frequency, even though such a correlation is often assumed in the literature. As explained above, the frequency decomposition analysis suggest that there is no quantum-chemical basis for such a correlation,

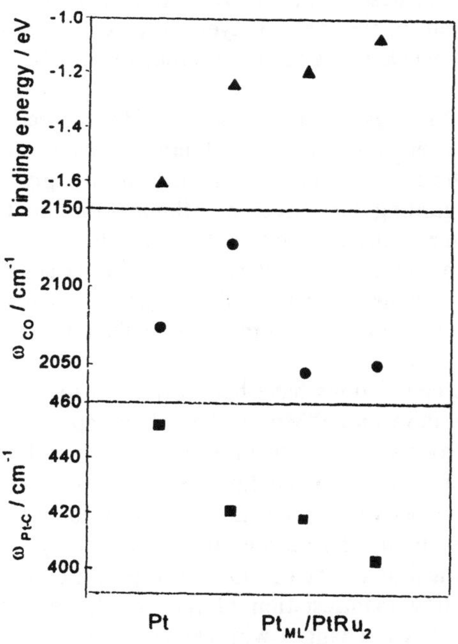

Figure 10. The binding energy, C-O stretching frequency, and Pt-C stretching frequency on four different surface, all with a Pt surface layer, but different bulk compositions (Pt, Pt_2Ru, $PtRu_2$, Ru). Reprinted with permission from M. M. T. Koper *et al.*, *J. Phys. Chem. B*, **106**, (2002) 686. Copyright 2002, American Chemical Society.

as both the back donation and steric contribution have an apparently equally strong influence of the C-O stretching frequency. On the other hand, the Pt-C stretching mode correlates well with the binding energy: a stronger bond causes the expected increase in the Pt-C frequency. We note that the weakening of Pt-CO bond on Pt/Ru(0001) in ultra-high vacuum was indeed observed experimentally by Buatier de Mongeot *et al.* using temperature-programmed desorption.[72]

The opposite effect is observed when a surface with a pure Ru overlayer is considered, and the bulk is stepwise enriched with Pt. The CO binding to the Ru overlayer is strengthened with a higher content of Pt in the bulk. In fact, the Ru/Pt(111) surface is the one that shows the strongest CO binding.[71] A similar effect is also observed when the top layer of a Pt(111) surface is mixed with Ru, leading to a surface alloy.[70] As the Ru content of the surface layer increases, the binding to the Pt surface sites is weakened and the binding to the Ru surface sites is strengthened.[70]

We note that cluster calculations of CO interacting with Pt-Ru surfaces have been reported in the literature using ab initio density functional methods.[73,74] These calculations do not give very clear-cut conclusions regarding the effect of alloying on the CO adsorption energy, and some of the cluster calculations in fact give results that are in conflict with the slab calculations. For the reasons mentioned in Section III.1, it is believed that the binding energy trends obtained from slab calculations should be more reliable than those obtained with cluster calculations.

The above calculations have been extended to a number of other alloys, notably PtSn and PtMo as these have special interest in CO oxidation electrocatalysis.[75] On PtSn, CO was found to interact only with the Pt sites, not with the Sn sites.[76] This leaves the Sn sites available for the activation of water necessary to oxidize the chemisorbed CO, explaining why PtSn is such an unusually good catalyst for the electrochemical oxidation of CO (at low potentials).[75]

An interesting manifestation of how on bimetallic surfaces the binding energy of CO changes with respect to pure Pt is illustrated in Figure 11.[76] In this figure, calculations are collected of CO adsorbed on a series of metals with a commensurate Pt overlayer. In all cases, the Pt-Pt distance in the overlayer was fixed at the lattice constant of the underlying substrate (it is realized that some of these systems are not stable in practice). When the change in binding energy compared to the binding energy on pure Pt(111) is plotted as a function of the Pt-Pt distance in the overlayer, a clear correlation is observed. An expanded

Pt overlayer, such as on Au, gives a stronger CO binding than pure Pt(111), whereas a contracted Pt overlayer, such as on Rh, leads to weaker CO binding energy. The solid line in Figure 11 represents results of a calculation of the CO binding to a single Pt layer with the Pt-Pt distance indicated on the figure's *x* axis (referred to the binding energy obtained for a Pt monolayer with equilibrium Pt-Pt *bulk* distance). The fact that the results for the overlayer systems follow closely those for the single monolayer illustrates that the CO binding energy is correlated primarily to the electronic properties of Pt overlayer, and that the underlying substrate plays only a secondary role. In terms of the Hammer-Nørskov model,[63] expanding (contracting) the Pt lattice leads to an upshift (downshift) of the d band, and hence a stronger (weaker) CO binding energy. A change in the Pt-Pt distance in the overlayer is alternative way to understand the binding energy trend shown in Figure 10. A related conclusion concerning the secondary role of the substrate was recently drawn on the basis of a combined spectro-electrochemical and density functional theory study of CO adsorption

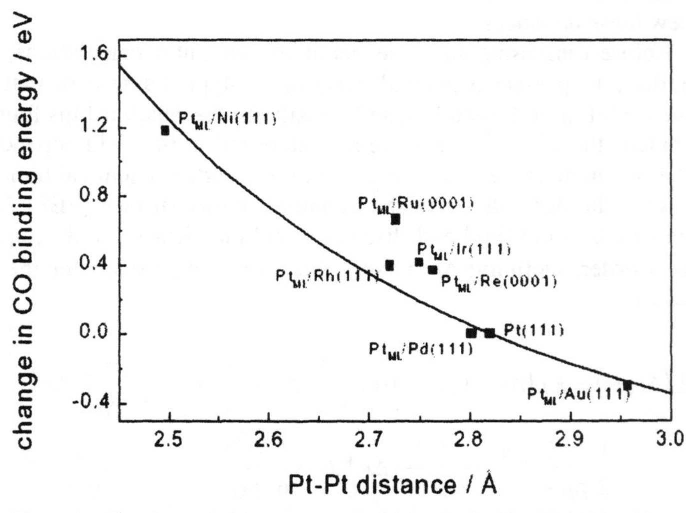

Figure 11. The change in binding energy of CO adsorbed in the most stable site of different Pt monolayer systems plotted vs. the Pt-Pt distance. CO on Pt(111) serves as a reference. The line is the result of a calculation of the CO binding in the hollow site of a 1-layer slab of Pt, with the Pt-Pt distance of 2.82 Å as an energy reference. Positive change in binding energy signifies weaker bond.

on Pd-modified Pt(111) by Gil *et al.*[77] The authors found that a monolayer of Pd on Pt(111) gives rise to essentially the same spectral characteristics of chemisorbed CO as a pure Pd(111) surface. This is expected since Pd and Pt have very similar lattice constants.

5, Field Dependent Chemisorption and the Interfacial Stark Effect: General Relationships

It is a natural tendency of electrochemists to be interested in the potential or electric field dependence of a chemisorption bond. In particular, electrochemists would like to know how the nature of the chemical bond, its coordination to the surface, the bond strength, and vibrational frequencies, both substrate-adsorbate and intramolecular, depend on the electrode potential. In recent years, some general rules concerning field-dependent chemisorbate bonding have emerged from extensive DFT calculations,[78-81] and it is the aim of this section to review these advances.

Before discussing in some detail the ab initio calculations, it is opportune to present a general perturbation-type framework that will prove useful in understanding and classifying the relationships found in the calculations.[80,82-85] Since we are interested in the field-dependence of the potential energy curve pertaining to a certain chemical bond, we can write the dependence of the potential energy surface (PES) $E(r,F)$ as an expansion in the bond distance r and the electric field F, up to a certain order. Defining $\Delta r = r - r_{eq}$, the formal expression for the PES reads as:

$$\begin{aligned} E(\Delta r, F) = E_b(\Delta r = 0, F = 0) + \frac{\partial E}{\partial \Delta r}\Delta r + \frac{\partial E}{\partial F}F + \frac{1}{2}\frac{\partial^2 E}{\partial \Delta r^2}\Delta r^2 \\ + \frac{1}{2}\frac{\partial^2 E}{\partial F^2}F^2 + \frac{\partial^2 E}{\partial \Delta r \partial F}\Delta r * F + \frac{1}{6}\frac{\partial^3 E}{\partial \Delta r^3}\Delta r^3 + \frac{1}{6}\frac{\partial^3 E}{\partial F^3}F^3 \\ + \frac{1}{2}\frac{\partial^3 E}{\partial \Delta r^2 \partial F}\Delta r^2 * F + \frac{1}{2}\frac{\partial^3 E}{\partial \Delta r \partial F^2}\Delta r * F^2 + O(4) \end{aligned} \tag{13}$$

where $O(4)$ signifies all terms higher than fourth order, and all partial derivatives are taken in $\Delta r=0$ and $F=0$ (making $\partial E/\partial\Delta r = 0$). Equation (13) can be rewritten as:

$$E(\Delta r, F) = E_b^0 + \frac{1}{2}K_0\Delta r^2 + G_0\Delta r^3 - \mu_s(F)F \\ - \mu_D(F)F * \Delta r - \alpha_D(F)F * \Delta r^2 + O(4) \tag{14}$$

with the different parameters defined as follows: $K_0 = \frac{\partial^2 E}{\partial \Delta r^2}$, $G_0 = \frac{1}{6}\frac{\partial^3 E}{\partial \Delta r^3}$, $\mu_s(F) = -\frac{\partial E}{\partial F} - \frac{1}{2}\frac{\partial^2 E}{\partial F^2}F - \frac{1}{6}\frac{\partial^3 E}{\partial F^3}F^2$, $\mu_D(F) = -\frac{\partial^2 E}{\partial \Delta r \partial F} - \frac{1}{2}\frac{\partial^3 E}{\partial \Delta r \partial F^2}F$, and $\alpha_D(F) = -\frac{1}{2}\frac{\partial^3 E}{\partial \Delta r^2 \partial F}$. Here, K_0 is the mode's force constant at zero field, G the mode anharmonicity at zero field, $\mu_s(F)$ the field-dependent static dipole moment, $\mu_D(F)$ the field-dependent dynamic dipole moment ($= \partial\mu_s/\partial\Delta r$), and $\alpha_D(F)$ a higher-order polarizability term. Note that the terms $\mu_s(0)$, $\mu_D(0)$, and $\alpha_D(0)$ are equal to the terms M_0, M_1, and M_2, respectively, in Eq. (7).

Equation (14) shows that the static dipole moment μ_s describes, to a first approximation, the field dependence of the bond strength, *i. e.*:

$$\frac{dE_b(F)}{dF} = -\mu_s(F) \tag{15}$$

Wasileski *et al.*[80] have checked the validity of this relationship by DFT-GGA cluster calculations. Figure 12 shows in the top and middle left-hand panels the field-dependent binding energy E_b and static dipole moment μ_s for four atomic adsorbates (Na, O, Cl, and I) at the hollow site of a 13-atom Pt(111) cluster. The overall adherence to Eq. (15) is clearly evident, for example, from the observation of uniformly positive dE_b/dF slopes and negative μ_s values for the electronegative O and Cl adsorbates on Pt(111), with the peaked E_b-F behavior for the more polarizable adsorbate iodine matching the sign alteration in μ_s. As expected, adsorbed sodium exhibits uniformly positive μ_s values, thereby yielding negative E_b-F slopes, *i. e.*, of opposite sign to the electron-withdrawing adsorbates. Consistent with polarizability expectations, oxygen exhibits almost field-independent μ_s values, with chlorine and especially iodine yielding μ_s values becoming more negative towards negative fields. Wasileski *et al.*[80] also report more negative μ_s values observed for these adsorbates on Ag(111) versus Pt(111), in harmony with the lower work function of the silver surface. However, as illustrated in Sec. III.3, the field-dependent μ_s values at actual electrode surfaces can be quite different from those in Figure 12, due to solvation and other effects not included in the DFT calculations.

It is important at this point to recall the difference between atom and ion adsorption. Figure 12 shows that chlorine and iodine adsorption become stronger towards negative fields, whereas it is known that halide adsorption on metal electrodes becomes more favorable towards more positive electrode potentials, and hence fields. This apparent contradiction is readily resolved by noting the different reference states involved (*i. e.*, atom and ion), leading to different E_b-F slopes as expressed in Eq. (11).

As binding energies, and hence their field dependence, are difficult to measure experimentally in an electrochemical cell, it is of interest to examine the field dependence of other fundamental binding characteristics, in particular vibrational properties of the chemisorbate. A fundamental equation for the field dependence of a vibrational frequency can be straightforwardly derived from Eq. (14). The force constant K_F at field F can be obtained by determining the minimum of Eq. (14) at field F and then inserting the resulting value for Δr in the expression for the second derivative of E with respect to Δr. This yields:

$$K_F = K_0 + \frac{6G_0\mu_D(0) - 2K_0\alpha_D(0)}{K_0} F \tag{16}$$

The harmonic vibrational frequency ν_H is then,

$$\nu_H(F) = \frac{1}{2\pi}\sqrt{\frac{K_F}{m*}} \tag{17}$$

where $m*$ is the reduced mass of the vibrational mode considered.

From a generalization of these equations, an expression for the slope of the force constant and the vibrational frequency at field F can be derived, *i. e.*

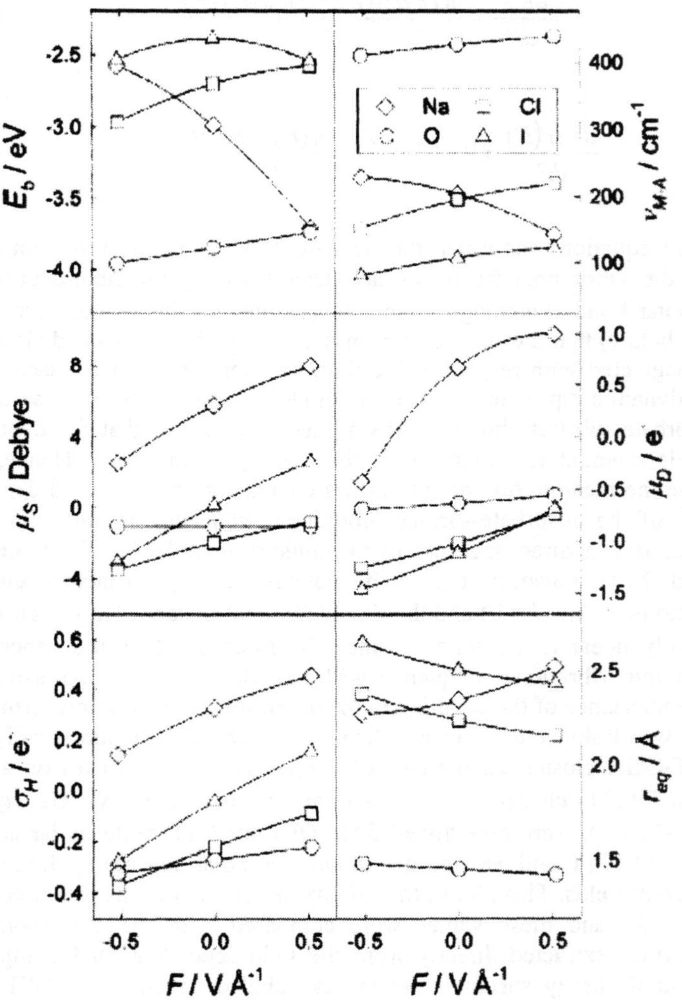

Figure 12. Plots of the binding energy E_b, anharmonic vibrational frequency $\nu_{M\text{-}A}$, static dipole moment μ_s, dynamic dipole moment μ_D, Hirshfeld charge σ_H, and equilibrium bond distance r_{eq}, as a function of the external field for hollow-site adsorbed Na (diamonds), O (circles), Cl (squares), and I (triangles) on 13-atom Pt clusters. Reprinted with permission from S. A. Wasileski *et al.*, *J. Chem. Phys.*, **115**, (2001) 8193. Copyright 2001, American Institute of Physics

$$\frac{dK_F}{dF} = \frac{6G_F\mu_D(F) - 2K_F\alpha_D(F)}{K_F} \tag{18}$$

and

$$\frac{d\nu_H(F)}{dF} = \nu_H(F)\frac{3G_F\mu_D(F) - K_F\alpha_D(F)}{K_F^2} \tag{19}$$

These equations show that the field dependence of the force constant and the vibrational frequency are determined by the field-dependent "dipolar bond stretching parameters" μ_D and α_D. In the common case (see below) that the second term in numerator of Eqs.(18) and (19) can be neglected with respect to the first, the slope is a direct measure of the dynamic dipole moment of the mode considered. In the case of an adsorbate-substrate bond, it was argued in Sec.III.3 that the dynamic dipole moment is a measure of the ionicity of the bond. Hence, we come the meaningful and interesting conclusion that the field dependence of the adsorbate-surface vibrational frequency is, under certain conditions, a direct measure of the ionicity or polarity of the surface bond. Note, however, that if the potential energy surface would be harmonic (*i. e.*, $G = 0$) and the dipole moment-distance (μ_s - r) curve is entirely linear (*i. e.*, $\alpha_D = 0$), then the force constant and hence the harmonic vibrational frequency will be field independent, illustrating the importance of the coupling to the higher-order anharmonic terms.

Wasileski *et al.*[80] have validated Eqs.18 and 19 numerically by DFT-GGA cluster calculations of simple atomic adsorbates on a 13-atom Pt(111) cluster. The four parameters of interest, K_F, G_F, $\mu_D(F)$, and $-2\alpha_D(F)$, were determined from DFT for four model polar adsorbates, O, Cl, I, and Na, and one model non-polar adsorbate, H, at three different fields. This allows the calculation of dK_F/dF as prescribed by Eq. (18), and these values were compared with the corresponding quantities extracted directly from the field-dependent DFT-computed potential energy surfaces themselves, labeled as dK_F/dF (DFT). In general, a good approximate correspondence (chiefly to within 10%) was observed between the values for dK_F/dF obtained from Eq. (18) and the dK_F/dF (DFT) values. This also leads to a similar correspondence between $d\nu_H/dF$ as obtained from Eq. (19) and $d\nu_H/dF$ (DFT). The latter quantity differs only marginally from the "real" $d\nu_{M\text{-}A}/dF$ values, which include the full anharmonicity of potential energy surface.

Wasileski *et al.*[80] extracted some interesting overall trends from their analysis. For most combinations of adsorbates and fields, the term $6G_F\mu_D(F)$ $/K_F$ dominates dK_F/dF, rather than the $-2\alpha_D(F)$ term in Eq. (18). Interestingly, the quantity G_F/K_F appears roughly system invariant, even among different adsorbates, making variations in the dynamic dipole moment μ_D the controlling quantity for variations in dK_F/dF. Furthermore, alterations in dK_F/dF with field are controlled primarily by the corresponding variations in μ_D. As expected, these trends are primarily valid for polar adsorbates. However, the α_D term was found to significantly influence dK_F/dF in some cases, where μ_D is small, as is anticipated for nonpolar metal-adsorbate bonds. Specifically, this situation was encountered for Na adsorbate on Pt(111) at negative field, and especially for H adsorbate in the hollow adsorption site on Pt(111), where the α_D term even dominates dK_F/dF.

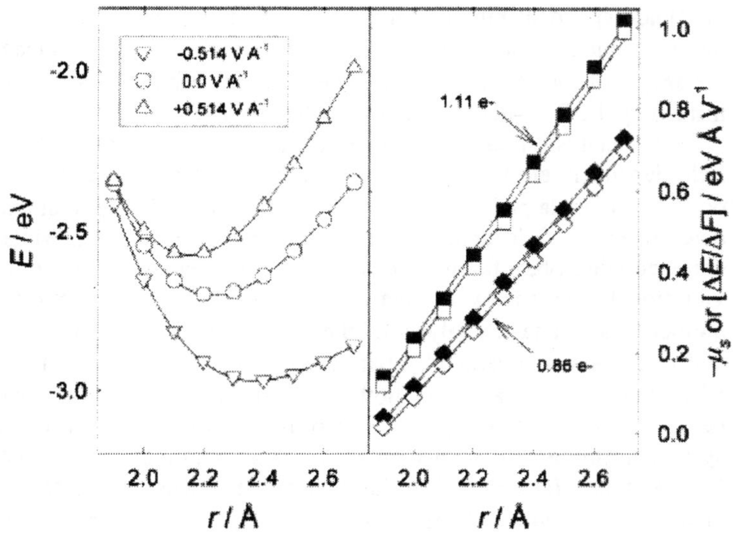

Figure 13. Left-hand portion: Plots of the metal-adsorbate interaction energies, $E(r)$ (i.e. PES) for Cl bound to hollow sites on a 13-atom Pt cluster at F=0 (circles) and $F = \pm$ 0.514 V Å^{-1} (upright and inverted triangles, respectively) *vs* bond length r. Right-hand portion: Plots of interaction energy displacements between $F = \pm$ 0.514 V Å^{-1} (open diamonds and squares, respectively) and F=0, normalized to the difference in field ($\Delta E/\Delta F$) for the PES given vs bond length; compared with the corresponding plots of static dipole moments at intermediate fields $F = \pm$ 0.257 V Å^{-1} (filled diamonds and squares, respectively). Reprinted with permission from S. A. Wasileski *et al.*, *J. Chem. Phys.*, **115**, (2001) 8193. Copyright 2001, American Institute of Physics

A graphical representation of the role of the field on the potential energy surface and the dipole parameters, is given in Figure13. The left-hand panel shows surfaces calculated for Cl in the hollow site of Pt(111) at $F = 0$ and at large positive and negative fields $F = \pm 0.514$ VÅ^{-1}. Despite the progressive decreases in equilibrium binding energy seen towards more positive fields, as already discussed, the equilibrium bond distance decreases and PES "well steepness" (and hence ν_{M-A}) increases under these conditions. The right-hand panel of Figure 13 shows the static dipole moment extracted from the field-dependence of the PES, which could be taken as an approximation to the static dipole moments at the intermediate fields $F = \pm 0.257$ V Å^{-1}. The slopes of these essentially linear curves are the dynamic dipole moments $\mu_D = \partial\mu_s/\partial r$at the respective fields, and are 0.86 and 1.11 e_0. Figure 13 illustrates the "splaying" of the PES caused by the applied electric field. The basic point to be gleaned from Figure 13 is that while the field-dependent binding energy E_b is determined by the static dipole moment μ_s at the equilibrium distance, the field dependence of the *shape* of the PES, and hence the ν_{M-A} - F behavior, is controlled instead by the field-dependent μ_s - r behavior, *i. e.*, μ_D.

The analysis presented above suggests some general features of field-dependent electrode-chemisorbate bonding that are worth pointing out.[81] As explained in detail above, the static dipole moment μ_s determines the field dependence of the binding energy E_b, while the corresponding Stark tuning behavior (*i. e.*, the field dependence of the vibrational frequencies) is controlled primarily by the dynamic dipole moment, μ_D. In fact, analysis of the field-dependent sensitivity of μ_s and μ_D leads to a general adsorbate classification.[81] For electronegative adsorbates, such O and Cl, both μ_s and μ_D are negative, the opposite being the case for electropositive adsorbates, such as Na. However, for systems forming dative-covalent bonds rather than ionic bonds, such as NH_3 and CO (see Sec III.6 for a detailed discussion of the field-dependent behavior of chemisorbed CO), μ_s and μ_D have opposite signs. The latter behavior arises from the diminishing metal-chemisorbate overlap, and hence the extent of charge polarization, as the bond is stretched, and is illustrated in cartoon fashion in Figure 14. Ionic bonds tend to retain or even enhance their ionic character as the bond is stretched, because of the strong driving force for electron flow either to (electronegative, anionic, class "A") or from (electropositive, cationic, class "C") the adsorbate. In simple energy terms, this is illustrated in Figure 15, showing that class A and C adsorbates have their main

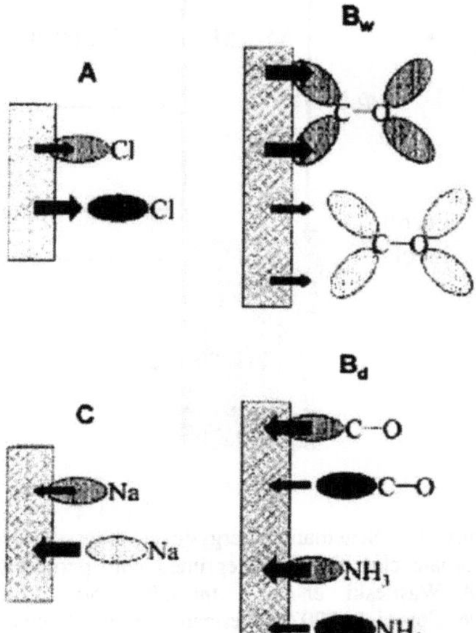

Figure 14. Schematic nature of the metal-chemisorbate charge polarization for bonding classifications A, B_w, B_d and C, as exemplified by orbital interactions for Cl, Na, CO and NH_3. Reprinted with permission from S. A. Wasileski *et al.*, *J. Am. Chem. Soc.*, **124**, (2002) 6796. Copyright 2002, American Chemical Society.

interacting energy level either far below or far above, respectively, the metal surface Fermi level. Covalent surface bonds (both withdrawing and drawing, classes "B_w" and " B_d"), on the other hand, exhibit an attenuation of the degree of metal-adsorbate charge polarization as the bond is stretched, as for these bonds the driving force for charge flow is the strength of the overlap between metal and adsorbate orbitals. These adsorbates will have their energy levels close to the metal's Fermi level, either just above (class B_d) or just below (class B_w). Figure 15 also illustrates the sensitivity of bonding character to the electrode

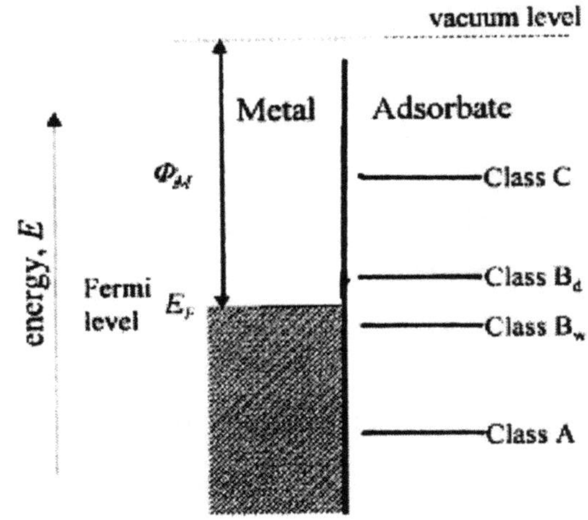

Figure 15. Schematic energy-level diagram for metal-adsorbate classification. Reprinted with permission from S. A. Wasileski *et al.*, *J. Am. Chem. Soc.*, **124**, (2002) 6796. Copyright 2002, American Chemical Society.

potential (electric field). Adjusting the electrode potential to more negative values corresponds to shifting the metal's energy levels (and hence the Fermi level) up with respect to the adsorbate levels, encouraging metal-to-adsorbate electron transfer. The opposite obviously happens at more positive fields. This will lead to corresponding changes in the dipole parameters μ_s and μ_D. Ultimately, by changing the electrode potential, one expects a transition from class A to B_w or B_d and even C with progressively more positive potential/field.

The final general situation concerning the field dependence of the dipole parameters μ_s and μ_D, the binding energy E_b, and the force constant $K_{M\text{-}A}$ is summarized schematically in Figure 16. At very negative field, both μ_s and μ_D start off as negative quantities. The solid lines illustrate a Class A-B_d-C transition for which, with progressively more positive field, μ_s changes sign before μ_D. The dashed lines illustrate a Class A-B_w-C transition for which, with progressively more positive field, μ_D changes sign before μ_s. The lower panels in the figure show how changes in the sign of μ_s and μ_D are accomponied by maxima in the $E_b - F$ and $K_{M\text{-}A} - F$ plots.

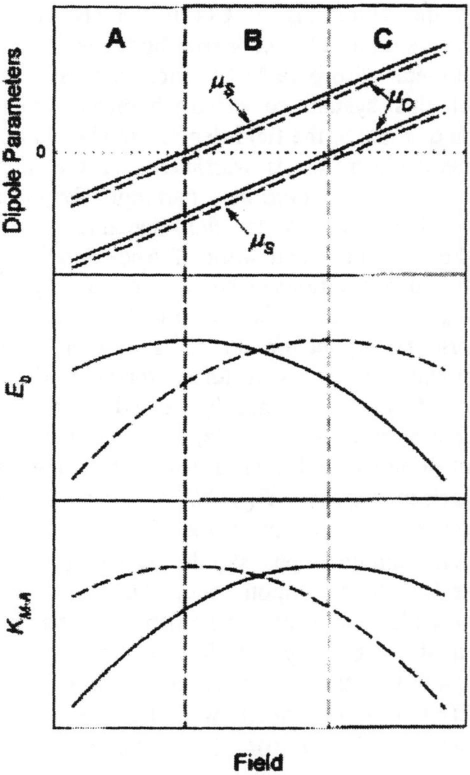

Figure 16. Top segment: schematic dependence of static (μ_s) and dynamic (μ_D) dipole moments on the applied field, showing field-induced transition expected for Class A/B_d/C and A/B_w/C metal-adsorbate bonds (solid and dashed lines, respectively). Middle segment: corresponding schematic plots metal-adsorbate binding versus field. (Note that increasing bond strength corresponds to lower, i.e. more negative, E_b values.) Bottom segment: corresponding schematic plots of the metal-adsorbate stretching force constant versus field (see text). Reprinted with permission from S. A. Wasileski *et al.*, *J. Am. Chem. Soc.*, **124**, (2002) 6796. Copyright 2002, American Chemical Society.

6. Field-Dependent Chemisorption of Carbon Monoxide

Ever since the first in situ Infrared spectral measurements of CO adsorbed on metal electrodes,[86] the experimentally observed potential dependence of the vibrations of CO in its chemisorbed state has attracted much attention. This concerns both the intramolecular C-O stretch and the metal-adsorbate M-CO vibration. Because of the great significance of this system to electrochemical surface science, a separate section devoted to the field-dependent chemisorption of CO on (transition-) metal electrodes is warranted. This is also the electrochemical system that has served as a paradigm for the application of quantum-chemical techniques to field-dependent electrochemistry, starting with the semi-empirical work of Anderson[87] and the ab initio Hartree-Fock based calculations of Bagus and coworkers.[88]

Bagus *et al.*[88,89] studied CO adsorbed on small Cu(100) and Pd(100) clusters and argued on the basis of their results that field-induced vibrational shifts observed for chemisorbed CO were mainly a physical Stark effect, *i. e.*, primarily caused by the changes in the electrostatic field across the molecule, and that chemical effects, *i. e.* field-dependent donation and back donation components, were small. They found Stark tuning rates $d\nu_{C\text{-}O}/dF$ of about 10^{-6} cm^{-1} (V/cm), in reasonable agreement with experiment. They also argued that the $d\nu_{M\text{-}CO}/dF$ Stark tuning slopes should be small due the non-ionic character of the CO chemisorption bond. The argument used by Bagus *et al.* to favor the physical Stark mechanism over the chemical mechanism (advocated, e.g., by Anderson in his semi-empirical calculations[87]) was that the self-consistent Hartree-Fock field-dependent energy $E(F)$ was described well by the perturbation-theory expression $E(F) = E(0) + \mu_s(0)F$. Since the zero-field static dipole moment $\mu_s(0)$ is a field-independent quantity, they concluded that the vibrational frequency shifts were caused only by a physical field-dipole interaction, and that chemical effects could be neglected.

Head-Gordon and Tully[90] reconciled these seemingly disagreeing points of view by noting that the static dipole moment is a first-order derivative, which indeed does not depend on the electric field, but that higher-order derivatives of the energy with respect to the field do implicitly incorporate changes in the wavefunction due to the field. This is basically because the field does not only change the energy, but also the geometry of the chemisorption bond. The perturbation theory coefficient that controls this behavior is precisely the dynamic dipole moment discussed extensively in the previous section, and the dynamic dipole moment does incorporate changes in the chemical bond due to field effects.

To illustrate the effect of the field on the CO chemisorption bond, Head-Gordon and Tully[90] carried out SCF-HF calculations (with neglect of electron correlation) for CO on 6- and 14-atom Cu clusters with a (100) surface orientation. Firstly, they looked at changes in the atomic Mulliken charges on the carbon and oxygen as a function of the field. These calculations clearly showed that as the field is made more negative, electrons flow from the metal to the chemisorbed CO. Decomposing the electron flow into its different symmetry contributions, it was found that the field effects occur almost exclusively through changes in the *E* HOMO, and in simplest qualitative terms this corresponds to changes in the extent of metal to $2\pi^*$ back donation due to the field. The field effects on the orbitals of A_1 symmetry, corresponding in part to changes in the donation component, were found to be much smaller. The fact that the field should primarily change the back donation contribution is expected on the basis of the simple energy diagram shown in Figure 7, since the $2\pi^*$ resonances are closest to the metal Fermi level.

Head-Gordon and Tully calculated Stark tuning slopes for the C-O stretch of *ca.* 2×10^{-6} $cm^{-1}/(V/cm)$ on both clusters, in reasonable agreement with the earlier HF calculations of Bagus *et al.* and experiment. However, the results for the Stark tuning slope of the Cu-CO stretch depended on the cluster: for the 6-atom Cu cluster, a negative Stark tuning slope was observed, in qualitative agreement with experiment, but for the larger 14-atom Cu cluster, non-monotonous behavior was observed (even though only three field strengths were considered). The latter result may in fact have reflected the existence of a maximum in the $\nu_{M\text{-}CO}$ - F relationship, as it is predicted to occur from the perturbation theory analysis in the previous section.

Further evidence for pronounced field effects on the bonding mechanism of chemisorbed CO came from the work of Illas, Ricart, and co-workers, who combined DFT calculations and SCF-HF calculations including electron correlation effects using the MP2 formalism.[91,92] These authors were also the first to study field effects of CO chemisorbed on a small 4-atom Pt(111) cluster, which is the electrode surface for which most of the available experimental results have been obtained. Using a projection operator technique, they showed that both the 5σ and $2\pi^*$ populations of the chemisorbed CO depend on the electric field. More electrons populate the $2\pi^*$ orbital with increasingly negative field, as expected from Figure 7, whereas the donation of electrons from the 5σ orbital to the metal orbitals decreases with negative field, also in accordance with Figure 7. The calculated Stark tuning slope was found to be somewhat higher for SCF-HF calculations compared to DFT-B3LYP results, but still in reasonable agreement with experiment. Interestingly, Illas *et al.* found a stronger CO chemisorption bond with increasingly negative field, in contrast to Head-Gordon and Tully[90] who found a stabilization of the CO chemisorption bond to their Cu(100) cluster at both positive and negative field. In retrospect, we can ascribe this to the two systems having their maximum in the E_b - F plot at different field strengths (see Figure 6). In later work, the Catalan group reported similar results for CO on a Pt(100) cluster,[92] for which they found a linear relationship between the field and the extent of $2\pi^*$ back donation, and also showed that the Stark tuning slope did not vary appreciably with Pt(111) cluster size.[93]

In collaboration with the Purdue group, we have carried out detailed analyses of the field-dependent chemisorption of CO on transition-metal (111) surfaces by DFT-GGA calculations on 13-atom clusters.[37,78,79,95] One of the objectives of our studies was to provide relationships between binding energetics, geometries, and vibrational properties, both of the intramolecular C-O and the metal-adsorbate bond.

We consider first the field-dependent binding energetics in the atop and hollow site of a 13-atom Pt(111) cluster. Figure 17 shows the total binding energy $E_b(F)$ (upper left panel) and its field-dependent decomposition into steric (lower left panel) and orbital components (right panels), where the latter is further decomposed into contributions of A_1 symmetry (donation contribution, upper right panel) and E symmetry (back donation contribution, lower right panel). The orbital components have been set to zero for vanishing electric field, as their absolute values are not of interest here. In addition to the calculations in which the C-O bond length was fully relaxed (solid lines), Figure 17 also shows the field-dependent binding parameters pertaining to a chemisorbed CO with its C-O bond length fixed at its uncoordinated vacuum value (dashed lines). This bond constraining is useful especially in analyzing the field-dependent steric contribution to the binding energy.

A significant result illustrated in Figure 17 is that atop coordination is preferred at positive fields, whereas at negative fields multifold hollow coordination is most favorable. This field-driven site switch on Pt(111) is in qualitative accordance with experiment.[65,66] It is seen that the back donation component gets more negative (*i. e.*, more stabilizing) towards negative field, whereas the opposite is the case for the donation component. This field-dependent donation and back donation effects are expected on the basis of Figure 7, and in agreement with earlier the ab initio calculations of Head-Gordon and Tully and Illas and co-workers. The back donation component is more strongly

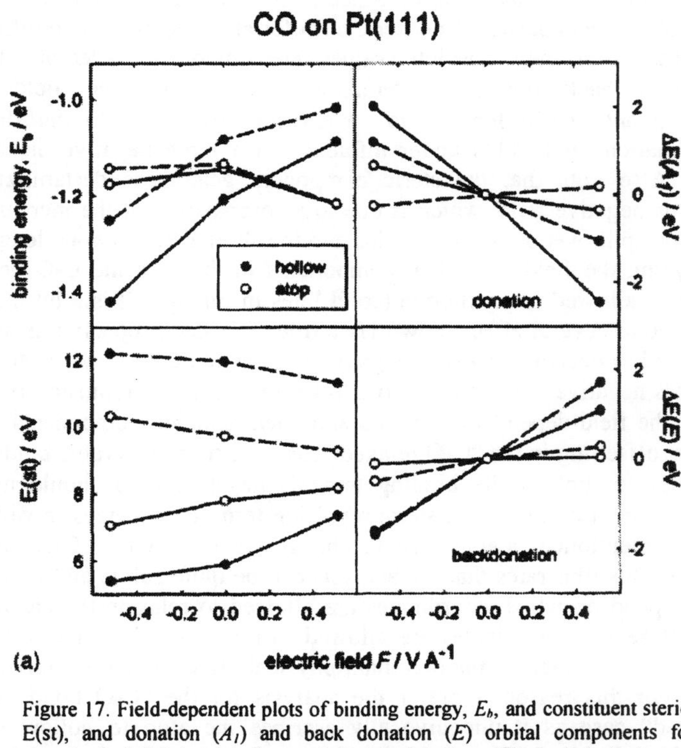

Figure 17. Field-dependent plots of binding energy, E_b, and constituent steric, E(st), and donation (A_1) and back donation (E) orbital components for chemisorbed CO in atop (open circles) and hollow (filled circles) sites on Pt(111) surface. Note that the orbital components are plotted as differences with respect to the zero-field values. Dashed plots refer to "bond-constrained" values, as described in the text. Reprinted with permission from M. T. Koper *et al.*, *J. Chem. Phys.*, **113**, (2000) 4392. Copyright 2000, American Institute of Physics.

field dependent for the multifold coordination, *i. e.* more stabilizing towards negative fields. This can be understood in terms of the bonding character of the back donation interaction, which will prefer to bind to as many surface atoms as possible. By contrast, the donation interaction mainly occurs through an antibonding resonance, and should therefore prefer to bind to as few surface atoms as possible, as already alluded to in Section III.4. This explains the more destabilizing effect of a more negative field on the back donation component for multifold coordination as compared to atop coordination. For the C-O "bond-constrained" case, the increase in the back donation interaction seen towards more negative fields is substantially larger than the opposite field dependence observed for the donation term. Since there is no significant difference in the field dependence of the steric term for atop and hollow coordination (although their absolute values are obviously different), the increased back donation interaction towards negative field is thedominant reason for CO to switch site from atop to multifold coordination on Pt(111) under influence of a more negative electric field. Also note that the steric component gets more destabilizing towards negative field, which is due to a combination of the increased electron spill over at negative fields and the shorter Pt-CO bond length. Analyzing the field-dependent components in the case the C-O bond length is allowed to equilibrate (solid lines in Figure 17) is a bit more involved as several effects now play a role. The steric repulsion is now observed to decrease towards negative field as the C-O bond stretches towards negative field, leading to a lower *overall* steric repulsion term. Since the field-dependent donation and back donation components are almost offsetting in the "C-O relaxed" case, the more favorable binding energies for hollow-site binding towards negative field should now formally be ascribed to the steric repulsion term, the changes in which are, as mentioned, mainly due to changes in the internal C-O bond length. This illustrates that the analysis can be quite subtle and may in fact depend on which coordinates are allowed to relax in the calculation. If several coordinates are allowed to relax, it will in general be difficult to disentangle the dominant physical effects in an unequivocal way. For this reason, I prefer the analysis for the "C-O bond constrained" case as it unequivocally ascribes the atop to hollow site switch under the influence of the electric field to the enhanced back donation contribution towards negative field.

Largely similar results were obtained for the other transition-metal surfaces, *i. e.* Pd, Ir, and Rh, although only on Rh a site switch is predicted to occur, similarly to Pt. On Ir and Pd, however, the energetic preference for atop and hollow-site CO binding, respectively, is sufficiently large to be unaffected by field alterations.[37]

We now turn our attention to the field-dependent C-O stretching frequency. Stark tuning slopes were calculated from the slope of the ν-F curve at $F = 0$, for both CO and NO (nitric oxide) on the four different transition-metal surfaces Rh, Ir, Pd, and Pt.[37] A comparison with the DFT-computed Stark tuning slopes with the experimental behavior[65,66] is shown in Figure 18. The experimental Stark tuning slopes $d\nu/d\mathrm{E}$, expressed in wavenumbers per Volt, were converted into field units, *i. e.* wavenumbers per Volt/Angstrom, by using the relation

$$\frac{d\nu}{dF} = d_{\mathrm{dl}} \frac{d\nu}{d\mathrm{E}} \tag{20}$$

where d_{dl} is the effective thickness of the double layer, taken to be *ca.* 3 Å.[96] The $d\nu_{DFT}/dF$ values are seen to be in rough accordance with the $d\nu_{exp}/dF$ estimates, although the variations of the former are smaller. These discrepancies are probably due to the role of co-adsorbed solvent in affecting the local electrostatic field. Nevertheless, the larger Stark tuning slopes computed for NO *vs.* CO are in accordance with experimental observations.

The field-dependent changes in the C-O and N-O frequencies were analyzed following the decomposition method described in Sec II.6. As reference state, the frequency at negative field F = -0.514 V/Å was taken, whereas the final state was the system at F = +0.514 V/Å. To the negative field PES the steric, donation, and back donation components calculated for the positive field were added sequentially, and the corresponding frequency changes calculated. Table 2 shows the results for CO adsorbed in the atop and hollow site of the Pt(111) cluster. Both the steric and the donation contributions are seen make a negative contribution (this was also found for CO and NO on other metals). The negative steric contribution likely arises from the lower metal electron surface electron density towards positive fields, diminishing the extent of steric repulsion with the chemisorbate electrons and hence yielding a shallower PES. The negative donation contribution is also readily understandable given that this orbital interaction should lessen towards more positive fields. Note, however, that the absolute values of this contribution are small, indicating an interesting insensitivity of the donation interaction to the electrostatic field. Table 2 also illustrates that the dominant contribution to the field-induced frequency changes

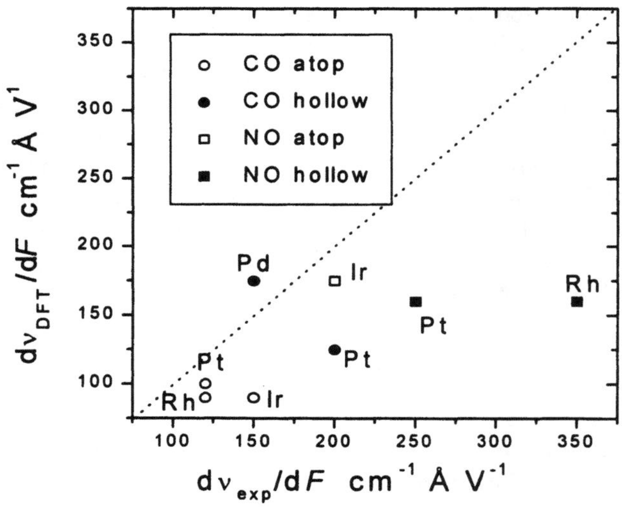

Figure 18. Comparison of the field-dependencies of ν_{CO} and ν_{NO} calculated by DFT, $d\nu_{DFT}/dF$, and experimental quantities, $d\nu_{exp}/dF$, as extracted from experimental potential-based values by using Eq. (20) (see text). Reprinted with permission from M. T. Koper *et al.*, *J. Chem. Phys.*, **113**, (2000) 4392. Copyright 2000, American Institute of Physics.

is the back donation component, being a large positive term, making the overall Stark tuning slope for the intramolecular vibration a positive quantity.

However, the decomposition analysis does not suggest any overriding trends that could be responsible for the variations observed in the Stark tuning slope as a function of adsorbate (CO or NO), coordination (atop *vs.* hollow) or metal substrate. For instance, Table 2 shows that the higher Stark tuning slope for CO bound in the hollow *vs.* the atop site (a result which itself is in agreement with experiment) cannot be attributed to a higher back donation component in the hollow site, as one might initially expect, at least not according to this method of analysis. The statement also holds for the higher Stark tuning slopes observed for chemisorbed NO *vs.* CO. The overall picture is rather one that invokes the offsetting influences of two or more components on the overall field-dependent behavior.

Table 2
Decomposition of Field-Induced Frequency Shift (in cm^{-1}) for CO Adsorbed in the Atop and Hollow Site on Pt(111).[a]

System	Steric	Donation	Back donation	Rest	Overall
Pt-CO atop	-131	- 7	+264	-16	+110
Pt-CO hollow	-72	-20	+201	+22	+131

[a] The far-left column is the cluster/binding site configuration. The far-right column is the frequency shift computed for a change in the applied field from –0.514 V/Å to +0.514 V/Å. The middle three columns give the corresponding field-induced frequency shift due to each contribution. The column Rest refers to the residual field-induced frequency shift not accounted for by the sum of these components.

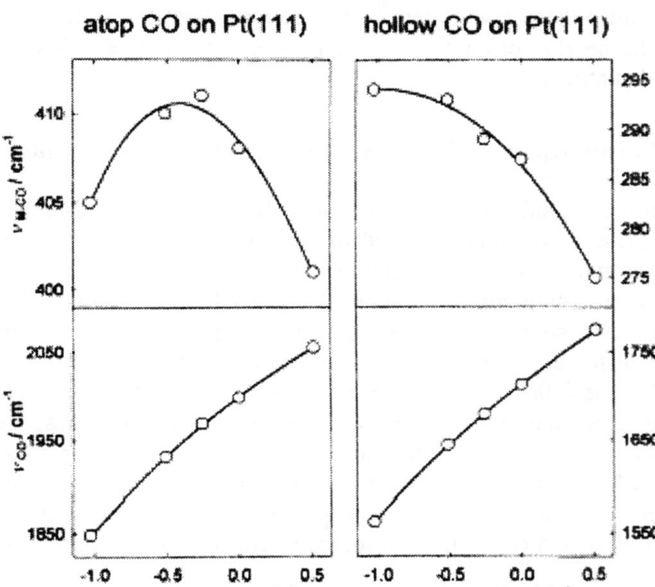

Figure 19. Stretching frequencies for metal-carbon ($\nu_{M\text{-}CO}$) and internal CO (ν_{CO}) bonds plotted against external field F for atop and hollow-site CO on Pt(111).

The field dependence of the metal-adsorbate $\nu_{M\text{-}CO}$ was also studied in some detail. Figure 19 shows the $\nu_{M\text{-}CO}$ - F curves compared to the $\nu_{C\text{-}O}$ - F for CO adsorbed at the atop and hollow site on a Pt(111) cluster. For CO in the hollow site and for CO in the atop site at sufficiently positive potential, the $\nu_{M\text{-}CO}$ - F curve has a negative Stark tuning slope, in agreement with Raman spectroscopic results on polycrystalline Pt.[97] However, the curve for atop CO shows a well-pronounced maximum, as at negative potentials/fields the $\nu_{M\text{-}CO}$ - F curve has a positive slope. Such behavior is indeed predicted to occur from the analysis presented in Section III.4 (see Figure 16), and is related, at least qualitatively, to a change in the sign of the dynamicdipole of the M-CO bond as a function of field. For hollow site CO, the maximum is not observed. This is most likely related to the much more negative chemisorption bond at negative fields for this geometry (see Figure 17), which favors progressively larger $\nu_{M\text{-}CO}$ values under these conditions.[79] The combination of the two effects yields the nearly constant $\nu_{M\text{-}CO}$ values seen at negative fields in the DFT calculations.

7. Chemisorption of Water and Water Dissociation Products

Most of the available quantum-chemical studies of water adsorption at metal surfaces have been limited to the interaction of a single water molecule with the metal surface. However, recently there have appeared a few publications presenting results for water at higher coverage (the so-called "bilayer"), and for "liquid water" at a metal surface using ab initio molecular dynamics. We will discuss these studies in the following. Apart from undissociated water, electrochemists are interested in the chemisorption of the primary dissociation products of water, namely chemisorbed hydrogen H_{ads} and chemisorbed OH_{ads}, which can be observed at most metal-liquid interfaces at sufficiently negative, respectively positive, electrode potential. Ab initio studies related to these adsorbates will also be discussed in this Section.

The chemisorption of a single water molecule has been studied mainly by ab initio cluster calculations.[98-105] Depending on the nature of the metal surface, the calculated adsorption energy is rather low, ranging from *ca.* 20 to 60 kJ/mol.[98-108] Practically all studies find that the water molecule prefers to adsorb on top of a surface metal atom, and also suggest that the most stable geometry for water adsorption is a tilted one, the tilt angle of the plane of the water molecule with the surface normal varying from 35 to 70°. (However, Nazmutdinov *et al.*[102] found that on a 7-atom mercury cluster using MP4 calculations water prefers a perpendicular orientation with the water plane parallel to the surface normal.)

Some of the earliest ab initio work on the interaction of water with metal surfaces was a DFT-LDA study by Müller and Harris,[98] who looked at a nine-atom Al(100) cluster with water adsorbed on top of a Al(100) surface atom. The binding energy in this state was *ca.* 50 kJ/mol and the tilt angle 55° with respect to the surface normal. They found that the bonding was accompanied by a small net donation of charge (*ca.* 0.1 electron) from the water molecule to unoccupied cluster levels. This donation leads to a strong reduction of the work function (which is known to occur experimentally) even when the intrinsic dipole of the adsorbed water is oriented (almost) parallel to the surface. The electron donation is accompanied by a slight expansion of the O-H bond length and the HOH bond angle. Müller and Harris interpret the binding in terms of the interaction of the two water lone pair states, *i. e.* the σ-like $3a_1$ and the π-like $1b_1$ orbitals, with empty *p*-type orbitals on the Al. For the on-top site the overlap of the $3a_1$ state with the metal orbitals is largest when the tilt angle with the surface normal is 0°, whereas the overlap of the $1b_1$ state with the metal orbitals is largest when the tilt angle with the surface normal is 90°. The competition between these two determines the final geometry. Very similar conclusions were reached by Ribarsky *et al.*[99] in an early DFT-LDA study of water on a five-atom Cu cluster.

Ignaczak and Gomes[105] presented a comparative DFT-GGA study of water adsorption on 12-atom (100) cluster models of Cu, Ag, and Au. They find that on these surfaces, there is no strong energetic preference for the atop or bridge site. When adsorbed on-top, the tilt angle with the surface normal was in the range 50-65°. Copper is found to be most hydrophilic metals of the three with a water adsorption energy of *ca.* 32 kJ/mol, followed by gold (*ca.* 30 kJ/mol) and silver (27 kJ/mol).

Izvekov and Voth[107] presented results of a slab calculation of a single water molecule with a three-layer Ag(111) slab with 16 surface atoms per unit cell. They find that the on-top adsorption site is strongly favored over the bridge and hollow sites, the binding energy in the on-top site being *ca.* 50 kJ/mol more favorable. The HH vector was found to be parallel to the surface, whereas the tilt angle was calculated to be 61°. As mentioned, a strong tilting is consistent with a picture in which the water molecule interacts with the metal surface through its lone pair $1b_1$ orbital, which derives mainly from the oxygen p_z orbital that is perpendicular to the plane of the water molecule. Very similar results were found by these authors in a related DFT-GGA slab study of water interacting with the Cu(110) surface,[106] although the interaction energy with the more open Cu(110) surface is found to be significantly larger, *ca.* 100 kJ/mol at the atop site and *ca.* 70 kJ/mol in the bridge position.

Vassilev *et al.*[108] studied water adsorption at different coverages on slabs of Rh(111) and Pt(111) using DFT-GGA theory. At low coverage, formally corresponding to 1/12 ML coverage, water was found to lie almost flat on both the Rh and the Pt surface, with Rh being slightly more hydrophilic (37 *vs.*. 30 kJ/mol adsorption energy). This suggest that the main binding interaction occurs through the water $1b_1$ orbital with empty metal orbitals, which is also corroborated by the broadening of the $1b_1$ orbital in a projected density-of-states analysis. No significant changes with respect to this low coverage are observed for a 1/3 ML coverage, corresponding to a $(\sqrt{3}x\sqrt{3})R30^{\circ}$ overlayer, but the effective adsorption energy per water molecule gets significantly more favorable, *ca.* 56 kJ/mol, for the 2/3 ML coverage. At this coverage, the water layer forms a so-called bilayer, the structure of which is illustrated in Figure 20(a), and in this structure strong hydrogen-bonding interactions exist between neighboring water molecules. Figure 20(b) shows how one can distinguish between different water molecules in the bilayer, the ones closest to the surface forming two hydrogen bonds with neighboring water molecules, and the ones further away from surface forming one hydrogen bond and having one O-H bond pointing away from the surface into the vacuum. Interestingly, when the binding energy of the entire bilayer is calculated, *i. e.* the energy of metal-bilayer system minus that of the metal surface and the isolated bilayer in vacuum (in a restricted geometry), it is found that the interaction with the metal surface is essentially vanishing, for both Pt and Rh. This is even true for an adsorbed bilayer that is flipped upside-down (see Figure 20c), *i. e.* with its "dangling" O-H bonds pointing towards the metal surface, which is found to be just as stable (or even slightly more stable) than the original bilayer with the dangling O-H bonds pointing to the vacuum. This suggests that in these calculations the structure of the water bilayer is dominated by the lateral water-water interactions and not influenced much by the water-metal interaction. From the lack of any orientational preference one also deduces that in a real electrochemical system the orientation of the interfacial water would be determined primarily by its interaction with the prevailing electric field,

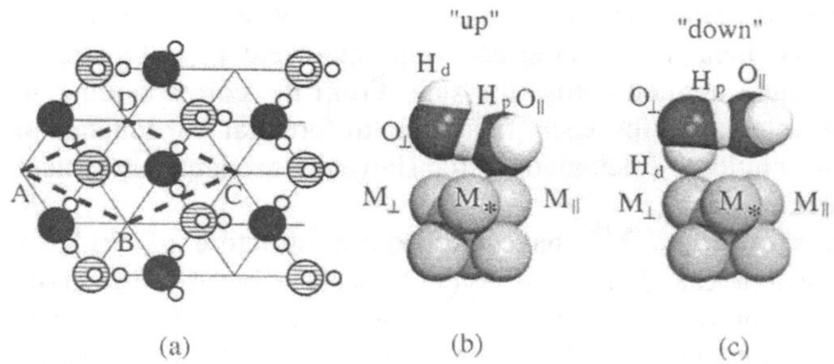

Figure 20. (a) Surface configuration of a perfect water bilayer and smallest possible unit cell ABCD used in the simulation. The substrate is represented through lines connecting the positions of the surface atoms; filled circles represent the oxygens of the first layer (closest to the surface), shaded circles represent the oxygens of the second layer, and open circles represent the hydrogens. (b) Side view of the "perfect" bilayer. (c) Side view of a configuration obtained by flipping the configuration shown in (a). $M_{\parallel}$ and $M_{\perp}$ are the two surface atoms with adsorbed water molecules, M• is the free surface atom.

and to a much smaller extent by its interaction with the metal. Two reasons may be given for this somewhat surprising lack of metal-water interaction at high water coverages. First, the small cell size may introduce artifacts; a larger unit cell may yield a lower-energy geometry with a more pronounced orientational preference. Secondly, at such low adsorption enthalpies, entropic effects may become important and hence finite-temperature effects may be relevant. To study the effect of temperature, Vassilev *et al.*[109] also carried out an ab initio molecular dynamics simulation of the water bilayer on Rh(111). The unit cell was larger in these studies, containing 8 water molecules and 12 surface metal atoms. The bilayer has a higher adsorption energy than in the smaller simulation cell, but on average there was no preferred orientation of the water in the bilayer.

Halley, Price and co-workers were the first to apply large-scale "very nearly" ab initio calculations to the study of the structure of liquid water-metal interface in a slab geometry.[12,110,111] The metal valence electrons were treated in the DFT formalism, but the water molecules were interacting via classical potentials. For the Cu(100)-

water interface they observed an effective repulsion between the Cu and the water in the sense that the first peak in the distribution function represented much less than a monolayer of water. The potential drop across the metal-water interface was calculated as a function of an electric field applied across the slab. From the corresponding surface charge calculated for each field, the differential capacitance of the interface could be calculated, which showed reasonable agreement with experiment.

Izvekov *et al.*[106,107] have studied the structure of "bulk" liquid water near a Cu(110) and a Ag(111) surface by ab initio molecular dynamics simulations using the Carr-Parrinello method. As all particles in the system were treated from first DFT principles, these are extremely time-consuming calculations that are at present only feasible for small system sizes and short simulation times. To realize the limitations, it is useful to quote the exact sizes and simulation times. For the Cu(110)-water interface, the unit cell consisted of a seven-layer metal slab, with nine atoms per layer, and the space between the slabs was filled with 24 water molecules so that a realistic water density was obtained. For the Ag(111)-water interface, the unit cell consisted of a five-layer metal slab, with sixteen atoms per layer, and the space between the slabs was filled with 48 water molecules. Total simulation times over which the statistical averaging occurred, was *ca.* 2.5 and 2.1 ps, respectively. Nevertheless, as Izvekov *et al.* showed, some insightful conclusions may still be drawn from such calculations. We will restrict our discussion here to the Ag(111)-water simulation, because this system is of great interest to many experimental electrochemists as some controversial Surface X-Ray Scattering data on this system were published by Toney *et al.*[112,113] Figure 21 shows the calculated Ag(111) wall-oxygen distribution function $g(z)$, z being the distance from the metal surface, and the running integration numbers (which give information about the number of water molecules in the water layers). The first relatively sharp peak corresponds to a water overlayer formed by *ca.* 9-10 water molecules (*i. e.* a density of around 0.63 water molecules per Ag surface atom), whereas the second peak is much broader and is formed by 5-6 water molecules on average. The simulation does not reproduce the extremely high surface water density in the first layer (*ca.* 1.1 to 1.8 water molecules per surface atom) extrapolated from the Surface X-Ray measurements. Izvekov *et al.* suggest that this may be due to an overlap of the first two peaks in the experimental distribution function. From an analysis of the trajectories of the water molecules in the first layer, Izvekov *et al.* find that their

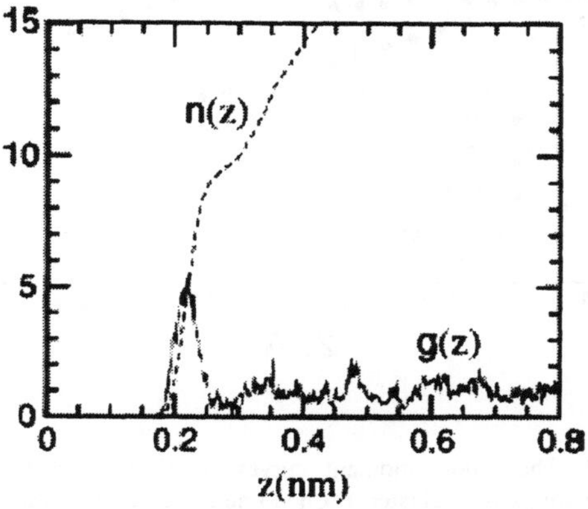

Figure 21. Wall-oxygen distribution function g(z) obtained from the AIMD simulation of the Ag(111)-water interface, and the corresponding running integration numbers n(z) as a function of the distance z from the Ag(111) surface. Reprinted with permission from S. Izvekov *et al.*, *J. Chem. Phys.*, **115**, (2001) 7196. Copyright 2001, American Institute of Physics.

positions are strongly localized, mainly residing on top of the surface metal atoms. This is linked to the strong preference of chemisorbed water for the atop adsorption site. The water molecules in the second layer are much more mobile and their projected position onto the surface may actually change over a lattice constant within the simula tion time. The preferred orientation of the water in the first layer is very close that of an isolated water molecule on the Ag(111) surface, with its HH vector parallel to the surface and the tilt angle being close to 60°. The orientational distribution in the second layer is much broader, though there is a tendency for the water molecules in this layer to have their dipole directed away from the metal surface. Izvekov *et al.* also looked at the changes in the electronic structure of the "solvated" Ag(111) surface compared to the clean Ag(111) surface. The electronic coupling of water molecules mainly occurs through the slab crystal electronic states, whereas the slab surface states remain almost uncoupled to the water. A similar result was found for the Cu(110)-water interface.

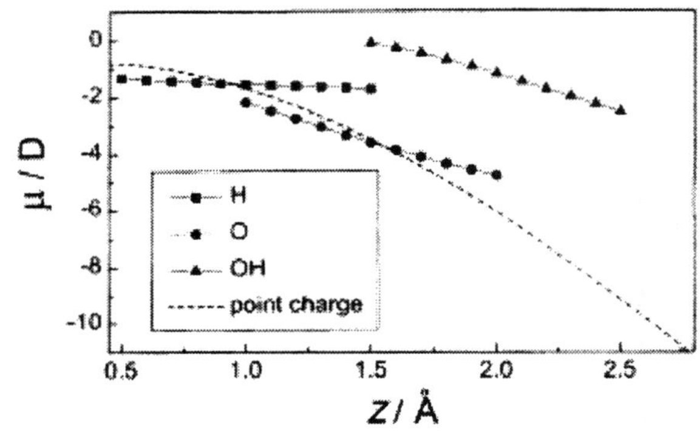

Figure 22. The dipole moment curves of H, O, and OH interacting with a Ag_{13} cluster. Dashed line is the corresponding dipole moment curve of a negative point charge interacting with the Ag_{13}^{+} cluster.

Apart from water being present as a solvent at the vast majority of electrochemical interfaces, it is also a reactant or the precursor to a reactant in many electrochemical reactions. The reactivity of water is usually related to its dissociation into adsorbed hydrogen at negative potentials, or into adsorbed hydroxyl or oxygen at positive potentials. The interaction of H, OH and O has therefore been studied in the electrochemical context in quite a few quantum-chemical studies.

Koper and van Santen[114] studied the adsorption of H, O, and OH on a series of metal clusters in the absence of water molecules. A key difference between these adsorbates is the type of bond they form with the metal surface. Atomic hydrogen forms an essentially covalent bond, whereas O and OH form much more polar bonds with the surface. Figure 22 illustrates this by showing the dipole moment curves for the three adsorbates interacting with the hcp hollow site of a 13-atom Ag(111) cluster, compared to the same curve for a negative point charge interacting with the Ag^{+} cluster. The lack of any distance dependence of the static dipole moment of the Ag(111)-H bond is suggestive of its covalent character. In agreement with experiment, the strength of the hydrogen bond with the transition-metal surface is rather metal independent, and amounts to *ca.* 240 kJ/mol. The adsorption on the coinage metals Cu, Ag, and Au is weaker by about 50 kJ/mol. This is consistent with the experimental observation that a strongly adsorbed underpotential deposited form of hydrogen exists on transition-metal electrodes, but not on the coinage-metal electrodes. The covalent character of the hydrogen-metal bond is also corroborated by the weak field dependence of its binding energy, the Mulliken and Hirshfeld charges on the hydrogen, the metal-hydrogen equilibrium distance, and the metal-hydrogen vibrational frequency.[114,115] An outstanding issue in the underpotential (upd) deposition of hydrogen on Pt(111) is the preferred adsorption site. In situ vibrational spectroscopy (Infrared and Sum Frequency generation) have given conflicting results as to whether upd hydrogen on Pt(111) is adsorbed in the atop, bridge or threefold hollow position.[116,117] Detailed slab DFT-GGA calculations of H on Pt(111) suggest that the atop site is energetically favored at a coverage of 0.25, but the difference with the other high-symmetry adsorption sites is not more than *ca.* 10 kJ/mol.[118] These calculations will have to be done in the presence of water to get more insight into the electrochemical situation.

Recently, Paredes Olivera *et al.*[119] presented an ab initio study of hydrated hydronium adsorption on Ag(111) using the SCF-HF MP2 formalism. They found that on a 28-atom Ag(111) cluster a single hydronium H_3O^+ ion prefers to coordinate with the oxygen above a fcc hollow site and the three hydrogens pointing towards atop silver atoms. The binding energy was *ca.* 230 kJ/mol and only weakly dependent on the adsorption site. Coordination through the oxygen was not favorable. Adding three water molecules to the system made the hydronium ion shift to a second layer, with the water molecules interacting directly with the silver surface, indicating that the hydronium ion does not specifically adsorb. At negative electric field, it was observed that the hydrogen bond of the hydronium ion with the water molecules shortens, suggesting the onset of proton transfer to the surface water. The surface waters also reorient with the electric field so that their dipoles screen the field more effectively.

The first product formed from water dissociation at positive electrode potentials is surface-bonded hydroxyl OH. The dipole moment curve shown in Figure 22, the calculated Mulliken en Hirshfeld charges (-0.37 and –0.27, respectively, for OH on both a 13-atom Ag(111) and a Pt(111) cluster), and the field dependence of the various binding characteristics, all suggest that OH forms a polar bond with the metal surface, and hence should be viewed upon as a surface hydroxide $OH^{\delta-}$.[114] This is due to the 1π orbital which is occupied by one electron in the uncoordinated OH, and whose energy lies below the Fermi level of the metal. Hence electronic charge is transferred from the metal to the OH upon adsorption. The bonding of OH is generally weaker than that of atomic O because of the lower degeneracy of the 1π orbital compared to the *2p* orbital on oxygen.[114]

The preferred binding site of OH on Pt(111) and Ru(0001) has been studied by DFT-GGA slab calculations by Michaelides and Hu[120] and Koper *et al.*[71] Both groups found that on Pt(111) the atop adsorption site is preferred, with a binding energy of *ca.* 223 kJ/mol. The binding geometries for the three different sites (atop, bridge, and hollow) are shown in Figure 23. Surface-bonded OH is tilted in the on-top and bridge sites, but adsorbs upright in the hollow site. These binding geometries in the different sites can be understood qualitatively by the sp^3 hydribization of the atomic orbitals on oxygen, which prefers a tetrahedral coordination. Contrary to Pt(111), OH prefers the upright position in the fcc hollow site on Ru(0001).[71] In agreement with the higher oxophilicity of Ru, the binding of OH to Ru(0001) is stronger than to Pt(111), *i. e.*, 340 *vs.* 223 kJ/mol.

Vassilev *et al.*[121] studied the chemisorption of OH on Rh(111) in the presence of co-adsorbed water molecules by ab initio molecular dynamics. The OH was "created" on the surface by starting with a configuration from a water bilayer on Rh(111), and then removing the two hydrogens from a water molecule in the first layer of the bilayer. The resulting oxygen species is not stable and reacts quickly in the simulation with a neighboring water molecule to produce two surface-bonded OH species. It was found that the OH species is highly mobile on the surface due to a fast proton hopping between neighboring water molecules and the OH, implying an OH hopping in the opposite direction, as illustrated in Figure 24. This is similar to the Grotthus mechanism for proton and hydroxyl mobility in bulk water. Interestingly, in the simulation the surface mobility of OH is higher than that calculated for OH^- in liquid water from similar simulations by Tuckerman *et al.*[122] Also, from the simulations it is concluded that surface-bonded OH should be considered as a $OH(H_2O)_2$ complex composed of the OH itself and two neighboring water molecules. A top view of the complex is shown in Figure 25. The complexing water molecules are characterized by shorter hydrogen bonds with the OH than with other water molecules.

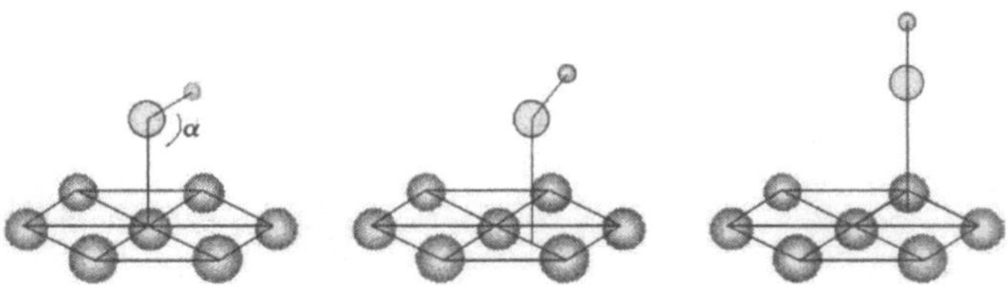

Figure 23. Typical orientations of surface-bonded OH in the atop, bridge, and hollow sites. Reprinted with permission from M. M. T. Koper *et al.*, *J. Phys. Chem. B*, **106**, (2002) 686. Copyright 2002, American Chemical Society.

8. Ab Initio Approaches to Modeling Electrode Reactions

Understanding and ultimately predicting chemical reactivity is one of the main goals of ab initio quantum chemistry. Electrochemical reactions are obviously one of the most challenging classes of chemical reactions to model, due the complexity of the electrode-electrolyte interface. It is the aim of this section to present a few approaches that have been taken in this direction. These approaches roughly divide into two different streams: one in which the potential energy surface including the effect of the solvent is studied as a function of a number of reaction coordinates at zero temperature, *i. e.* as a series of electronic structure calculations, and one in which the dynamics of the reacting system is calculated from first principles using AIMD. Both approaches are very time consuming and may in fact sample only a limited part of the available configuration space. From the statistical-mechanical point-of-view, only classical molecular dynamics can at present provide a statistically satisfactory description of the role of the solvent for selected reactions, though even this approach is computationally very demanding. Classical molecular dynamics simulations of electrode reactions have been described in a number of recent papers.[123-127]

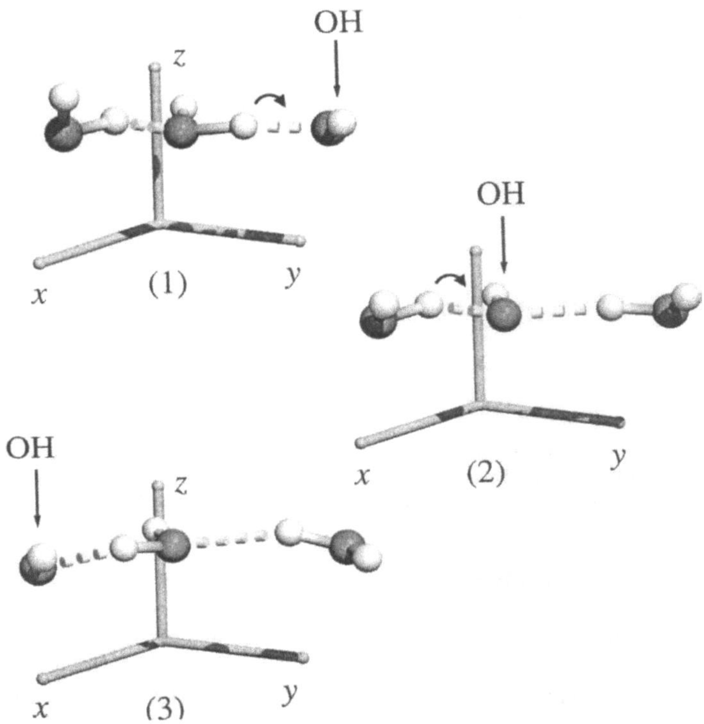

Figure 24. Subsequent configurations in the chain of proton exchange reactions between neighboring OH and H_2O molecules leading to an effective "shift" in the position of the surface hydroxyl. The Rh(111) surface is in the *x-y* plane.

Kuznetsov and Lorenz[128,129] were the first to apply ab initio quantum-chemical methods to study one of the most elementary reactions in electrochemistry, the hydrogen evolution reaction. Using SCF-HF calculations they computed potential energy surfaces for the approach of a hydrogen from a solvated hydronium (*i. e.* $H_3O(H_2O)_2$) to

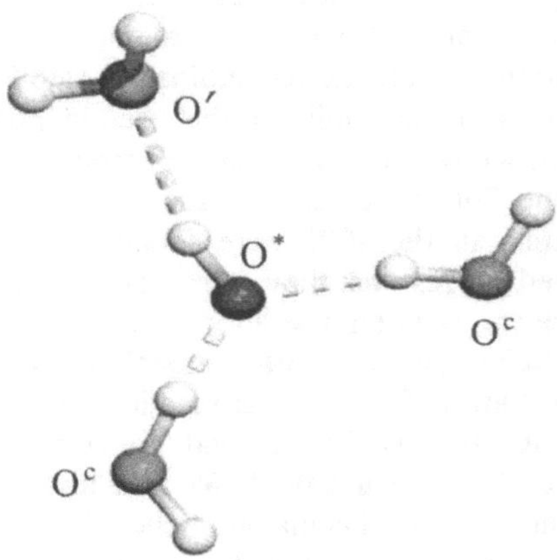

Figure 25. Top view of the local environment of the surface hydroxyl species. The metal surface is in the plane of view. O* is the oxygen of the hydroxyl, O^c are the oxygens of the complexing water molecules, O' is the oxygen of the nearest non-complexing water.

a Cu(100) cluster surface, to model the Volmer reaction $H_3O^+ + e \rightarrow H_{ad} + H_2O$. One of the OH bonds of the hydronium was pointing towards the hollow site of the cluster surface, and the potential energy as a function the distance of this hydrogen was calculated for different distance of the hydronium's oxygen from the surface, keeping the rest of the system to a fixed C_{2v} symmetry. For sufficiently large distances of the oxygen from the surface, the PES posseses two minima, the one corresponding to the left-hand side of the reaction, the other to the right-hand side of the reaction. The structure of the hydronium is only weakly distorted due to the presence of the metal. In the calculation, the barrier for the proton transfer is quite small and the overall reaction quite exothermic, meaning that the right-hand side of the reaction equation is more stable. A complete geometry optimization suggests a structure in which the water from the hydronium directs one its hydrogens to the surface, leading the a H(HOH) structure reminiscent of a H_3O^- anion. Indeed, the charges (presumably Mulliken charges) calculated suggest a negatively charged H_3O species, which becomes positively charged with increasing distance of the oxygen from the surface. Kuznetsov and Lorenz also studied the next step in the hydrogen evolution, namely the recombination step $H_3O^+ + H_{ad} + e \rightarrow H_2 + H_2O$, using a structure similar to that used in modeling the Volmer reaction, *i. e.* a structure that has a hydrogen atom already present in the hollow site of the Cu(100) cluster. In spite of the restricted geometry (C_{2v}), two minima in the PES were found, and their charges and geometries indeed suggest that they correspond to a hydronium ion and molecular hydrogen. The demonstration that it is feasible to study this reaction using such a simple ab initio model is quite significant. Very similar ab initio studies of the Volmer reaction on Pt(111) and Ag(111) were recently published by Ohwaki and Yamashita,[130] who explicitly included the role of the electric field. We note also that for the Volmer reaction, a recent model by Pecina and Schmickler[131] suggests that the activation energy for proton transfer to a non-catalytic metal is determined by solvent reorientation, which can be studied by classical simulation methods, rather than by the proton shuttling itself.

Anderson and co-workers[132-135] used similar extensive ab initio DFT-GGA electronic structure calculations to study an equally important electrochemical reaction, namely oxygen reduction. The oxygen reduction was supposed to take place in the following four electron transfer steps:

$$O_2 + H^+(aq) + e^-(U) \rightarrow HO_2^\bullet(aq) \tag{21}$$

$$HO_2^\bullet + H^+(aq) + e^-(U) \rightarrow H_2O_2(aq) \tag{22}$$

$$H_2O_2 + H^+(aq) + e^-(U) \rightarrow HO^\bullet(aq) + H_2O(aq) \tag{23}$$

$$HO^\bullet + H^+(aq) + e^-(U) \rightarrow H_2O(aq) \tag{24}$$

where U is the electrode potential. The solvation structure of the proton was taken into account by modeling it as a solvated hydronium ion $H_3O^+(H_2O)_2$, whereas the solvation of the other species was not considered. The potential energy surface was calculated in a restricted

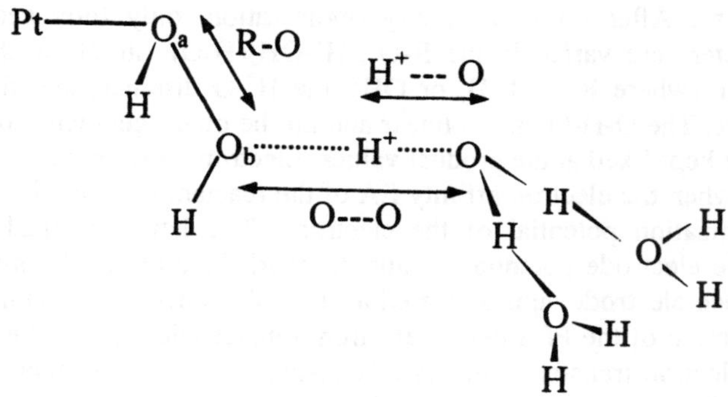

Figure 26. Structure and parameter definitions of the reaction complex illustrated schematically for the case of H_2O_2 reacting with a Pt atom. Reprinted with permission from A. Anderson *et al.*, *J. Electrochem. Soc.*, **147**, (2000) 4229. Reproduced by permission of The Electrochemical Society, Inc.

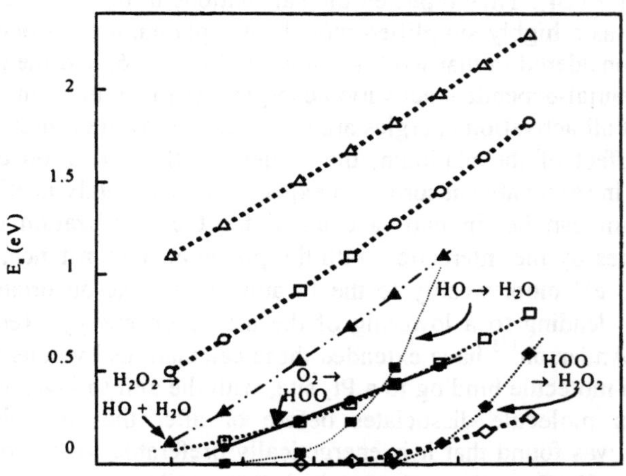

Figure 27. Activation energy, E_a, for the four steps in the oxygen reduction to water, reactions (21)-(24), as a function of the electrode potential *U*. Heavy lines connect points with species undergoing reduction bonded to a platinum atom; dotted lines connect points with no bonding to platinum. The same key applies to both sets of curves. Reprinted with permission from A. Anderson *et al.*, *J. Electrochem. Soc.*, **147**, (2000) 4229. Reproduced by permission of The Electrochemical Society, Inc.

geometry. After initial geometry optimization, only three geometric parameters are varied in the R-O...H^+-$OH_2(H_2O)_2$ structure, the R-O distance (where R is O, H, or OH), the H^+-O distance, and the O-O distance. The O-H-O part is linear and all the other geometric parameters are kept fixed at the product values. Electron transfer is assumed to occur when the electron affinity EA of the reactant complex is equal to the ionization potential of the electrode. The latter is equal to the variable electrode potential U plus the work function of the hydrogen reference electrode, approximated as 4.6 eV. Hence, one examines the intersection of the PES before electron transfer with that of the system after electron transfer at points of constant EA = U, and then use the one of lowest energy to specify the transition state parameters. Using this procedure, one can calculate the activation energy for each of the reactions as a function of the electrode potential.

From this type of calculation, Anderson and Albu[132,133] found that the activation energies for reactions 21-24 decrease in the order (23) > (21) > (22) > (24). They repeated the calculations in the presence of a Pt atom,[134] as a highly simplified model for a platinum electrode. The structure considered in that work is shown in Figure 26, and the results for the potential-dependent activation energies are given in Figure 27. It is seen that all activation energies are significantly lowered through the catalytic effect of the platinum, the values for the activation energy now being in reasonable accord with experiment. The catalytic effect of the platinum can be, in part, accounted for the stabilization of the intermediates by the interaction with the platinum. For instance, O_2 is bonded 1.3 eV more weakly to the Pt atom in an end-on orientation than HOO, leading to a lowering of the activation energy. Recently, Sidik and Anderson[135] have extended these calculations by considering the oxygen molecule binding to a Pt_2 site, with the aim to investigate if the oxygen molecule dissociates before or after the first electron transfer. It was found that it is energetically preferable for the oxygen to first form HOO on the surface, which then easily dissociates into O and OH on the surface. In this model, the first electron step to form HOO on the surface has the highest activation energy, in agreement with kinetic models from the literature (see, e.g., the review by Markovic and Ross[75]).

Anderson *et al.*[136] used a similar approach based on ab initio SCF-HF MP2 theory to study the formation of surface-bonded OH and O from water on one- and two-atom Pt electrode models. They conclude that OH will not be formed by the dissociation of water into surface-bonded OH and H, but rather by a transfer of the releasing proton to neighboring water. Specifically, Anderson *et al.* considered the reaction:

$$Pt\text{-}HOH...OH_2(OH_2)_2 \rightarrow Pt\text{-}OH...H^+\text{-}OH_2(OH_2)_2 + e^- \qquad (25)$$

and conclude from their calculations that this reaction should be thermodynamically possible at electrode potentials of 0.6 V (NHE) and higher, in good agreement with the potential at which the formation of OH_{ad} is presumed to take place on Pt(111).[75] The low activation energy found for the process is also in agreement with the highly reversible nature of OH_{ad} formation as usually assumed in the experimental literature.[75] Similar results are obtained for the formation of atomic oxygen from surface-bonded OH.[136]

A very similar result was obtained recently in a DFT-GGA study by Desai and Neurock,[137] investigating the role of the solvent in the dissociation of water over a Pt-Ru alloy. The PtRu system was modeled in the slab geometry, with a Pt:Ru ratio of 2:1 in $\sqrt{3}x\sqrt{3}$ arrangement and a slab thickness of three layers. The unit cell contained 9 surface atoms and 23 water molecules. In the absence of water, the authors found the $H_2O \rightarrow OH_{ad} + H_{ad}$ to be endothermic by 53 kJ/mol, having an activation barrier of 105 kJ/mol. In the presence of water, the same reaction has a reaction energy of 43 kJ/mol, and an activation barrier of 80 kJ/mol. This lowering of the energies is due to a stabilization of the charge separation in the OH and H adsorbates owing to the interaction with the water. However, the most favorable reaction is the one similar to reaction 25, *i. e.* dissociation into OH_{ad} and a solvated proton, with a reaction energy of 26 kJ/mol and a barrier of only 27 kJ/mol. The hydroxyl intermediates are most stable on top of Ru sites, whereas the proton is solvated as an $H_5O_2^+$ species, with only one water molecule separating it from the surface-bonded OH. Using ab initio molecular dynamics simulations, the authors found that the surface-bonded OH may migrate from Ru to Pt by a mechanism equivalent to the OH migration mechanism on Rh(111)[121] described in the previous section. The intriguing simulation by Desai and Neurock suggests that the presence of water on Ru promotes the formation of OH on a neighboring Pt site, as normally an isolated OH on Pt (even in a PtRu alloy) is much less stable than an isolated OH on Ru.[71]

The formation of surface-bonded OH is of great interest for another important electrocatalytic process, namely the oxidation of CO. It is usually assumed by experimentalists that the surface-bonded OH reacts with CO to form CO_2 in the reaction scheme:[75]

$$H_2O \rightarrow OH_{ad} + H^+ + e^- \tag{26a}$$

$$CO_{ad} + OH_{ad} \rightarrow CO_2 + H^+ + e^- \tag{26b}$$

A good CO oxidation catalyst should combine a high OH_{ad} coverage with a low CO binding energy, as this would lead to CO oxidation at a low potential. For this purpose, Pt-based Pt-M bimetallic catalysts are often used, where the OH forms on the more hydrophilic M site, but would preferably also lead to a lowering of the binding energy of CO on the Pt sites. The latter is often referred to as the electronic effect. Many authors have calculated CO and OH adsorption energetics on both clusters and slabs, confirming that there is indeed an electronic effect on for instance PtRu (as discussed already in section III.4). Ishikawa *et al.*[138] calculated the reaction and activation energies for the $H_2O \rightarrow OH_{ad} + H_{ad}$ reaction on 10-atom Pt-M mixed metal clusters by DFT-GGA calculations to study possible electrocatalysts. As mentioned, this is not the relevant reaction, but since the H adsorption energies vary less than the OH adsorption energies, the results could still be useful, though all reaction and activation energies will be too high. They estimate the activation energy from the bond energies using an analytical approach suggested by Shustorovich.[139] They find that PtRe and PtW should have lower reaction and activation energies for the $H_2O \rightarrow OH_{ad} + H_{ad}$ reaction than PtRu. They used a similar approach for the CO oxidation reaction, assuming that this reaction takes place through the formation of a COOH intermediate, *i. e.* $CO_{ad} + OH_{ad} \rightarrow COOH_{ad}$. They calculate this reaction to be intrinsically the fastest on Pt, although the calculated activation energy (*ca.* 0 kJ/mol) seems somewhat improbable from the experimental point-of-view. It is also good to mention that the overall reaction rate does not only depend on reaction energetics but also on site availability. The lack of adsorption of CO on the Sn of a PtSn alloy surface is presumably the main reason why PtSn is such a good CO oxidation catalyst, as CO does not block OH adsorption sites.[75,76]

Desai and Neurock[140] studied the mechanism of CO oxidation on PtRu in the absence and presence of water molecules using DFT-GGA slab calculations. The computational setup was similar to that used in their simulations of the water dissociation. In the vapor phase, they calculate the reaction $CO_{ad} + OH_{ad} \rightarrow CO_2 + H_{ad}$, with CO on Pt and OH on Ru to be endothermic by 11 kJ/mol with an activation barrier of 71 kJ/mol. The same reaction in the presence of water is less favorable, having a reaction energy of +24 kJ/mol and an activation energy of 90 kJ/mol. This is expected since there is an uncompensated loss in solvation energy of the OH_{ad} reactant, which has a polar bond with the surface (as discussed in Section III.7). The most favorable reaction in water, however, is obviously the one in which the hydrogen ends up as a proton in water, $CO_{ad} + OH_{ad} \rightarrow CO_2 + H^+ + e^-$, having a reaction energy of -6 kJ/mol and an activation barrier of 60 kJ/mol. The importance of the coupling with the solvent is illustrated in Figure 28, which shows the transition state associated with the reaction. The hydrogen begins to dissociate from the hydroxyl intermediate and starts to form a covalent bond with the water molecules to which it was hydrogen bonded. The water molecule, in turn, begins to transfer one of its internal hydrogens to the next adjacent water molecule. The chain of proton transfer reactions occurs by a concerted movement of hydrogen atoms along the hydrogen bonds of water. Note that the calculated activation barrier for the CO oxidation (60 kJ/mol) is higher than that for the previously calculated water activation (27 kJ/mol), suggesting that the former reaction would be rate determining. However, no field effects were as yet included in these calculations, and one should also bear in mind that all the reaction paths were calculated at zero temperature, and hence do not include any finite temperature entropic contributions.

Desai *et al.*[141] also simulated the effect of water on the dissociation of acetic acid on Pd(111) using DFT-GGA slab calculations. They found that in the presence of water, the heterolytic dissociation of acetic acid into solvated acetate and a solvated proton is preferred over homolytic dissociation into an adsorbed acetate radical an adsorbed hydrogen, which is obviously the preferred reaction in the gas phase. Even though in the calculations the acetate ion is more stable in the solution than at the Pd(111) surface, there appears to be a barrier for it to desorb from the metal surface.

Izvekov *et al.*[142] studied the event of an atom/ion leaving a cube-shaped 35-atom Li cluster embedded in water solvent. The dynamics of the cluster, the atom/ion leaving it, and the electrons was solved by the Carr-Parrinello method. The surrounding water molecules and the Na^+ and Cl^- ions dissolved in it were described by classical potentials and classical molecular dynamics. This is not a purely ab initio calculation but an interesting result obtained with this method is the free energy profile of the leaving metal ion/atom as leaves the metal and dissolves into the solution. This free energy is calculated by integrating the mean force obtained at specific distances of the ion from the surface. The resulting curve is shown in Figure 29. In agreement with the result

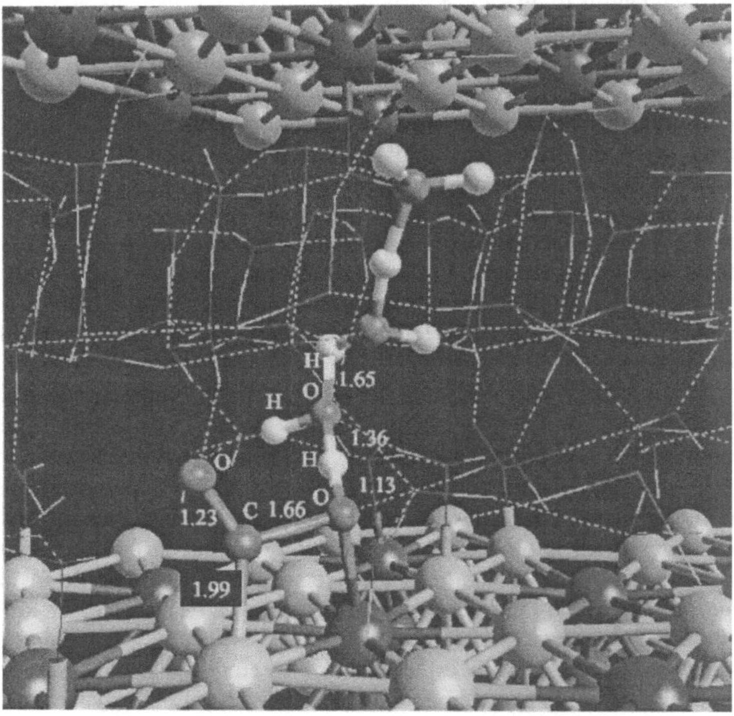

Figure 28. Transition state structure for the reaction of CO and OH over the Pt-Ru alloy surface. The O-H bond is activated to release hydrogen into the solvent in the form of a proton.

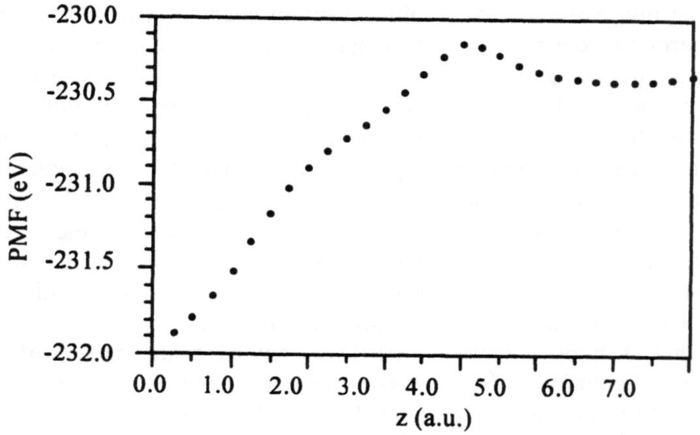

Figure 29. The ion potential of mean force (PMF) vs. distance, *z*, of the ion to the equilibrium position in the cluster. Positive *z* represents the solution part of the simulation cell. Reprinted with permission from S. Izvekov *et al.*, *J. Electrochem. Soc.*, **147**, (2000) 2273. Reproduced by permission of The Electrochemical Society, Inc..

obtained by Desai *et al.*[141] there appear to be two minima, one corresponding to the adsorbed state and the other to the solvated state. In this case, the adsorbed state is more stable. The barrier for adsorption is related to the fact that before the ion can adsorb, it has to lose part of its solvation shell. In desorption terms, the barrier exists because the ion needs to build up a full solvation shell, for which it needs to be sufficiently far away from the surface.

IV. OUTLOOK

In this review, I have discussed some recent applications of computational ab initio quantum chemistry to electrochemistry. My selection of examples and references is incomplete and to some extent based on my personal involvement. For instance, I did not touch upon ab initio quantum-chemical calculations associated with the metal contribution to the double layer capacity,[143,144] underpotential deposition phenomena,[145-147] and the adsorption of fuel cell relevant molecules such as oxygen[148] and methanol.[149] At this point, it may also be useful to point out that my restriction to ab initio quantum chemical calculations was to a certain extent a convenient delineation of the vast field of computational (quantum) chemistry. By no means do I consider "ab initio" synonymous to "better". In many cases, it will better, preferable, and more useful to study the problem at hand by other computational methods, such semi-empirical quantum chemistry, classical molecular dynamics, or Monte Carlo simulations. This depends primarily on the time-scale and size of the system to be modeled, but also on the type of physical or chemical insight that one is aiming at. Certainly when the "non-local" role of the solvent is concerned, classical simulations methods are preferable as they allow a much better statistics. Having said that, I hope I have nonetheless convinced the reader of the enormous potential of ab initio quantum chemistry in studying problems relevant to electrochemistry. The future of applied quantum chemistry lies in a further utilization of ab initio methods to study chemical binding and chemical reactions, especially in combination with ab initio molecular dynamics methods. As far as electrochemistry is concerned, these studies should aim at studying and understanding the role of the field and the solvent on the structure and reactivity of the electrode-electrolyte interface. The studies described in this paper are only the first stages towards a new field of electrochemistry, namely that of computational ab initio electrochemistry.

V. ACKNOWLEDGEMENTS

This review is dedicated to the memory of Michael J. Weaver, in collaboration with whom part of the work described herein was carried out. Working and discussing with Mike was always fun, exciting, and fruitful, and his impact and contribution as "an experimentalist with an interest in theory" to the field of ab initio electrochemistry will be greatly missed. I am grateful to Rutger van Santen for his continuous and generous support of my work at Eindhoven, for our many discussions, and for encouraging me to apply ab initio methods to electrochemistry. I would also like to thank the PhD students and post-docs Tatyana Shubina, Peter Vassilev, Sally Wasileski, and Steven Mitchell for their contributions and many discussions. Matthew Neurock was so kind to send me preprints of his work. Finally, I would like to acknowledge financial support of different parts of my ab initio work by the Royal Netherlands Academy of Sciences (KNAW), the Netherlands Organization for Scientific Research (NWO), and the Energy research Center of the Netherlands (ECN).

REFERENCES

[1] P. A.M.Dirac, *Proc.Roy.Soc.(London) A* **123** (1929) 714.
[2] J.A. Pople, *Angew.Chem.Int.Ed.* **38** (1999) 1894.
[3] W.Kohn, *Rev. Mod. Phys.* **71** (1999) 1253.
[4] J. L.Whitten, H.Yang, *Surf. Sci. Rep.* 218 (1996) 55
[5] R.A. van Santen and M.Neurock, *Catal. Rev. Sci. Eng.* **37** (1995) 357.
[6] B. Hammer and J. K.Nørksov, *Adv. Catal.* **45** (2001) 71.
[7] J. Greeley, J. K. Nørksov, and M. Mavrikakis, *Annu. Rev. Phys. Chem.* **53** (2002) 319.
[8] R. R. Nazmutdinov and M. S.Shapnik, *Electrochim. Acta* **41** (1996) 2253.
[9] W. Schmickler, *Chem. Rev.* **96** (1996) 3177.
[10] W. Schmickler, *Annu. Rep. Prog. Chem., Sect.C* **95** (1999) 117.
[11] J. W. Halley, P. Schelling, and Y. Duan, *Electrochim. Acta* **46** (2000) 239.
[12] J. W. Halley, S. Walbran, and D. L. Price, *Adv.Chem.Phys.* **116** (2001) 337.
[13] M. T. M. Koper, R .A. van Santen, and M. Neurock, in: *Catalysis and Electrocatalysis at Nanoparticle Surfaces*, Ed. by E. R. Savinova, C. G.Vayenas, and A. Wieckowski, Marcel Dekker, New York, in press.
[14] J. P. Lowe, *Quantum Chemistry*, Academic Press, San Diego, 1993.
[15] F. Jensen, *Introduction to Computational Chemistry*, John Wiley & Sons, Chicester, 1999.
[16] P. Hohenberg, W. Kohn, *Phys. Rev.* **136** (1964) B864.
[17] W. Kohn, L. J. Sham, *Phys. Rev.* **140** (1965) A1133.
[18] J. P. Perdew, J. A. Chevary, S. H. Vosko, K. A. Jackson, M. R. Pederson, D. J. Singh, and C. Fiolhais, *Phys. Rev. B* **46** (1992) 6671.
[19] J. P. Perdew, K. Burke, and M. Ernzerhof, *Phys.Rev.Lett.* **77** (1996) 3865.
[20] B. Hammer, L. B. Hansen, and J. K. Nørskov, *Phys. Rev. B* **59** (1999) 7413.
[21] M. Sprik, J. Hutter, M. Parrinello, *J. Chem. Phys.* **105** (1996) 1142.
[22] P. Vassilev, C. Hartnig, M. T. M. Koper, F. Frechard, and R. A. van Santen, *J. Chem. Phys.* **115** (2001) 9815.
[23] M. C. Payne, M. P. Teter, D. C. Allen, T. C. Arias, and J. D. Joannopoulos, *Rev. Mod. Phys.* **64** (1992) 1045.
[24] D. Vanderbilt, *Phys.Rev.B* **41** (1992) 7892.
[25] F. L. Hirshfeld, *Theor. Chem. Acta* **44** (1977) 129.
[26] K. B. Wiberg and P. R. Rablen. *J. Comp. Chem.* **14** (1993) 1504.
[27] P. S. Bagus, K. Hermann, and C. W. Bauschlicher, Jr., *J. Chem. Phys.* **80** (1984) 4378.
[28] T. Ziegler and A. Rauk, *Theor. Chim. Acta* **46** (1977) 1.
[29] R. Hoffmann, *Solids and Surface: A Chemist's View of Bonding in Extended Structures*, VCH, Weinheim, 1988.
[30] R. Carr and M. Parrinello, *Phys. Rev. Lett.* **55** (1985) 2471.
[31] P. L. Silvestrelli, M. Parrinello, *J. Chem. Phys.* **111** (1999) 3572.
[32] D. Marx, M. E. Tuckerman, J. Hutter, M. Parrinello, *Nature* **397** (1999) 601.
[33] B. Ensing, E. J. Meijer, P. E. Blöchl, and E. J. Baerends, *J. Phys. Chem. A* **105** (2001) 3300.

[34]C.Henry, *Surf. Sci. Rep.* **31**, 231, 1998.
[35]K. Kinoshita, in: *Modern Aspects of Electrochemistry*, Ed. by J. O'M. Bockris, B. E. Conway, and R. E. White, Vol.14, New York, Plenum Press, 1982, p.557.
[36]A. Y. Lozovoi, A. Alavi, J. Kohanoff, and R. M. Lynden-Bell, *J. Chem. Phys.* **115** (2001) 1661.
[37]M. T. M. Koper, R. A. van Santen, and S. A. Wasileski, M. J. Weaver, *J. Chem. Phys.* **113** (2000) 4392.
[38]C. Bureau and G. Lécayon, *J. Chem. Phys.* **106** (1997) 8821.
[39]X. Crispin, V. M. Geskin, C. Bureau, R. Lazzaroni, W. Schmickler, and J. L. Brédas, *J. Chem. Phys.* **115** (2001) 10493.
[40]I. Tavernelli, R. Vuilleumier, M. Sprik, *Phys. Rev. Lett.* **88** (2002) 213002.
[41]P. S. Bagus, G. Pacchioni, and M. R. Philpott, *J.Chem.Phys.* **90** (1989) 4287.
[42]M. Blanco, J. M. Ricart, J. Rubio, and F. Illas, *J. Electroanal. Chem.* **267** (1989) 243.
[43]J. Rubio, J. M. Ricart, J. Casanovas, and F. Illas, *J.Electroanal.Chem.* **359** (1993) 105.
[44]P. S. Bagus, G. Pacchioni, *Electrochim.Acta* **36** (1991) 1669.
[45]A. Ignaczak, J. A. N. F. Gomes, *J.Electroanal.Chem.***420** (1997) 71.
[46]M. T. M. Koper and R. A. van Santen, *Surf.Sci.* **422** (1999) 118.
[47]H. Sellers, E. M. Patrito, and P. Paredes Olivera, *Surf. Sci.***356** (1996) 222.
[47]P. Paredes Olivera, E. M. Patrito, H. Sellers, and E. Shustorovich, *J. Mol. Catal. A* **119** (1997) 275.
[49]K. Doll, N. M. Harrison, *Phys. Rev. B* **63** (2001) 165410.
[50]E. Bertel, F. P. Netzer, *Surf. Sci.* **97** (1980) 409.
[51]S. Wang and P. A. Rikvold, *Phys. Rev. B* **65** (2002) 155406.
[52]S. J. Mitchell, M. T. M. Koper, in preparation.
[53]J. Seitz-Beywl, M. Poxleitner, M. M. Probst, and K. Heinzinger, *Int. J. Quantum. Chem.* **42** (1992) 1141.
[54]G. Toth, E. Spohr, and K. Heinzinger, *Chem. Phys.* **200** (1995) 347.
[55]R. R. Nazmutdinov and E. Spohr. *J. Phys. Chem.* **98** (1994) 5956.
[56]A. Ignaczak, J. A. N. F. Gomes, and S. Romanowski, *J. Electroanal. Chem.* **450** (1998) 175.
[57]V. Kairys and J. D.Head, *Surf. Sci.* **440** (1999) 169.
[58]W. Lorenz, J. Reinhold, R. Siegemund, and C. Krebs, *Pol. J. Chem.* **52** (1978) 2431.
[59]An. M. Kuznetsov, J. Reinhold, and W. Lorenz, *Electrochim. Acta* **29** (1984) 801.
[60]S. A. Wasileski and M. J. Weaver, *J. Phys. Chem. B* **106** (2002) 4782.
[61]G. Blyholder, *J.Phys.Chem.* **68** (1964) 2772.
[62]B. Hammer, Y. Morikawa, and J. K. Nørskov, *Phys.Rev.Lett.* **76** (1996) 2141.
[63]B. Hammer and J. K. Nørskov, *Surf. Sci.* **343** (1995) 211.
[64]A. Ruban, B. Hammer, P. Stoltze, H. L. Skiver, and J. K. Nørskov, *J. Mol. Catal.A* **115** (1997) 421.
[65]M. J. Weaver, S. Zou, C. Tang, *J.Chem.Phys.* **111** (1999) 368.
[66]M. J. Weaver, *Surf. Sci.* **437** (1999) 215.
[67]F. Illas, S. Zurita, J. Rubio, A. M. Márquez, *Phys. Rev. B* **52** (1995) 12372.
[68]F. Illas, S. Zurita, A. M. Márquez, and J. Rubio, *Surf. Sci.* **376** (1997) 279.
[69]S. A. Wasileski and M. J.Weaver, *Faraday Disc.* (2002), in press.
[70]Q,Ge, S.Desai, M.Neurock, K.Kourtakis, *J.Phys.Chem.B* **105** (2001) 9533.
[71]M. T. M. Koper, T. E. Shubina, and R. A.van Santen, *J. Phys. Chem. B* **106** (2002) 686.
[72]F. Buatier de Mongeot, M. Scherer, B. Gleich, E. Kopatzki, and R. J .Behm, *Surf. Sci.* **411** (1998) 249.
[73]M.- S.Liao, C. R. Cabrera, and Y. Ishikawa, *Surf. Sci.* **445** (2000) 267.

[74]R. C. Binning, M.-S.Liao, C. R. Cabrera, Y. Ishikawa, H. Iddir, R. Liu, E. S .Smotkin, A. Aldykiewicz, and D. J. Myers, *Int. J. Quantum. Chem.* **77** (2000) 589.
[75]N. M. Markovic and P. N. Ross, Jr., *Surf. Sci. Rep.* **45** (2002) 117.
[76]T. E. Shubina, M. T. M. Koper, *Electrochim.Acta* (2002), in press.
[77]A. Gil, A. Clotet, J. M. Ricart, F. Illas, B. Alvarez, A. Rodes, and J. M. Feliu, *J.Phys.Chem.B* **105**(2001) 7263.
[78]S. A. Wasileski, M. J. Weaver, M. T. M. Koper, *J. Electroanal. Chem.* **500** (2001) 344.
[79]S. A. Wasileski, M. T. M. Koper, and M. J. Weaver, *J. Phys. Chem. B* **105** (2001) 3518.
[80]S. A. Wasileski, M. T. M.Koper, and M. J. Weaver, *J. Chem. Phys.* **115** (2001) 8193.
[81]S. A. Wasileski, M. T. M. Koper, and M. J. Weaver, *J. Am. Chem. Soc.* **124** (2002) 2796.
[82]D. K. Lambert, *J. Chem. Phys.* **89** (1988) 3847.
[83]D. K. Lambert, *Electrochim.Acta* **41** (1996) 623.
[84]J. Marti and D. M. Bishop, *J. Chem. Phys.* **99** (1993) 3860.
[85]J. R. Reimers and N. S. Hush, *J. Phys. Chem. A* **103** (1999) 10580.
[86]B. Beden, C. Lamy, A. Bewick, and K. Kunimatsu, *J. Electroanal. Chem.* **121** (1981) 343.
[87]S. P. Mehandru and A. B. Anderson, *J.Phys.Chem.* **93** (1989) 2044.
[88]P. S. Bagus, C. J. Nelin, K. Hermann, and M. R. Philpott, *Phys. Rev. Lett.* **36** (1987) 8169.
[89]P. S. Bagus, G. Pacchioni, *Surf. Sci.* **236** (1990) 233.
[90]M. Head-Gordon, J. C. Tully, *Chem. Phys.* **175** (1993) 37.
[91]F. Illas, F. Mele, D. Curulla, and A. Clotet, *Electrochim. Acta* **44** (1998) 1213.
[92]D. Curulla, A. Clotet, J. M. Ricart, and F. Illas, *Electrochim.Acta* **45** (1999) 639.
[93]D. Curulla, A. Clotet, J. M. Ricart, and F. Illas, *J. Phys. Chem. B* **103** (1999) 5246
[94]M. García-Hernández, D. Curulla, A. Clotet, and F. Illas, *J. Chem. Phys.* **113** (2000) 364.
[95]M. T. M. Koper and R. A. van Santen, *J. Electroanal. Chem.* **476** (1999) 64.
[96]M. J. Weaver, *Appl. Surf. Sci.* **67** (1993) 147.
[97]P. Gao, and M. J. Weaver, *J. Phys. Chem.* **90** (186) 4057.
[98]J. E. Müller and J. Harris, *Phys. Rev. Lett.* **53** (1984) 2493.
[99]M. W. Ribarsky, W. D. Luedtke, and U. Landman, *Phys. Rev. B* **32** (1985) 1430.
[100]C. W. Bauschlicher, Jr., *J. Chem. Phys.* **83** (1985) 3129.
[101]I. I. Zakharov, V. I. Andreev, and G. M. Zhidomirov, *Surf. Sci.* **277** (1992) 407.
[102]R. R. Nazmutdinov, M. Probst, and K. Heinzinger, *J. Electroanal. Chem.* **369** (1994) 227.
[103]S. Jin, J. D. Head, *Surf. Sci.* **318** (1994) 204.
[104]M. D. Calvin, J. D. Head, and S. Jin, *Surf. Sci.* **345** (1996) 161.
[105]A. Ignaczak and J. A. N. F. Gomes, *J. Electroanal. Chem.* **420** (1997) 209.
[106]S. Izvekov, A. Mazzolo, K. van Opdorp, and G. A. Voth, *J. Chem. Phys.* **114** (2001) 3248.
[107]S. Izvekov and G. A. Voth, *J. Chem. Phys.* **115** (2001) 7196.
[108]P. Vassilev, M. T. M.Koper, and R. A. van Santen, *Phys. Rev. B*, submitted.
[109]P. Vassilev, M. T. M. Koper, and R. A. van Santen, in preparation.
[110]D. L. Price and J. W. Halley, *J. Chem. Phys.* **102** (1995) 6003.
[111]D. L. Price, *J. Chem. Phys.* **112** (2000) 2973.
[112]M. F. Toney, J. N. Howard, J. Richer, G. L. Borges, J. G.Gordon, O. R. Melroy, D. G. Wiester, D. Yee, and L. B.Sorensen, *Nature* (*London*) **368** (1994) 444.
[113]M. F.Toney, J. N. Howard, J. Richer, G. L. Borges, J. G. Gordon, O. R. Melroy, D. G. Wiester, D. Yee, and L. B. Sorensen, *Surf. Sci.* **335** (1995) 326.

[114]M. T. M. Koper and R. A. van Santen, *J.Electroanal.Chem.* **472** (1999) 126
[115]X. Xu, D. Y. Wu, B. Ren, H. Xian, and Z.-Q.Tian, *Chem. Phys. Lett.* **311** (1999) 193.
[116]R. J. Nichols and A. Bewick, *J. Electroanal. Chem.* **243** (1988) 445.
[117]A. Peremans, A. Tadjeddine, *Phys. Rev. Lett.* **73** (1994) 3010.
[118]R. A. Olsen, G. J. Kroes, and E. J. Baerends, *J. Chem. Phys.* **111** (1999) 11155.
[119]P. Paredes Olivera, A. Ferral, E. M. Patrito, *J. Phys. Chem. B* **105** (2001) 105.
[120]A. Michaelides and P. Hu, *J. Chem. Phys.* **114** (2001) 513.
[121]P. Vassilev, M. T. M. Koper, and R. A. van Santen, *Chem. Phys. Lett.* **359** (2002) 337.
[122]M. Tuckerman, K. Laasonen, M. Sprik, and M. Parrinello, *J. Chem. Phys.* **103** (1995) 150.
[123]D. A. Rose and I. Benjamin, *J. Chem. Phys.* **100** (1993) 3545.
[124]J. B. Straus, A. Calhoun, and G. A. Voth, *J. Chem. Phys.* **102** (1995) 529.
[125]A. Calhoun, M. T. M. Koper, and G. A. Voth, *J. Phys. Chem. B* **103** (1999) 3442.
[126]C. Hartnig and M. T. M. Koper, *J. Chem. Phys.* **115** (2001) 8540.
[127]C. Hartnig and M. T. M. Koper, *J. Electroanal. Chem.* (2002) in press.
[128]An. M. Kuznetsov and W. Lorenz, *Chem. Phys.* **183** (1994) 73.
[129]An. M. Kuznetsov and W.Lorenz, *Chem. Phys.* **185** (1994) 333.
[130]T. Ohwaki and K. Yamashita, *J. Electroanal. Chem.* **504** (2001) 71.
[131]O. Pecina and W. Schmickler, *Chem. Phys.* **228** (1998) 265.
[132]A. B. Anderson and T. V. Albu, *Electrochem. Comm.* **1** (1999) 203.
[133]A. B. Anderson and T. V. Albu, *J. Am. Chem. Soc.* **121** (1999) 11855.
[134]A. B. Anderson and T. V. Albu, *J. Electrochem. Soc.* **147** (2000) 4229.
[135]R. A. Sidik and A. B. Anderson, *J.Electroanal.Chem.* **528** (2002) 69.
[136]A. B. Anderson, N. M. Neshev, R. A. Sidik, and P. Shiller, *Electrochim. Acta* **47** (2002) 2999.
[137]S. Desai and M. Neurock, *Phys. Rev. B*, submitted.
[138]Y. Ishikawa, M.-S. Liao, and C. R. Cabrera, *Surf. Sci.* **513** (2002) 98.
[139]E. Shustorovich, *Adv. Catal.* **37** (1990) 101.
[140]S. Desai and M. Neurock, in preparation; S. Desai, Ph. D. Thesis, University of Virginia, 2002.
[141]S. Desai, V. Pallassana, and M. Neurock, *J. Phys. Chem. B* **105** (2001) 9171.
[142]S. Izvekov, M. R. Philpott, and R. I. Eglitis, *J. Electrochem. Soc.* **147** (2000) 2273.
[143]W. H. Mulder, J. H. Sluyters, and J. H. van Lenthe, *J. Electroanal. Chem.* **261** (1989) 273.
[144]R. R. Nazmutdinov, M. Probst, and K. Heinzinger, *Chem. Phys. Lett.* **221** (1994) 224.
[145]C. G. Sanchez and E. P. M. Leiva, *Electrochim. Acta* **45** (1999) 691.
[146]M. I. Rojas, M. G. Del Popolo, and E. P. M. Leiva, *Langmuir* **16** (2000) 9539.
[147]C. G. Sanchez, E. P. M. Leiva, and J. Kohanoff, *Langmuir* 17 (2001) 2219.
[148]T. Li, P. B. Balbuena, *J. Phys. Chem. B* **105** (2001) 9943.
[149]S. K. Desai, M. Neurock, and K. Kourtakis, *J. Phys. Chem. B* **106** (2002) 2559 .

3

Macroscopic and Molecular Models of Adsorption Phenomena on Electrode Surfaces

P. Nikitas

Laboratory of Physical Chemistry, Department of Chemistry, Aristotle University of Thessaloniki, 54124 Thessaloniki, Greece

I. INTRODUCTION

The experimental study of the adsorption of organic compounds on electrodes began in the first decade of the previous century with Gouy's electrocapillary work.[1] Since then it has attracted considerable attention, mainly because it affects the mechanism of most of the processes occurring on electrodes. The first attempts to present a theoretical description of the effect of the electric field on adsorption appeared in 1925 and 1926 by Frumkin,[2, 3] who formulated the macroscopic model of condensers in parallel. The interpretation of the electrosorption of organic compounds at a molecular level was initiated by Butler[4] in 1929, but it was the work of Bockris and co-workers in 1963 that put the bases of the contemporary microscopic modelling.[5] The main contribution by Bockris *et al.* was the introduction of the concept of the competition between solvent and adsorbate molecules for adsorption and the reorientation of the adsorbed molecules on the electrode upon the variation of the electric field.[5,6]

In this Chapter we present critically the most important models

Modern Aspects of Electrochemistry, Number 36, edited by C. G. Vayenas, B. E. Conway, and R. E. White, Kluwer Academic / Plenum Publishers, New York, 2003

developed to describe adsorption phenomena on electrodes. The discussion is restricted to the reversible adsorption of neutral organic molecules, *i.e.*, when the adsorbed molecules retain their chemical individuality and equilibrium is established between adsorbed and non-adsorbed molecules. The accomplishments as well as the difficulties in the development of a fully satisfactory, quantitative theory are indicated and discussed.

II. FEATURES OF ELECTROSORPTION AND FACTORS AFFECTING THEM

The simplest of the adsorption processes on electrodes is the monolayer reversible adsorption of a neutral organic compound on an energetically homogeneous electrode surface, like the surface of a Hg electrode. The typical features of this process when the solute molecules do -not reorientate on the electrode are the following:[7-9]

1. The maximum in adsorption occurs at a constant value of the applied potential E, the potential drop across the adsorbed monolayer $\Delta\phi$, and the electrode charge density σ^M. This property leads to the appearance of a common intersection point in the plots of σ^M *vs.* E at constant solute mole fractions x in the electrolyte solution.
2. The adsorption is accompanied by a decrease in the differential capacity C in the region of E_{max}, the potential of maximum in adsorption. At potentials more positive or more negative than E_{max} the organic adsorbate is desorbed giving rise to typical adsorption-desorption peaks.
3. The adsorption at constant E or $\Delta\phi$ or σ^M is described by the *Frumkin* isotherm

$$\beta x = \frac{\theta}{1-\theta}\exp(-2A^{AS}\theta) \tag{1}$$

where θ is the surface coverage by the organic solute, $\beta = \exp(-\Delta G_{ads}/RT)$, ΔG_{ads} is the standard Gibbs energy of adsorption at zero coverage, and $A^{AS} = z\{w_{AS} - (w_{AA} + w_{SS})/2\}/kT$. Here, the subscript S denotes a solvent molecule, A a solute molecule, w_{ij} is the lateral interaction energy between pairs of adsorbed i and j molecules, and z is the coordination number of the lattice structure of the adsorbed layer.

4. The standard Gibbs energy of adsorption, ΔG_{ads}, depends quadratically upon the electrical variable X (= E, $\Delta\phi$ or σ^M).
5. Depending on the system the interaction parameter A^{AS} may or may not vary with X. If A^{AS} is independent of E (or $\Delta\phi$), the isotherms are called *congruent* with respect to E (or $\Delta\phi$) and the plots of σ^M *vs..* θ at constant E (or $\Delta\phi$) are linear. Similarly, if A^{AS} is independent of σ^M, the isotherms are congruent with respect to σ^M and the plots of E *vs..* θ at constant σ^M are linear.

Deviations from the above properties have been observed attributed to various factors. For example, the reorientation of the solute molecules on the electrode surface is a factor that may disturb seriously the above picture, especially when the area covered by a solute molecule on the electrode surface changes upon reorientation.[7,10,11] In this case the maximum in adsorption is concentration dependent and for this reason the plots of σ^M *vs..* E do not exhibit a common intersection point, the capacitance plots show characteristic humps in the region between the adsorption-desorption peaks, and the validity of the Frumkin isotherm is questionable.

The properties of the adsorbed layer also change when a surface phase transition occurs[12-16] or when polylayers[17,18] or surface aggregates[19-27] are formed. The properties of the adsorbed layer in these cases are discussed in the relevant sections below.

The adsorption of neutral organic compounds on energetically homogenous electrode surfaces is determined by a great number of specific effects, which may be divided into the following three categories: a) size effects, b) effects determining the "effective" field acting on each adsorbed particle, and c) effects due to short-range interactions. Electron spillover from the electrode to the electrolyte solution, partial charge transfer and direct correlations between the solvent molecules in the adsorbed layer and non-adsorbed solvent molecules of the adjacent solution may also affect the properties of the adsorbed layer.

The size effects come from the fact that the adsorbed solute molecules have greater dimensions than those of the solvent molecules. The adsorption from solution is in fact a replacement process, during which a solute molecule A from the bulk electrolyte solution replaces r pre-adsorbed solvent S molecules from the adsorbed layer. Since the dimensions of A are in general greater than those of S, we expect that r should be greater than unity.

However, we have endless results that $r = 1$ in the vast majority of the experimental systems. Thus the wide applicability of the Frumkin isotherm is an indirect evidence that $r = 1$. Indeed, if we adopt $r = 1$, assume an adsorbed layer with a two-dimensional lattice structure and a random distribution of the adsorbed molecules, then statistical mechanics readily yields the Frumkin isotherm, Eq. (1).[28-30] A direct indication that r is close to unity comes from the thermodynamic method we have proposed for the determination of r.[31-33] This method is applicable to monolayers composed of solvent and constant orientated solute molecules. These conditions can be safely detected experimentally and if they are fulfilled, r can be obtained from surface pressure data by means of two extrapolations. Due to these extrapolations the method is extremely sensitive to experimental errors and for this reason it is applicable only to air / solution and liquid / liquid interfaces. All applications of this method gave the value $r = 1 \pm 0.2$, despite the considerable differences in the size of the adsorbates used.[31-35]

The value $r \approx 1$ may be explained by the work of Guidelli *et al.*[36] which shows that a strict statistical mechanical treatment of an adsorbed layer composed of adsorbate and solvent molecules distributed over a two-dimensional lattice yields roughly linear plots of $F = \ln\{\theta/x(1-\theta)\}$ *vs.*. θ even when the dimensions of the solute molecules are greater than those of the solvent molecules, provided that H-bonding among the solvent molecules is taken into account. Note that the linearity of the $F = \ln\{\theta/x(1-\theta)\}$ *vs.*. θ plots shows the validity of the Frumkin isotherm, which in turn indicates that $r \approx 1$.

Alternatively the value $r \approx 1$ may be explained if we assume that the solvent molecules associate on the electrode surface to form clusters S_c with dimensions quite greater than those of the solute molecules. When a solute molecule A is adsorbed, it cuts from S_c a smaller cluster S_A of solvent molecules, which has necessarily equivalent to A dimensions. The S_A cluster is displaced towards the bulk solution, where it is disintegrated into g monomeric units. Thus the adsorbed layer behaves as if it were composed of adsorbate A molecules and solvent clusters S_A with dimensions equal to A and the adsorption process may be expressed, at least phenomenologically, as[33,37,38]

$$A(\text{bulk}) + S_A(\text{ads}) \Leftrightarrow A(\text{ads}) + gS(\text{bulk}) \tag{2}$$

It is seen that the peculiar behavior of the adsorbed layer to exhibit $r \approx 1$ can be explained by the assumption of surface clusters. However, surface clusters can be formed only if strong hydrogen bonds are formed among the adsorbed solvent molecules. It is seen that Guidelli's findings about the role and the effect of H-bonding are in line with an adsorption mechanism represented by Eq. (2). Therefore, there are at least two approaches to take size effects into account: a direct approach based on Guidelli's *et al.* work and an indirect approach based on Eq. (2).

The factors that contribute to the "effective" field acting on each adsorbed particle are:

a) the thickness of the adsorbed layer,
b) the permanent and induced dipole moments of the adsorbed particles,
c) the distribution of the adsorbed particles, and
d) image effects due to the metal surface and possibly to the bulk electrolyte solution.

The existence of hydrogen bonds among the solvent molecules is also expected to affect the "effective" field. Finally, the various short-range interactions are affected by the existence of hydrogen bonds and in general the distribution of the interacting molecules.

We should point out that although we know the factors that determine electrosorption, we are not fully aware of the extent that each factor contributes to the properties of the adsorbed layer. In addition, the introduction of the various factors and effects in a theoretical treatment is not straightforward and for this reason idealizations and assumptions are necessary. These idealizations differentiate the various models.

III. MACROSCOPIC MODELS

The macroscopic description of the adsorption on electrodes is characterised by the development of models based on classical thermodynamics and the electrostatic theory. Within the frames of these theories we can distinguish two approaches. The first approach, originated from Frumkin's work on the parallel condensers (PC) model,[2,3] attempts to determine the dependence of σ^M upon the applied potential E based on the Gibbs adsorption equation. From the relationship $\sigma^M = g(E)$, the surface tension γ and the differential capacity C can be obtained as a function of E by simple mathematical transformations and they can be further compared with experimental data. The second approach denoted as STE (simple thermodynamic-electrostatic approach) has been developed in our laboratory,[38,39] and it is based on the determination of analytical expressions for the chemical potentials of the constituents of the adsorbed layer. If these expressions are known, the equilibrium properties of the adsorbed layer are derived from the equilibrium equations among the chemical potentials. Note that the relationship $\sigma^M = g(E)$, between σ^M and E, is also needed for this approach to express the equilibrium properties in terms of either σ^M or E. Here, this relationship is determined by means of the Gauss theorem of electrostatics.

1. PC Approach

The first model of this approach is the two parallel condenser model (TPC), which visualizes the adsorbed layer as two capacitors connected in parallel.[2,3] Only water (solvent) molecules are present between the plates of one of these capacitors and only solute molecules between the plates of the other. The model has been progressively generalized and extended to take account of the reorientation of the solute and solvent molecules, the heterogeneity of the surface of solid electrodes as well as variations in the adsorbed layer thickness upon adsorption.[7,40-44]

As pointed out above, the TPC model is based on the determination of the $\sigma^M = g(E)$ relationship via the Gibbs adsorption equation[7,44,45]

$$-d\gamma = \sigma^M dE + RT\Gamma_{max}\theta d\ln x \tag{3}$$

where Γ_{max} is the maximum solute surface concentration. Using cross differentiation and simple mathematical transformations, Eq. (3) yields

$$\left(\frac{\partial \sigma^M}{\partial \theta}\right)_E = -RT\Gamma_{max}\left(\frac{\partial \ln x}{\partial E}\right)_\theta \tag{4}$$

Up to this point the treatment is strictly thermodynamic. However, the next steps necessarily demand the use of extra-thermodynamic assumptions. The first of them is the choice of an adsorption isotherm, which in general may be written as $\beta x = f(E, \theta)$. The most commonly adopted isotherm is the Frumkin isotherm, Eq. (1), assuming arbitrarily a linear or a quadratic dependence of A^{AS} upon E. The next necessary assumption concerns the dependence of β upon E, for which the following expression is usually adopted: $\beta = \beta_{max}\exp\{-b(E-E_{max})^2\}$, where b is a constant. Thus if a certain adsorption isotherm $\beta x = f(E, \theta)$ is selected, the partial derivative of ln x with respect to E at constant θ can be calculated and therefore Eq. (4) after integration yields the relationship $\sigma^M = g(E)$. From this relationship the dependence of γ upon E is obtained by integration and the differential capacity C is calculated from $C = d\sigma^M/dE$.

The TPC model has been reviewed several times and for this reason we briefly point out that, despite its simplicity and the fact that it was formulated in the third decade of the previous century, this model and its extensions are still in use. In fact they are by far the models having the most applications. The potentialities of these models to reproduce γ *vs..* E and C *vs..* E experimental data are quite satisfactory. However, this does not mean that they represent equally satisfactorily the physical reality of the adsorbed layer. Their main drawbacks are:

a) the choice of the adsorption isotherm is arbitrary;
b) when the Frumkin isotherm is used, there is no physical explanation about the field dependence of A^{AS};
c) the use of $\beta = \beta_{max}\exp\{-b(E-E_{max})^2\}$ is based on the assumption that the *effective* field acting on each adsorbed particle is proportional to the potential E,[7] which is a rough approximation; and
d) the relationship $\sigma^M = g(E)$ determined by means of Eq. (4) does not coincide with the corresponding expression based on the Gauss theorem of electrostatics.[46, 47]

Finally, the TPC model and its extensions have become progressively unattractive due to the lack of any correlation to the molecular structure and properties of the adsorbed layer.

An approach similar to PC has been proposed by Mohilner *et al.* (MNM theory).[48-52] Their basic idea was to treat the adsorbed layer as a two-component non-electrolyte solution called the *surface solution*.[48,49] The field effect as well as any correlation to molecular or structural properties of the surface solution are missing from the original MNM theory. At this stage this theory differs a little from a curve fitting procedure. In subsequent papers the introduction of the field effect has been attempted following the TPC approach.[50-52] Thus the MNM theory and its extensions do not offer a real alternative approach to the theoretical description of adsorption on electrodes.

2. STE Approach

The adsorbed layer is assumed to be a non-electrolyte surface solution composed of N neutral species, which may be the solvent molecules, the molecules of one or more solutes, or even different states of the solvent molecules due to different orientations of their permanent dipole moment. The location of the adsorbed layer is between two perfectly conducting walls, which are the electrode surface on one side and the electrolyte solution on the other side. Thus the Gibbs free energy of this layer is given by

$$dG^{ads} = -SdT + Vdp + \gamma dA - A\sigma^M d\Delta\phi + \sum_{i=1}^{N} \mu_i^{ads} dN_i \tag{5}$$

where N_i is the number of molecules of the i-th component, μ_i^{ads} is its chemical potential, and A is the area of the adsorbed layer. The area A may be written as $A = \sum A_i N_i$, where A_i is the partial area of the i-th species. Cross-differentiation of Eq. (5) followed by integration along a path of constant pressure, temperature and composition results in.[38,39]

$$\mu_i^{ads} = \mu_i^* - \gamma A_i - \int_0^{\Delta\varphi} \frac{\partial(A\sigma^M)}{\partial N_i} d\Delta\phi \tag{6}$$

where μ_i^* is the chemical potential of i in the absence of field and surface effects. Therefore, we may write $\mu_i^* = \mu_i^0 + kT \ln f_i\theta_i$, where μ_i^0 is the value of μ_i^* in its standard state, f_i is an activity coefficient, and θ_i is the surface coverage of i. The activity coefficients depend upon the surface composition and analytical expressions of this dependence may be obtained by means of either statistical mechanics or thermodynamics.[38,53-55]

It is seen that the calculation of the chemical potentials of the adsorbed species and consequently the study of the equilibrium properties of the adsorbed layer by means of the STE approach necessarily require knowledge of the dependence of σ^M upon the adsorbed layer composition and the potential drop $\Delta\phi$. This dependence is determined by means of the Gauss theorem of electrostatics, which readily yields[38,39]

$$\varepsilon_0 \Delta\phi / \ell = \sigma^M + \sigma^{ind} = \sigma^M + \sum_{i=1}^{N} (P_i - \alpha_i \Delta\phi / \ell) N_i / \ell A \tag{7}$$

Here, l is the thickness of the adsorbed layer, σ^{ind} is the induced charge density on the surface of the adsorbed layer which is adjacent to the electrode surface, P_i is the permanent dipole moment normal to the electrode surface of the i-th species, α_i is its polarizability, and ε_0 is the permittivity of vacuum.

Now Eqs. (6) and (7) result in the following simple expression for the chemical potential of the adsorbed species i:

$$\mu_i^{ads} / kT = \mu_i^0 / kT - \gamma A_i / kT + \ln f_i \theta_i + b_i \Delta\phi - c_i \Delta\phi^2 \tag{8}$$

where $b_i = P_i/lkT$, $c_i = \varepsilon_0 \varepsilon_i A_i / 2lkT$ and $\varepsilon_i = 1 + \alpha_i/\varepsilon_0 A_i l$. The equilibrium properties of the adsorbed layer can be determined by means of Eq. (8) as a function of the potential drop $\Delta\phi$ provided that we have defined the adsorption processes precisely in terms of chemical potentials.

It is interesting to note that Eq. (8) can be derived by strict statistical mechanical arguments as a limiting case of a more general molecular model, which is discussed below. For this reason we do not present here comparisons between model predictions and experimental data. Such comparisons may be found in [37,39,56]. All the comparisons show that the model, despite its great simplicity, predicts at least qualitatively all the experimental features of both the stable and the transition region. In addition, it can be straightforwardly extended to account for many adsorption phenomena occurring in an adsorbed layer.[18,25,57,58]

The most drastic approximation adopted in the STE approach is the assumption that the effective field acting on the adsorbed species is proportional to $\Delta\phi$. Note that in the PC approach this field is assumed to be proportional to E. Both approximations arise from the uniform smearing of the charges of the dipoles over the conducting plates, between which the adsorbed layer is located. That is, the macroscopic models disregard discreteness of charge effects and this is their most serious shortcoming. This defect cannot be effectively overcome within the frames of the macroscopic approach, therefore it becomes necessary to utilise molecular arguments.

IV. MOLECULAR MODELS

The molecular models adopt a statistical mechanical treatment of the adsorbed layer. In most cases a lattice structure is assumed and the differences of the various models lie in the effects on which the emphasis is put. There are two main molecular approaches: one has been developed by Guidelli and his colleagues[9,29,36,59-64] and the other is based on the LBS theory.[65] Guidelli's approach emphasizes local order and hydrogen bonding among adsorbed water (solvent) molecules, whereas the models based on the LBS theory disregard local order and focus their attention on the polarizability of the adsorbed molecules.

1. Guidelli's Approach

Guidelli was the first to develop a strict statistical mechanical model for the adsorption of neutral organic compounds on electrodes and he continued with his colleagues to present some novel ideas in this area.[9,29,36,59-64] In the first models the adsorbed layer is treated as a two-dimensional lattice sheet consisting of H-bonded solvent molecules and solute molecules occupying r sites.[29,36,59-61] Later Guidelli and Aloisi included in their treatment a three-dimensional lattice extending from the electrode surface into the bulk electrolyte solution.[9,62-64]

(i) Two-Dimensional Lattice Model

The adsorbed layer is idealized as a two-dimensional lattice consisting of solvent molecules adsorbed in a given number of orientations and neutral solute molecules which occupy *r* sites and behave like flexible, open polymeric molecules. The canonical partition function Q for the adsorbed monolayer may be expressed as[60]

$$Q=\left(\prod_i q_i^{N_i}\right)\sum_{\{^xN_{jk}\}} g(\{N_i\},\{^xN_{jk}\})\exp(-\sum_{x,j,k} {}^xN_{jk}\,{}^xw_{jk}/kT) \quad (9)$$

where N_i is the number of the i-th adsorbed species per unit area, q_i is the partition function of i, ${}^xN_{jk}$ is the number of nearest-neighboring pairs of j and k species which follow each other in the positive direction of the x-axis, ${}^xw_{jk}$ is the lateral interaction energy of the above j-k pair, and g is the number of configurations corresponding to a given set of $\{N_i\}$, $\{{}^xN_{jk}\}$.

The i-th adsorbed species may be a solute r-mer or a solvent monomer molecule in a certain orientation. The solvent molecules are assumed to be in 25 distinct orientations. These are chosen as the maximum number of single H-bond formed per adsorbed solvent molecule within the adsorbed layer. Thus the contributions to the ${}^xw_{jk}$ energy come from hydrogen bonds and dipole-dipole interactions. Short-range interactions are assumed to have a negligible contribution. For the calculation of the dipole-dipole and dipole-field interactions that contribute to q_i, all the adsorbed species are considered non-polarizable point dipoles embedded in a fictitious continuum of dielectric constant ε, which accounts for the distortional polarizability of the adsorbed molecules. Image effects are treated explicitly. Due to the hydrogen bonds and the strong dipole-dipole interactions local order is expected to play a dominant role and for this reason, the function g in Eq. (9) is calculated by means of the quasi-chemical approximation. Using this approximation and standard statistical mechanical techniques, the equilibrium properties of the adsorbed layer are calculated as a function of σ^M.

The model attempts to consider rigorously as many specific effects as possible. However, this approach has its own cost. The use of the quasi-chemical approximation results in extremely complicated

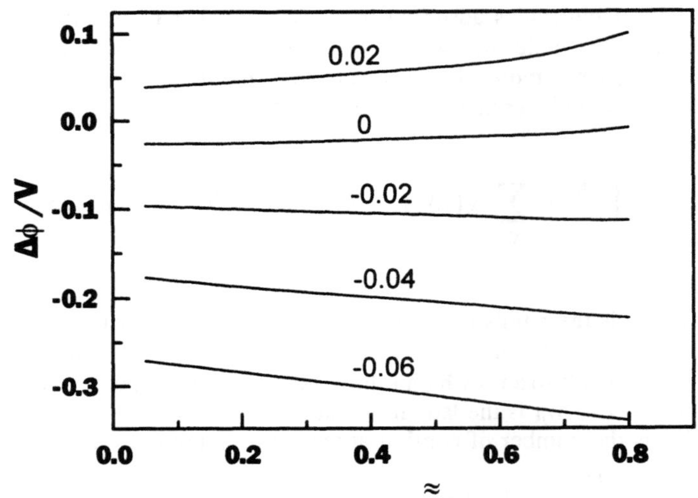

Figure 1. Plot of $\Delta\phi$ *vs.*. θ at constant σ^M = 0.02, 0, -0.02, -0.04, -0.06 C m^{-2} (from top to bottom) predicted by Guidelli's *et al.* two-dimensional model when r = 3 and ε = 5.5. (Reprinted from *J. Electroanal. Chem.*, **197**, R. Guidelli, M. L. Foresti, Statistical-Mechanical Treatment of an Adsorbed Monolayer with Local Order. Part III. p. 103, Copyright ©1986, with permission from Elsevier Science).

equilibrium equations making any direct comparison with experiment difficult if not impossible. In addition, the assumption of an *effective* dielectric constant ε independent of the adsorbed layer composition is not realistic. Despite these weaknesses, the model accounts satisfactorily for the shape of the $\Delta\varphi$ *vs.*. θ curves at constant σ^M (Figure 1), predicts Frumkin type adsorption isotherms with an attractive interaction parameter (Figure 2), and a quadratic dependence of the Gibbs free energy of adsorption upon σ^M. Note that the plots of Figure 2 are satisfactorily linear in the region $0 < \theta < 0.7$, despite the fact that $r = 3$. The reason is that due to the hydrogen bond interactions,

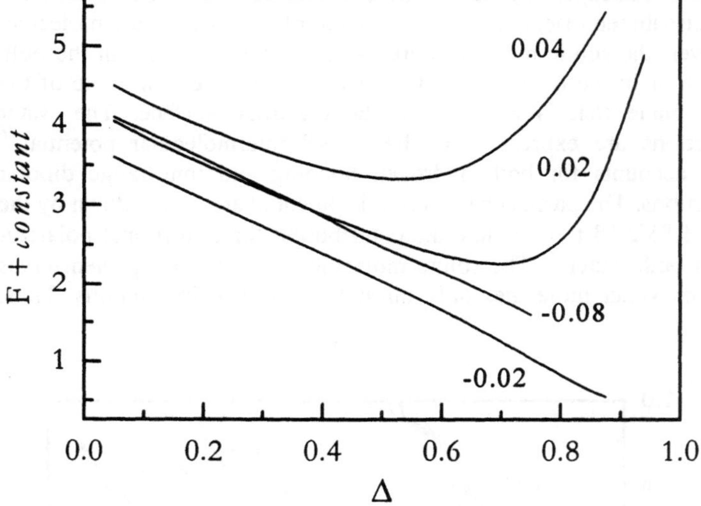

Figure 2. Plot of $F = \ln\{x(1-\theta)/\theta\}$ *vs.*. θ at constant σ^M (indicated by each curve in C m^{-2}) predicted by Guidelli's *et al.* two-dimensional model when $r = 3$ and $\varepsilon = 5.5$. (Reprinted from *J. Electroanal. Chem.*, **197**, R. Guidelli, M. L. Foresti, Statistical-Mechanical Treatment of an Adsorbed Monolayer with Local Order. Part III, p. 103, Copyright ©1986, with permission from Elsevier Science).

the water molecules tend to aggregate on the electrode surface. This increases the probability of a solute molecule to find a cluster of *r* water molecules and it tends to become equal to (1 - θ), as in the Frumkin isotherm.

(ii) Three-Dimensional Lattice Model

The two-dimensional lattice model disregards completely any direct correlation between the adsorbed water molecules and the adjacent water molecules of the solution phase. In order to overcome this weakness Guidelli and Aloisi have developed a three-dimensional lattice model,[9,62-64] the main features of which are the described next.

The interface is represented by a body-centered cubic lattice extending from the electrode surface to the bulk of the solution phase. Each site is occupied by either a water molecule in any of 24 symmetrically distributed orientations or a segment of an r-mer solute molecule. However, the solute molecules are assumed to exist either in the bulk solution or in the lattice but with at least one segment in a site of the lattice plane that is adjacent to the electrode surface. The water interactions are expressed by the TIP4P intermolecular potential,[66] which accounts for both hydrogen bonding and long-range dipolar interactions. The partial charges of this potential are scaled down by the factor 1.85/2.18 to eliminate the contribution for distortional polarization in bulk water.[9] The solute molecules interact among themselves and with water molecules only through coulombic interactions. Thus,

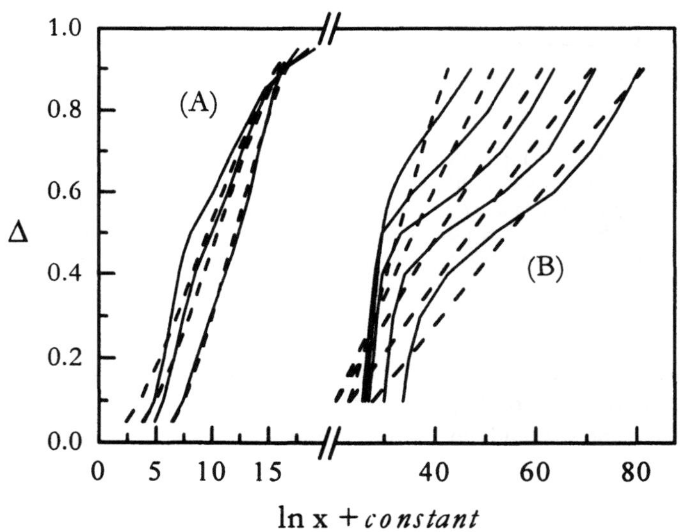

Figure 3. Adsorption isotherms predicted by Guidelli's *et al.* three-dimensional model (—) for (A) a polar dimeric solute adsorbed vertical or flat at σ^M = 0, 0.05, 0.1 C m^{-2} (from left to right), and (B) a polar monomeric solute at σ^M = 0.04, 0, -0.04, -0.08, -0.12 C m^{-2} (from left to right). Broken lines represent the best Frumkin's isotherms. (Solid lines were reprinted from *J. Electroanal. Chem.*, **329**, R. Guidelli, and G..Aloisi, A three Dimensional Lattice Model of TIP4P Water Molecules and of Polar Monomeric and Dimeric Solute Molecules Against a Charged Wall.. p.39, Copyright ©1992, with permission from Elsevier Science).

short-range, London type interactions are disregarded. The model takes into account the water polarizability but it is not clear whether it considers the solute polarizability as well.

The statistical mechanical treatment is carried out on the basis of Barker's self-consistent field theory. This treatment accounts for local order and yields a set of non linear equations, which are solved numerically by iterations. The most interesting point of this model is that it does not use adjustable parameters. However, the predictions are rather poor.

Up to now the model has been applied with monomeric, dimeric and trimeric solute molecules. Although the study of these cases is not complete, possibly due to computational difficulties, it seems that some of the adsorption features are satisfactorily predicted only in the case of non-polar monomeric and polar dimeric solute molecules, provided that the latter exhibit certain orientations on the electrode surface.[9,62,64] In the case of polar monomeric and dimeric molecules that may adsorb either vertically or flat, the model does not give satisfactory predictions. This is shown in Figure 3 where the solid lines represent adsorption isotherms predicted by the model and the dotted lines represent the best Frumkin's isotherms that describe them. In the case of the trimeric solutes, the model predicts the existence of a surface phase transition.[63] However, the transition properties, due to the use of an inappropriate statistical mechanical treatment, contradict thermodynamic and experimental data.[67] Thus, despite its novelty the three-dimensional lattice approach has not given the expected results yet.

2. Models Based on the LBS Approach

The models of this category are based on the pioneering work by Levine, Bell and Smith[65] in 1969, who modeled the inner layer of a charged interface in the absence of solute molecules as a two-dimensional sheet of polarizable point dipoles situated *in vacuo*. The dipoles are randomly distributed over the sites of a hexagonal lattice. This model has been extended to describe adsorption of solute molecules on electrode surfaces by Sangaranarayanan, Rangarajan and their colleagues,[68-75] and later by the present author to study many adsorption phenomena on electrodes.[16,23,37,76-84] Note that the adsorbed species are polarizable point dipoles with a random distribution, whereas in the two-dimensional model by Guidelli the adsorbed molecules are non-polarizable point dipoles exhibiting a local order. Details on the statistical mechanical treatment of the models based on the LBS approach may be found in the original papers.[65,68-73] Here, we present a different and more general mathematical treatment.

(*i*) *The Mathematical Treatment*

The adsorbed layer is modeled as a two-dimensional sheet with a hexagonal lattice structure composed of adsorbate and solvent molecules, which behave as polarizable point dipoles. The solvent molecules in the form of either monomers or clusters may be in different polarization states (orientations) treated as independent species. Each adsorbed species i may occupy r_i lattice sites. In addition, the thickness of the adsorbed layer may vary upon adsorption. For this reason we denote by l_i the thickness of the adsorbed layer when it is composed exclusively of the i-th species.

If $\Delta\phi'/l_i$ is the effective electric field acting on each adsorbed species i, then the electrical energy of the monolayer U^{el} is equal to the energy required to charge the monolayer *in vacuo*, *i.e.*, to increase $\Delta\phi'$ from zero to its final value, plus the work needed to bring the adsorbed species from an infinite distance outside the interface into the adsorbed layer when it is subject to a constant potential difference $\Delta\phi'$, plus the work needed to produce induced dipoles. Therefore, U^{el} may be expressed as[37]

$$U^{el} = -\varepsilon_0(\Delta\phi')^2\sum A_i N_i / 2\ell_i + \sum P_i N_i \Delta\phi' / \ell_i - \sum \alpha_i N_i (\Delta\phi')^2 / \ell_i^2 + \sum \alpha_i N_i (\Delta\phi')^2 / 2\ell_i^2 \qquad (10)$$

Indeed, the first term of Eq. (10) expresses the electrical energy enclosed in the monolayer when $\Delta\phi'$ is increased *in vacuo* from zero to its final value, the second term represents the interaction energy of the permanent dipoles with the electric field, the third term is the interaction energy of the induced dipoles with the electric field, and the last term expresses the work needed to produce induced dipoles.

Differentiation of Eq. (10) with respect to N_i at constant $\Delta\phi'$ yields the electrical contribution to the chemical potential of i

$$\mu_i^{el} / kT = \mu_i^{0,el} / kT + b_i \Delta\phi' - c_i \Delta\phi'^2 \qquad (11)$$

where again $b_i = P_i/l_ikT$, $c_i = \varepsilon_0\varepsilon_iA_i/2l_ikT$ and $\varepsilon_i = 1 + \alpha_i/\varepsilon_0A_il_i$. The contribution arising from the entropy of mixing and the various short-range interactions may be expressed as[84-87]

$$\mu_i^{rest}/kT = \mu_i^{0,rest}/kT + \ln f_i\theta_i + r_i\sum(r_j - 1)\theta_j/r_j \tag{12}$$

provided that the size effects are taken into account by means of Flory's approximation.[84-87] In Eq. (12) the surface coverage θ_i is defined from $\theta_i = r_iN_i/M$, M being the total number of lattice (adsorption) sites. Therefore, the chemical potential of i in the adsorbed layer and in the absence of surface effects may be written as

$$\mu_i^*/kT = \mu_i^0/kT + \ln f_i\theta_i + b_i\Delta\phi' - c_i\Delta\phi'^2 + r_i\sum(r_j - 1)\theta_j/r_j \tag{13}$$

Note that this very simple and elegant expression for the chemical potentials in the absence of surface effects is related to the true surface chemical potential of i, μ_i^{ads}, via the relationship[88]

$$\mu_i^{ads} = \mu_i^* - \gamma A_i \tag{14}$$

However, in order for either μ_i^{ads} or μ_i^* to be useful

a) analytical expressions for the activity coefficients f_i are needed, and

b) $\Delta\phi'$ should be expressed in terms of either $\Delta\phi$ or E.

The expression of $\ln f_i$ may be derived directly from lattice statistics. Thus under Flory's approximation ($q_i = r_i$), we readily obtain from Eq. (35) in [77]

$$\ln f_i = r_i\sum_{j\neq i}\theta_jA^{ij} - r_i\sum_{k<j}\theta_k\theta_jA^{kj} \tag{15}$$

In what concerns the relationship between $\Delta\phi'$ and $\Delta\phi$, it may be derived as follows. If $\Delta\phi'/l_i$ is the effective electric field acting on each adsorbed species i, Eq. (7) is still valid but with the following modifications. First, $\Delta\phi$ on the right-hand side of this equation should be replaced by $\Delta\phi'$. Next, if we consider that the adsorbed layer is composed exclusively of i species, Eq. (7) is reduced to $\varepsilon_0\Delta\phi/\ell_i = \sigma^M + P_i M / A\ell_i r_i - \alpha_i M\Delta\phi'/A\ell_i^2 r_i$. Now if this equation is multiplied by θ_i and summed over all i, we obtain

$$\varepsilon_0\Delta\phi\sum\theta_i/\ell_i = \sigma^M + M\sum(P_i - \alpha_i\Delta\phi'/\ell_i)\theta_i/A\ell_i r_i \tag{16}$$

In addition, if again the adsorbed layer is composed exclusively of i species and X_i is the field due to the surrounding dipoles acting on species i, we have $\Delta\phi'/\ell_i = \sigma^M/\varepsilon_0 + X_i$, which yields

$$\varepsilon_0\Delta\phi'\sum\theta_i/\ell_i = \sigma^M + \varepsilon_0\sum X_i\theta_i = \sigma^M + \varepsilon_0 X \tag{17}$$

Here, X is a mean field calculated on the basis of the dipole-dipole interaction energy U^{dd}, because[69,76]

$$U^{dd} = \frac{1}{4\pi\varepsilon_0 d^3}\sum_{i,j}\overline{N}_{ij}(P_i^t/r_i)(P_j^t/r_j) = (X/2)\sum_i P_i^t N_i \tag{18}$$

where $P_i^t = P_i - \alpha_i(\sigma^M/\varepsilon_0 + X)$ is the total dipole moment of the i-th species, d is the intermolecular distance, and $\overline{N}_{ij}$ is the average number of the neighboring contacts between i and j species. The adsorbed species are assumed randomly distributed over the lattice sites and therefore

$$\overline{N}_{ij} = z_e r_i N_i r_j N_j / M \quad \text{and} \quad \overline{N}_{ii} = z_e r_i^2 N_i^2 / 2M \tag{19}$$

where z_e is an effective coordination number, which accounts indirectly for both image effects and non-nearest neighboring interactions.[65,89] From Eqs. (18) and (19) we obtain

$$X = z_e\sum P_i^t\theta_i / 4\pi\varepsilon_0 d^3 r_i \tag{20}$$

which, in combination with Eq. (17), results in

$$\varepsilon_0 \Delta\phi' \sum \theta_i / \ell_i = \sigma^M + z_e \sum (P_i - \alpha_i \Delta\phi' / \ell_i)\theta_i / 4\pi d^3 r_i \qquad (21)$$

It is seen that Eqs. (16) and (21) allow for the calculation of $\Delta\phi'$ from $\Delta\phi$ and *vice versa,* provided that the composition of the adsorbed layer is known. This composition is determined from the equilibrium equations, which are discussed below.

Note that Eq. (13) is in fact an extension of Eq. (8), which accounts for size effects and uses $\Delta\phi'$ instead of $\Delta\phi$. It is easy to show that these two potentials become identical if we assume that the charge of the poles of the dipoles is smeared uniformly over the plates of the condenser that includes the adsorbed layer. From a mathematical point of view this is equivalent to the validity of: $l = l_i = 4\pi d^3(M/A)/z_e$. If we consider that $d = \sqrt{2/(M/A)\sqrt{3}}$, z_e is close to 10.6,[89] and l lies between d and $1.5d$,[57] then $l \approx 4\pi d^3(M/A)/z_e$ and therefore $\Delta\phi' \approx \Delta\phi$. This explains the validity of the STE model in many experimental systems.

(ii) Equilibrium Properties

The great flexibility of the present model as well as of the STE model lies in the simple expression of the chemical potentials, Eqs. (8) and (13). In order to investigate the equilibrium properties of an adsorbed layer, we may write down the equilibrium equations in terms of chemical potentials and apply Eqs. (13) and (14). Thus, for the single adsorption of a solute A that is described by Eq. (2), the equilibrium equations in terms of chemical potentials may be expressed as

$$\mu_A^{ads} - \mu_1^{ads} = \mu_A^b - g\mu_S^b \qquad (22a)$$

$$\mu_j^{ads} = \mu_1^{ads} \qquad (22b)$$

where $j = 1, 2, \ldots,$ m denotes solvent clusters S_A which are in different polarization states. Eqs. (22) in combination with Eqs. (13), (14), and (15) result in the following adsorption isotherm

$$\ln\frac{\theta}{1-\theta} + A^{AS}(1-2\theta) = \ln(\beta_A x) - \ln B - (b_A - b_1)\Delta\phi' + (c_A - c_1)\Delta\phi'^2 \qquad (23)$$

where $\theta = \theta_A$, $\ln \beta_A = -\Delta G_{ads}/RT$ and B is equal to 1 when $m = 1$, otherwise it is given by

$$B = 1 + \sum_{i=2}^{m} \beta_{i1} \exp\{-(b_i - b_1)\Delta\phi' + (c_i - c_1)\Delta\phi'^2\} \tag{24}$$

Note that Eq. (15) yields $f_l/f_l = 1$ and $\ln (f_A/f_l) = A^{AS}(1 - 2\theta)$.

Extensive tests show that this simple model predicts all the adsorption features at least qualitatively. As an example, Figures 4 and 5 show calculated and experimental plots of θ *vs.*. σ^M, $\delta(\Delta G_{ads})/RT$ *vs.*.

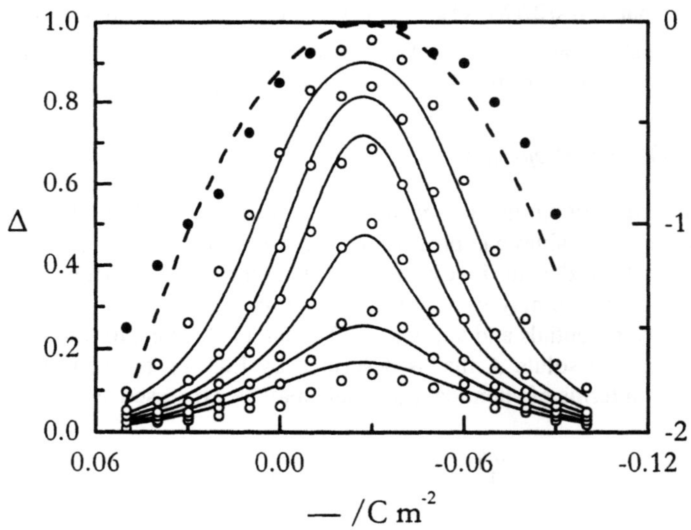

Figure 4. Plots of θ *vs.*. σ^M (o o o and —) and $\delta(\Delta G_{ads})/RT$ *vs.*. σ^M (• • • and - - -) due to triethylamine adsorption on a Hg electrode at concentrations 0.002, 0.00126, 0.001, 0.000794, 0.000631, and 0.0005 mol dm^{-3} (from top to bottom). Points are experimental data reprinted from *J. Electroanal. Chem.*, **255**, M. L. Foresti *et al.*, Adsorption of Triethylamine at the Mercury/Water Interphase from Charge and Interfacial Tension Measurements. p. 267, Copyright ©1988, with permission from Elsevier Science. Curves were calculated from Eqs. (16), (21), and (23) using the following parameters: $r_A = r_l = 1$, m = 1, $b_A = 2.02$ V^{-1}, $b_l = -0.57$ V^{-1}, $c_A = 2.25$ V^{-2}, $c_l = 7.44$ V^{-2}, $l_A = l_l = 0.8$ nm, $M/A = 1/A_A = 2.17\times10^{18}$ molecules/m^2, $A^{AS} = 1.7$, $z_e = 12.8$, and $\ln \beta_A = 6.82$.

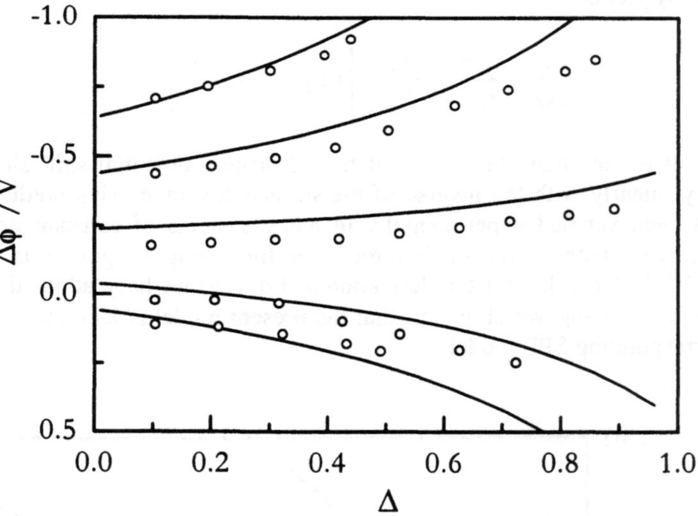

Figure 5. Plots of $\Delta\phi$ *vs.*. θ at constant σ^M values equal to -0.12, -0.08, -0.04, 0, 0.02 C m^{-2} (from top to bottom). Points are experimental data reprinted from *J. Electroanal. Chem.*, **255**, M. L. Foresti *et al.*, Adsorption of Triethylamine at the Mercury/Water Interphase from Charge and Interfacial Tension Measurements, p. 267, Copyright ©1988, with permission from Elsevier Science. Curves were calculated from Eqs. (16), (21), and (23) using the parameters of Figure 4.

σ^M and $\Delta\phi$ *vs.*. θ due to triethylamine adsorption on a Hg electrode.[90] Here, $\delta(\Delta G_{ads}) = \Delta G_{ads} - \Delta G_{ads,max}$, where $\Delta G_{ads,max}$ is the maximum value of ΔG_{ads}. More examples are given below in the section Complicated Adsorption Phenomena, whereas variations of this model are examined in.[23,37,56, 57,68,72]

An interesting adsorption feature, which is also described by this simple model in the limiting case when $m = 1$ and $\Delta\phi' = \Delta\phi$, is the surface potential shift upon adsorption, $\Delta\chi$, defined from

$$\Delta\chi = \Delta\phi(\sigma^M = 0, \theta = 0) - \Delta\phi(\sigma^M = 0, \theta \neq 0) \qquad (25)$$

For an adsorbed layer composed of constant orientated adsorbate molecules and solvent molecules in one polarization state, Eq. (21) readily yields[56]

$$\frac{1}{\Delta\chi} = \frac{\varepsilon_0 \varepsilon_S A_S}{P_S \varepsilon_A - P_A \varepsilon_S}\left[(\varepsilon_A - \varepsilon_S) + \frac{\varepsilon_S}{\theta}\right] \tag{26}$$

It is seen that the inverse of the adsorption potential shift should vary linearly with the inverse of the surface coverage. This prediction has been verified experimentally in many systems of constant orientated adsorbates in the stable region, see for example Figure 6 as well as [56,91-93]. Note that for the derivation of Eq. (26) we have adopted m = 1 and $\Delta\phi' = \Delta\phi$, which means that the present model is identical to the corresponding SPE model.

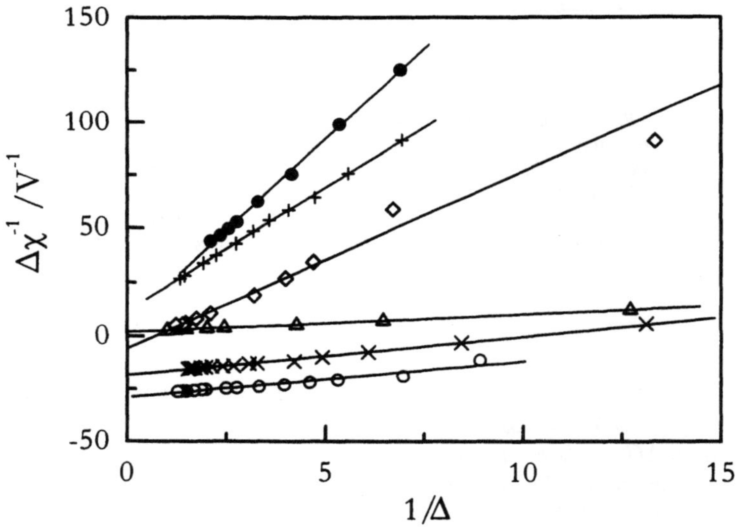

Figure 6. Plots of $1/\Delta\chi$ *vs.*. $1/\theta$ for the adsorption of acetonitrile (•, o), propionitrile (+, ×) and butyronitrile (◊, Δ) at the uncharged Hg | solution (•, +, ◊) and air | solution (o, ×, Δ) interfaces. The straight lines were calculated using a least squares procedure. Data were reprinted from *J. Electroanal. Chem.*, **385**, P. Nikitas and A. Pappa-Louisis, The Surface Potential Shift upon Adsorption at the Air/Solution and the Uncharged Hg/Solution Interfaces. p. 257, Copyright ©1995, with permission from Elsevier Science.

V. COMPLICATED ADSORPTION PHENOMENA

The adsorbed layer is the scene of various interesting phenomena, like re-orientation of the adsorbate molecules, co-adsorption, polylayer formation, surface aggregation, adsorption of oligomers and, finally, surface phase transformations. All these phenomena can be treated within the frames of either the STE model or the models based on the LBS approach in precisely the same way: We express the equilibrium equations first in terms of chemical potentials and next we introduce into these equations the expressions of the chemical potentials given by Eqs. (13) and (14). In some cases certain modifications are needed, as discussed below.

1. Co-Adsorption and Reorientation

(*i*) *Co-Adsorption of Neutral Solutes*

Consider the co-adsorption process of two neutral solutes A and B on an electrode surface. If $r = A_B/A_A$, the whole adsorption process may be expressed as[37]

$$\mu_A^{ads} - \mu_1^{ads} = \mu_A^b - g_A\mu_S^b \tag{27a}$$

$$\mu_B^{ads} - r\mu_1^{ads} = \mu_B^b - g_B\mu_S^b \tag{27b}$$

$$\mu_j^{ads} = \mu_1^{ads} \tag{27c}$$

where again j = 1, 2, ..., m. Note that a B molecule replaces from the adsorbed layer a cluster with dimension equal to B and chemical potential $\mu_{1B}^{ads} = r\mu_1^{ads}$. The view expressed in[39] that the solvent clusters replaced by A and B molecules have the same surface coverage θ_S is not correct, because in such a case the equality $\mu_{1B}^{ads} = r\mu_1^{ads}$ would not be valid even in the case where these clusters existed as independent entities. From Eqs. (13), (14), and (27) we obtain

$$\frac{\theta_A}{1-\theta_A-\theta_B}\frac{f_A}{f_1}=A_A \tag{28a}$$

and

$$\frac{\theta_B}{(1-\theta_A-\theta_B)^r}\frac{f_B}{(f_1)^r}=A_B \tag{28b}$$

where

$$A_A=\beta_A x_A \exp\{-(b_A-b_1)\Delta\phi'+(c_A-c_1)\Delta\phi'^2\}/B \tag{29}$$

$$A_B=\beta_B x_B \exp\{-(b_B-rb_1)\Delta\phi'+(c_B-rc_1)\Delta\phi'^2\}/B^r \tag{30}$$

The activity coefficients may be calculated from Eq. (15), which yields

$$\ln(f_i/f_1^r)=r(1-2\theta_i-\theta_k)A^{iS}+r\theta_k(A^{AB}-A^{kS}) \tag{31}$$

where $k \neq i$ and i, $k = A$ or B.

We observe that the surface composition of an adsorbed layer composed of solvent S and two solute molecules A and B in terms of $\Delta\phi'$ can be calculated from Eqs. (28)–(31) and in combination with Eqs. (16) and (21) the properties of the adsorbed layer can be expressed as a function of either $\Delta\phi$ or σ^M. Comparisons of model predictions with experimental data when $m = 1$, $\Delta\phi' = \Delta\phi$ and $A^{ij} = 0$ show that this simple model can provide a satisfactory description of the co-adsorption of two organic adsorbates, without using additional adjustable parameters, but from the known properties of their single adsorption.[37]

Two limiting cases of the co-adsorption are of particular interest. One is the reorientation of a solute on the electrode surface and another the effect of the specific adsorption of anions on the adsorption features of a neutral solute.

(*a*) *Reorientation of a neutral solute*

The reorientation of the solute molecules on an electrode surface from the flat position (B) to the normal one (A) may be treated as co-adsorption of two solutes, A and B. Therefore, the reorientation process can be described by Eqs. (28)-(31), provided that we use $x = x_A = x_B$, $A^{AS} = A^{BS}$ and $A^{AB} = 0$. Then the model predicts at least qualitatively all the adsorption features that characterize reorientation of solute molecules on electrodes.[37] For example, Figures 7 and 8 show calculated and experimental plots of a) the surface pressure Π *vs.*. $\Delta\phi$, and b) σ^M *vs.*. ln x at constant $\Delta\phi$ values due to 4-methylpyridine adsorption on a Hg electrode. From the experimental study of 4-methylpyridine we know that this compound reorientates on the Hg electrode from the flat state to the normal one with the N atom facing the solution.[10] For the application of the model we used $\Delta\phi = \Delta\phi'$ and the following parameters: $m = 1$, $b_A - b_l = 4$ V^{-1}, $c_A - c_l = -8$ V^{-2}, $b_B - rb_l = -0.3$ V^{-1}, $c_B - rc_l = -12$ V^{-2}, $r = 1.6$, $\varepsilon_A = 2.5$, $P_S = -1\times10^{-30}$ C m, $l_A = l_B = l_l = 0.5$ nm, $M/A = 3.33\times10^{18}$ molecules/m^2, $A^{AS} = 0.1$, ln $\beta_A = 3.9$, ln $\beta_B = 4.4$, and $RT\Gamma_{max} = 13.7$. The surface pressure was calculated from[94]

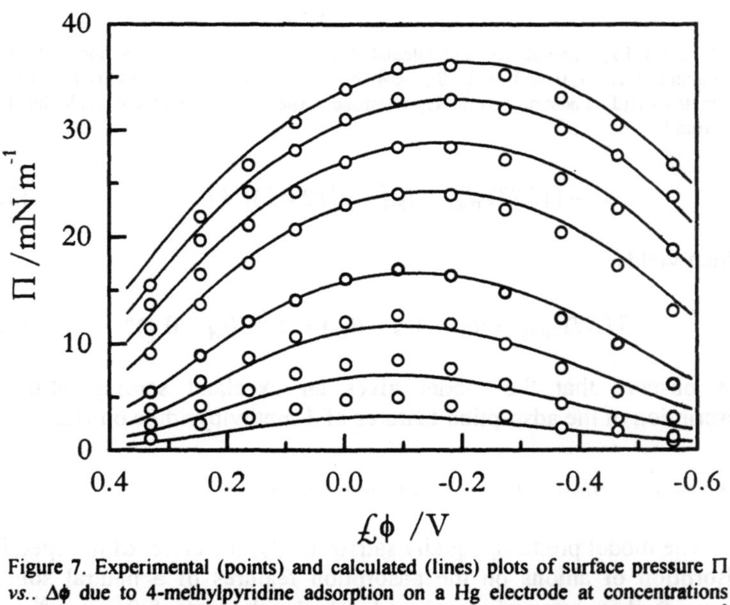

Figure 7. Experimental (points) and calculated (lines) plots of surface pressure Π *vs.*. $\Delta\phi$ due to 4-methylpyridine adsorption on a Hg electrode at concentrations 0.1237, 0.09262, 0.06552, 0.04653, 0.02083, 0.012, 0.0063, and 0.0026 mol dm^{-3} (from top to bottom). Experimental data were recalculated from [10]. Lines were calculated from Eqs. (16), (21), and (28)-(33) using the parameters given in text.

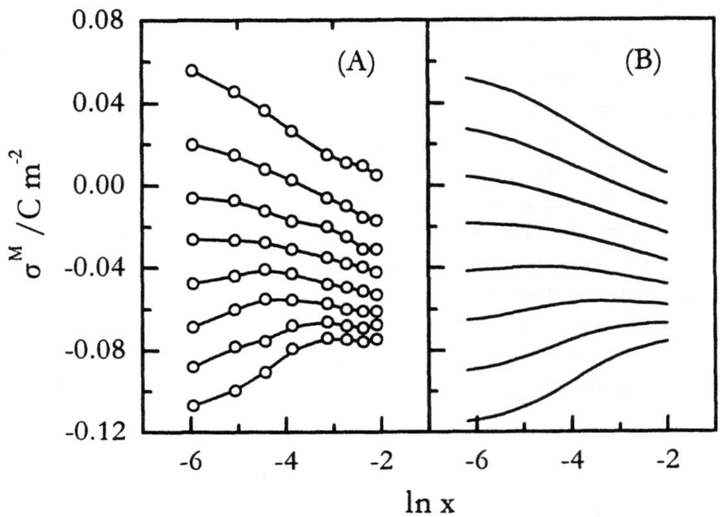

Figure 8. Experimental (A) and calculated (B) plots of σ^M *vs.*. ln x at constant $\Delta\phi$ equal to 0.2, 0.1, 0, -0.1, -0.2, -0.3, -0.4 and –0.5 V (from top to bottom) due to 4-methylpyridine adsorption on a Hg electrode. Experimental data were recalculated from [10].

$$-\Pi / RT\Gamma_{max} = [\mu_1^* - \mu_1^*(\theta = 0)] / kT \tag{32}$$

which yields

$$-\Pi / RT\Gamma_{max} = \ln(1 - \theta_A - \theta_B) + A^{AS}(\theta_A + \theta_B)^2 \tag{33}$$

We observe that the model gives an excellent semi-quantitative description of the adsorption features of 4-methylpyridine on Hg.

(b) Co-adsorption of neutral and ionic species

The model predicts, equally satisfactorily, the effect of the specific adsorption of anions on the adsorption features of a neutral solute. Figure 9 shows experimental and calculated plots of θ *vs.*. σ^M and $\delta(\Delta G_{ads})/RT$ *vs.*. σ^M due to ethylene glycol adsorption on a Hg electrode from aqueous 0.1 M NaF solutions, *i.e.*, in the absence of specific adsorption.[95] The model predictions were calculated from Eqs. (16), (21), and (23) using $m = 1$, $r_A = r_l = 1$, $b_A = 0.73\ \text{V}^{-1}$, $b_l = -0.23\ \text{V}^{-1}$, $c_A = 6.62\ \text{V}^{-2}$, $c_l = 10.65\ \text{V}^{-2}$, $l_A = l_1 = 0.6$ nm, $M/A = 3.23\times10^{18}$ molecules/m^2, $A^{AS} = 0$, $z_e = 14.49$, and $\ln \beta_A = -0.54$.

In order to calculate the effect of the co-adsorption of anions, Eqs. (28)–(31) should be modified as follows. First, we should use in Eq. (27) the electrochemical potential of the anions B instead of the chemical potential. The electrochemical potential of an inorganic ion B is given by

$$\begin{aligned}\bar{\mu}_B^* / kT &= \mu_B^0 / kT + \ln f_B \theta_B + z_B e^0 \phi^d / kT \\ &+ z_B e^0 \rho_B \Delta\phi' / \ell_B kT - c_B \Delta\phi'^2 + r_B \sum (r_j - 1)\theta_j / r_j\end{aligned} \tag{34}$$

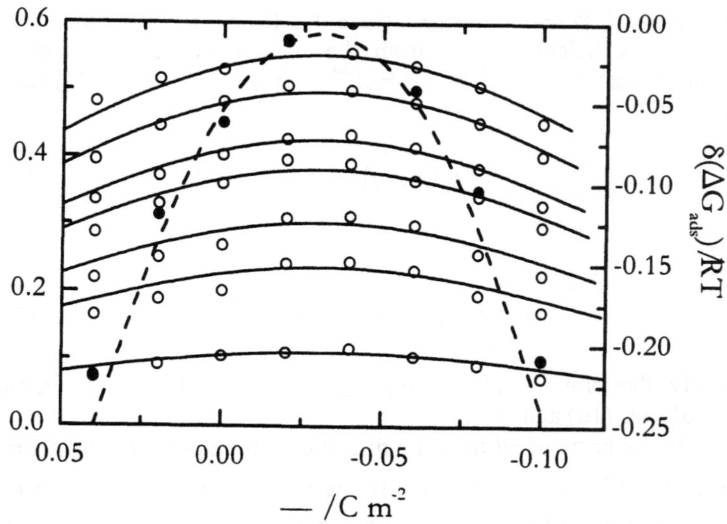

Figure 9. Plots of θ *vs.*. σ^M (o o o and —) and $\delta(\Delta G_{ads})/RT$ *vs.*. σ^M (• • • and - - -) due to ethylene glycol adsorption on a Hg electrode at concentrations 2, 1.6, 1.2, 1.0, 0.7, 0.5 and 0.2 mol dm^{-3} (from top to bottom). Points are experimental data reprinted from *J. Electroanal. Chem.*, **28**, S. Trasatti, Effect of the Nature of the Supporting Electrolyte on the Thermodynamic Analysis of the Adsorption of Organic Substances on Mercury. Adsorption of Ethylene Glycol form 0.1 m Aqueous Solutions of Halides, p. 257, Copyright ©1970, with permission from Elsevier Science. Curves were calculated from Eqs. (16), (21), and (23) using the parameters given in text.

where e^0 is the proton charge, ϕ^d is the inner potential at the OHP calculated from the Gouy-Chapman theory, and ρ_B is the distance of the center of the ion from the OHP. Equation (34) arises from Eq. (13) if we add the equality $z_B e^0 \phi^i = z_B e^0 \phi^d + z_B e^0 \rho_B \Delta\phi'/l_B$. In this equality $z_B e^0 \phi^i$ is the total work required to bring the ion with charge number z_B from an infinite distance to its final position inside the adsorbed layer, $z_B e^0 \phi^d$ is the work needed to bring the ion from an infinite distance through the diffuse layer to a site at the boundaries of the adsorbed layer, and finally the work required to bring the same ion from this site to its final position inside the adsorbed layer is $z_B e^0 \rho_B \Delta\phi'/l_B$. In this case, Eqs. (28) and (29) are still valid but Eq. (30) is modified to

$$A_B = \beta_B a_B \exp\{rb_1\Delta\phi' + (c_B - rc_1)\Delta\phi'^2 + e^0(\phi^d + \rho_B\Delta\phi'/\ell_B)/kT\}/B^r \tag{35}$$

In addition, it is reasonable to assume that the main contribution to the activity coefficients comes from the repulsive forces between the anions. Then $A^{AS} = 0$, $A^{AB} = A^{BS} = A^{BB} = -zw_{BB}/2kT$ and Eq. (31) results in

$$\ln(f_A / f_1) = 0 \tag{36a}$$

and

$$\ln(f_B / f_1^r) = r(1 - 2\theta_B)A^{BB} \tag{36b}$$

Finally, the term $-z_B e^0 (M/A)\rho_B\theta_B/r_B l_B$ should be added to the left-hand side of Eqs. (16) and (21).[39]

On the basis of all these modifications, we calculated the dependence of $\delta(\Delta G_{ads})/RT$ upon the difference $(\sigma^M - \sigma^M_{\max})$ when the bulk concentration c of a specifically anion varies from 0 to 1 mol dm^{-3}. Figure 10 compares the predicted behavior with the experimental dependence of $\delta(\Delta G_{ads})/RT$ upon $(\sigma^M - \sigma^M_{\max})$ in the presence of NaF, KCl, KBr and KI. The predicted curves were calculated from Eqs. (28), (29), (35), and (36) and the modified Eqs. (16) and (21) using the parameters of Figure 9 plus $r = 0.43$, $\varepsilon_B = 25$, $rA^{BB} = -4$, $\rho_B = 0.3$ nm and $\ln \beta_B = -1.5$. Note that the increase in the concentration c of the

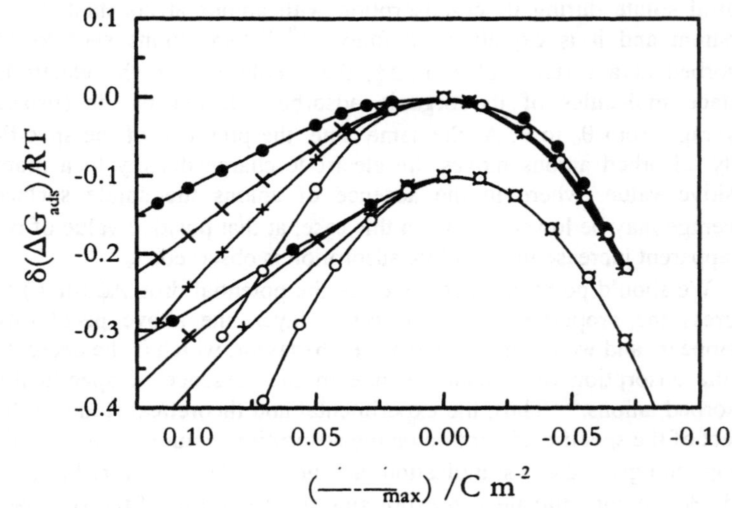

Figure 10. Plots of $\delta(\Delta G_{ads})/RT$ *vs.* $\sigma^M - \sigma^M_{max}$ due to ethylene glycol adsorption on a Hg electrode in the presence of NaF (o), KCl (+), KBr (×) and KI (•) (upper curves), and the same plots calculated from the model using the parameters given in text plus $c = 0$ (o), 0.01 (+), 0.1 (×) and 1 (•) mol dm^{-3} (lower curves). The experimental data were recalculated from [95]. The calculated data are shifted by –0.1 along the y-axis.

electrolyte in the bulk solution is equivalent to the use of constant c but different electrolytes in the order F^-, Cl^-, Br^-, I^-. We observe that the model gives a very satisfactory description of the effect of anion co-adsorption on the ethylene glycol electrosorption.

However, the most interesting feature of anion co-adsorption depicted in Figure 10 is the increase in the adsorption of a neutral solute in the presence of co-adsorbed anions when σ^M is used as the independent electrical variable. This is a very peculiar property, because the co-adsorption always leads to a decrease in the adsorption of each co-adsorbed species. Note that although this phenomenon was observed long ago, no satisfactory explanation had been given by the previous adsorption theories.[95-97] In contrast, the present model gives predictions that are in complete agreement with the experimental behavior. According to this model, the increase in the adsorption of a neutral solute during its co-adsorption with anions at constant σ^M is apparent and it is explained as follows.[39] When anions start to be adsorbed at a certain value of $\Delta\phi$, they replace from the electrode surface molecules of the organic adsorbate decreasing its surface coverage from θ_0 to θ. At the same time, the presence of the specifically adsorbed anions moves the electrode charge density to a more positive value, where in the absence of anions the solute surface coverage may be lower than θ. In this case, at that positive value of σ^M an apparent increase in the solute adsorption is observed.

We should point out that if we use the potential drop $\Delta\phi$ (or E) to express the properties of the adsorbed layer, the above peculiarity disappears and we observe the expected behavior, which is the decrease in the adsorption of a neutral solute in the presence of specifically adsorbed anions.[39] Thus, the experimental and theoretical study of the effect of the specific adsorption on the adsorption features of an organic compound gives the first indication that the two electrical variables, σ^M and $\Delta\phi$, are not equivalent in expressing the properties of the adsorbed layer and that the use of σ^M as the independent electrical variable should be handled with care. This issue of the equivalence of the two electrical variables is also discussed from another point of view in the section Polarization Catastrophe and Other Artifacts below.

(*ii*) *Co-Adsorption Phenomena in the Electrochemically Modulated Liquid Chromatography*

The co-adsorption of neutral and/or ionic species plays a dominant role in the retention characteristics of solutes in electrochemically modulated chromatographic columns. The electrochemically modulated liquid chromatography is a new and promising technique, which uses conductive stationary phases and the whole column is configured as an electrochemical cell.[98-105] If the mobile phase consists of the polar solvent S, an inert electrolyte, the organic modifier B and the eluite A, the capacity factor, k, for eluite A is given by[106-108]

$$\ln k = \lim_{n_A^b \to 0} \ln \frac{n_A^{ads}}{n_A^b} = \ln k^* + \lim_{\theta_A \to 0} \ln \frac{\theta_A}{x_A} \tag{37}$$

where n_A^{ads}, n_A^b are the numbers of moles of A on the stationary phase and in the mobile phase inside the chromatographic column, respectively. Since the retention on the stationary phase is governed by the co-adsorption of the eluite A and the organic modifier B, the ratio θ_A/x_A may be easily calculated from Eqs. (28)-(31) using the condition $\theta_A \to 0$.

The expression of the capacity factor in terms of $\Delta\phi$ is significantly simplified in the limiting case where $r \approx 1$, m = 1, $\Delta\phi' = \Delta\phi$ and $A^{ij} \approx 0$. Then we readily obtain that

$$\ln k = \ln k_0 - \ln(1 + A_B) - (b_A - b_1)\Delta\phi + (c_A - c_1)\Delta\phi^2 \tag{38}$$

where A_B is given by Eq. (30) with $r = 1$, $\Delta\phi' = \Delta\phi$ and $B = 1$. Equation (38) describes the majority of the experimental data reported up to now, because it predicts either a linear or a quadratic dependence of ln k upon the applied potential E over a certain range of potentials,[108] as found in many experimental studies.[102-105]

2. Polylayer Formation

The formation of polylayers on electrode surfaces is a rather rare phenomenon and for this reason the corresponding publications are few.[17, 18] Our previous studies on the formation of polylayers have shown that this process may be described by the following equilibrium equations[18,109]

$$\mu_{A_1}^{ads} - \mu_1^{ads} = \mu_A^b - \mu_S^b \tag{39a}$$

$$\mu_{A_i}^{ads} - \mu_{A_{i-1}}^{ads} = \mu_A^b \tag{39b}$$

$$\mu_j^{ads} = \mu_1^{ads} \tag{39c}$$

According to the above scheme, first a monomer solute molecule is adsorbed on the electrode surface by replacing a solvent molecule (or cluster), next dimers normal to the adsorbing surface are formed by the adsorption of monomers on top of previously adsorbed solute molecules, trimers from the adsorption of monomers on top of dimers, and so on.

The polylayer formation is such a complicated phenomenon that, at least as a first approximation, we may use $\Delta\phi' = \Delta\phi$ and $m = 1$. Then, substitution of Eqs. (13) and (14) into Eqs. (39) results in the following equilibrium equations[18]

$$\frac{\theta_{A_1}}{1-\theta_{A_1}-\theta_{A_2}-\ldots-\theta_{A_n}}\frac{f_{A_1}}{f_1} = A_A \tag{40a}$$

$$\frac{\theta_{A_i}}{\theta_{A_{i-1}}}\frac{f_{A_i}}{f_{A_{i-1}}} = K_i \tag{40b}$$

Here, A_A is given by Eq. (29) with $B = 1$ and $\Delta\phi' = \Delta\phi$, and K_i may be written as

$$K_i = \beta_i x_A \exp\{-(b_{A_i} - b_{A_{i-1}})\Delta\phi + (c_{A_i} - c_{A_{i-1}})\Delta\phi^2\} \tag{41}$$

where i ranges from 2 to n, n being the number of sublayers of the adsorbed film.

The treatment of the polylayer formation is significantly simplified if we assume that the short-range interactions have a negligible effect on the equilibrium properties. This yields $f_{A_1}/f_1 = 1$ and $f_{A_i}/f_{A_{i-1}} = 1$. A further simplification arises if we assume that $\beta_2 = \beta_3 = \ldots = \beta_n = \beta$, an approximation that is usually adopted in studies of multiplayer adsorption.[109-111] In addition, we may observe that $\varepsilon_{A_i} \approx \varepsilon_A$ for every i, which yields $c_{A_i}\ell_i = c_{A_1}\ell_1$, and thus, $c_{A_i} = c_{A_1}/i$. Finally, the parameters b_{A_i} depend upon the orientation of the adsorbed molecules. For example, if the orientation of the solute molecules in the first sublayer is opposite to that in the higher sublayers, we have $b_{A_i} = -(i-2)P_{A_1}/i\ell_1 kT = -b_{A_1}(i-2)/i$.

The equilibrium properties of a polylayer film can be determined from Eqs. (40) and (41), which allow for the calculation of the surface composition in terms of $\Delta\phi$. Next the charge density may be calculated from Eq. (16) and finally the differential capacity is obtained via numerical differentiation of the σ^M *vs..* $\Delta\phi$ data. Figure 11 shows calculated C *vs..* $\Delta\phi$ plots at different values of n when the orientation of the molecules in the first sublayer is opposite to that in the higher sublayers. For the calculations we used the following parameters: $b_{Al} - b_l = 5.5$ V^{-1}, $c_{Al} - c_l = -6$ V^{-2}, $\varepsilon_A = 3$, $P_S = -5\times10^{-30}$ C m, $l_1 = 0.6$ nm, $M/A = 3\times10^{18}$ molecules/m^2, $\ln \beta_A x = 2$ and $\ln \beta = 9$.

For comparison, the inset of Figure 11 shows experimental C *vs.* E plots taken from [17]. It is seen that both plots exhibit the same basic features, *i.e.*, a characteristic dip at positive polarizations and a plateau in the negative region. The lack of capacitance peaks in the experimental plots is likely due to the presence of the supporting electrolyte. A more thorough investigation of the polylayer formation is given in [18].

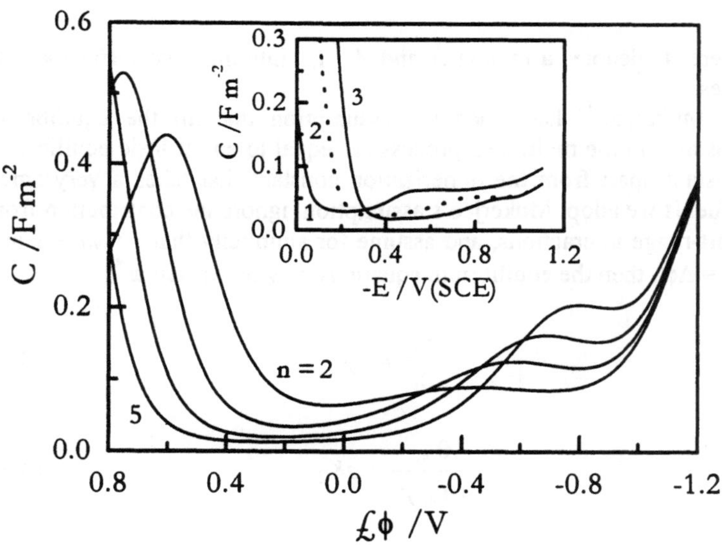

Figure 11. Plots of C *vs..* $\Delta\phi$ for polylayer films of different thickness predicted by the model using Eqs. (40) and (41) and the parameters given in text. Inset: Experimental C *vs.* E plots due to the adsorption on a Hg electrode from (1) 0.05 M H_2SO_4 solution saturated with n-hexanol, (2) 0.1 M KCl solution saturated with n-hexanoic acid, and (3) 0.1 M HCl solution saturated with n-heptanoic acid. Experimental data reprinted from *J. Colloid and Interface Science*, **105**, K. G. Baikerikar and R. S. Hansen, p. 143, Copyright ©1985, with permission from Academic Press, an imprint of Elsevier Science.

3. Surface Aggregation

In contrast to the scarcity of polylayer formation on electrodes, surface aggregation processes have been observed in many cases.[19-27] In order to model them we may adopt the single step process, $rA_1 \rightarrow A_r$,[25] or the more realistic multi-step process, $A_1 + A_{i-1} \rightarrow A_i$.[84,112] The latter process may be expressed in terms of chemical potentials as

$$\mu_{A_1}^{ads} - \mu_1^{ads} = \mu_A^b - \mu_S^b \tag{42a}$$

$$\mu_{A_{i-1}}^{ads} + \mu_{A_1}^{ads} = \mu_{A_i}^{ads} \tag{42b}$$

$$\mu_j^{ads} = \mu_1^{ads} \tag{42c}$$

where A_1 denotes a monomer and A_i, A_{i-1} intermediary surface aggregates.

Mukerjee[113] has made the assumption that all the equilibrium constants in the multi-step process are equal to an intrinsic equilibrium constant apart from the dimerization constant that takes a very small value. If we adopt Mukerjee's assumption, ignore the contribution from short-range interactions, and assume for simplicity that $j = m = 1$ and $\Delta\phi = \Delta\phi'$, then the equilibrium equations may be written as[84]

$$\frac{\theta_{A_1}}{\theta_1} = A_A \tag{43a}$$

$$\frac{\theta_{A_2}}{(\theta_{A_1})^2} = gK_2 \tag{43b}$$

$$\frac{\theta_{A_i}}{\theta_{A_{i-1}}\theta_{A_1}} = K_i \tag{43c}$$

where $g \ll 1$, $\theta_1 = 1 - \theta_{A_1} - \theta_{A_2} - \cdots - \theta_{A_r}$, A_A is given by Eq. (29) with $B = 1$ and $\Delta\phi' = \Delta\phi$, and K_i including K_2 may be expressed as

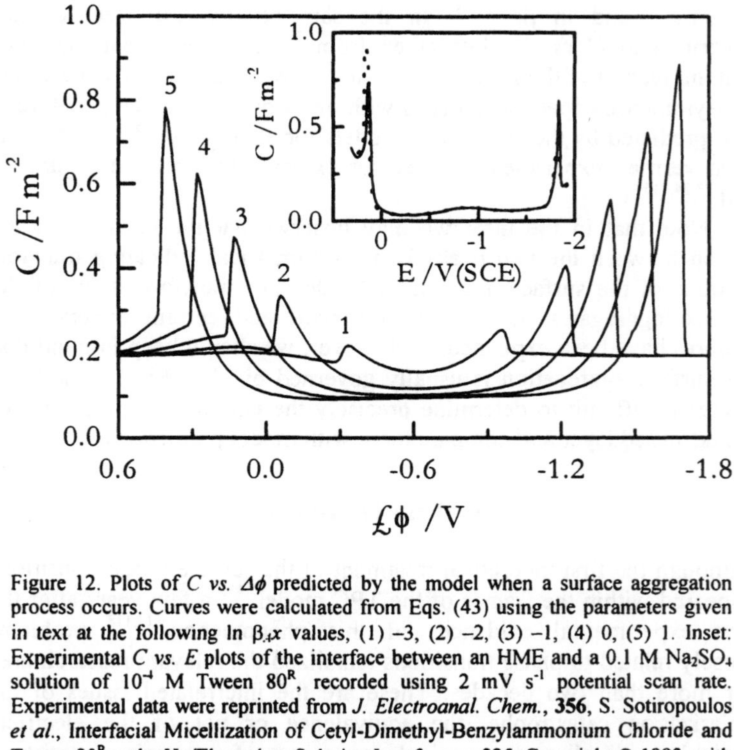

Figure 12. Plots of C *vs.* $\Delta\phi$ predicted by the model when a surface aggregation process occurs. Curves were calculated from Eqs. (43) using the parameters given in text at the following ln $\beta_A x$ values, (1) –3, (2) –2, (3) –1, (4) 0, (5) 1. Inset: Experimental C *vs.* E plots of the interface between an HME and a 0.1 M Na_2SO_4 solution of 10^{-4} M Tween 80^R, recorded using 2 mV s^{-1} potential scan rate. Experimental data were reprinted from *J. Electroanal. Chem.*, **356**, S. Sotiropoulos *et al.*, Interfacial Micellization of Cetyl-Dimethyl-Benzylammonium Chloride and Tween-80^R at the Hg/Electrolyte Solution Interface, p. 225, Copyright © 1993, with permission from of Elsevier Science.

$$K_i = \beta_r \exp\{-(b_r - b_{A_1})\Delta\phi + (c_r - c_{A_1})\Delta\phi^2\} \tag{44}$$

because Mukerjee's assumption necessarily imposes the equalities: $\beta_i = \beta_r$, $b_{A_i} - b_{A_{i-1}} = b_r$ and $c_{A_i} - c_{A_{i-1}} = c_r$ for every i ≥ 2.

Equation (43) can be used for the calculation of the surface composition in terms of $\Delta\phi$, and if the surface composition is known, the charge density is again calculated from Eq. (16) and the differential capacity C is obtained by numerical differentiation of the σ^M *vs..* $\Delta\phi$ data. Figure 12 shows C *vs..* $\Delta\phi$ plots predicted by the model when r =

100, $b_{A l} - b_l = 5$ V^{-1}, $c_{A l} - c_l = -6$ V^{-2}, $b_r - b_l = 0$ V^{-1}, $c_r - c_l = 2$ V^{-2}, ε_A = 3, g = 0.00001, $P_S = 1\times10^{-30}$ C m, l = 0.6 nm, $M/A = 3\times10^{18}$ molecules/m^2, and ln β_r = 2. In the same figure, the inset depicts experimental C *vs.* E plots taken from [24]. It is seen that the most striking feature of the formation of surface aggregates is the appearance of asymmetric capacitance peaks with one side vertical. This feature is also predicted by the single-step models for aggregation[25,114] and it has been verified experimentally, see for example the inset of Figure 12 and [23,24,115].

Note that in the inset we may observe that the capacity in the region between the two peaks is not symmetrical indicating transformations of the surface aggregates. Models for transformations of the surface aggregates have been developed and certain experimental features have been explained.[25] However, we should keep in mind that the surface aggregation is usually governed by slow kinetics and this makes it difficult to determine precisely the equilibrium properties of the adsorbed layer and compare them with model predictions.

4. Phase Transitions

Although the first theoretical treatment of the surface phase transitions appeared within the frames of the TPC model[7,116] a few years after the first experimental evidences of this phenomenon,[117,118] problems closely related to phase transitions tantalized the relevant research area for more than two decades. These are the interrelated issues of the polarization catastrophe, the equivalence or not of the electrical variables as well as the equivalence or not of the various statistical mechanical treatments. Due to their significance, these issues are discussed separately in the section below. Here, we focus our attention on the types and properties of the phase transitions predicted by the models for electrosorption.

The molecular models may predict at least two types of phase transitions: Transitions leading to (i) two immiscible surface solutions, and (ii) surface precipitation.

(i) Transitions Leading to Two Immiscible Surface Solutions

This is the conventional type of phase transitions predicted by both the macroscopic and molecular (two-dimensional) models, and they are closely related to the magnitude of the short-range interactions. For a first-order phase transition at the critical point we always have[14,16,119]

$$\left(\frac{\partial \ln(\beta_A x)}{\partial \theta}\right)_{T,V,\Delta\phi} = 0 \tag{45a}$$

$$\left(\frac{\partial^2 \ln(\beta_A x)}{\partial \theta^2}\right)_{T,V,\Delta\phi} = 0 \tag{45b}$$

Thus, the critical properties depend upon the adsorption processes that take place on the electrode surface. For example, if the adsorption is described by the equilibrium Eq. (23) with $m = 1$ and assume for simplicity that $\Delta\phi = \Delta\phi'$, then the critical values of A^{AS} and θ are: $A_c^{AS} = 2$ and $\theta_c = 0.5$.

Therefore, the adsorbed layer undergoes a phase separation transition when A^{AS} takes values greater than 2. The typical feature of this type of transition is the appearance of transition loops at the plots of θ *vs.*. $\Delta\phi$ and σ^M *vs.*. $\Delta\phi$. These loops may be replaced by vertical steps if we use the generalized ensemble Δ introduced in[16,67,82] or make use of the equal-areas theorem.[83] In this case the capacitance peaks in the C *vs.* $\Delta\phi$ plots become needle like and pits are getting formed (Figure 13). In addition, the model predicts that the square of the pit width is a linear function of $\ln x$ when T is constant and varies linear with T at constant x.[25] All these features have been experimentally observed with E in place of $\Delta\phi$.[12]

If $\Delta\phi'$ varies significantly from $\Delta\phi$, then the critical properties A_c^{AS} and θ_c depend upon $\Delta\phi$ but the general features of the transition region are as described above.[16,75] Similar transition properties are also predicted in the case of re-orientated adsorbates.[25,83]

Although the model predicts qualitatively all the properties of phase transitions, it does not shed light on the relation between the solute structure and the critical value of A^{AS}. If we model the adsorbed layer slightly differently by assuming that the solvent molecules occupy one lattice site and the solute molecules r sites, then, instead of Eq. (23) we will obtain a Flory-Huggins type adsorption isotherm with A_c^{AS} decreasing by increasing r. This is roughly in agreement with experiment.

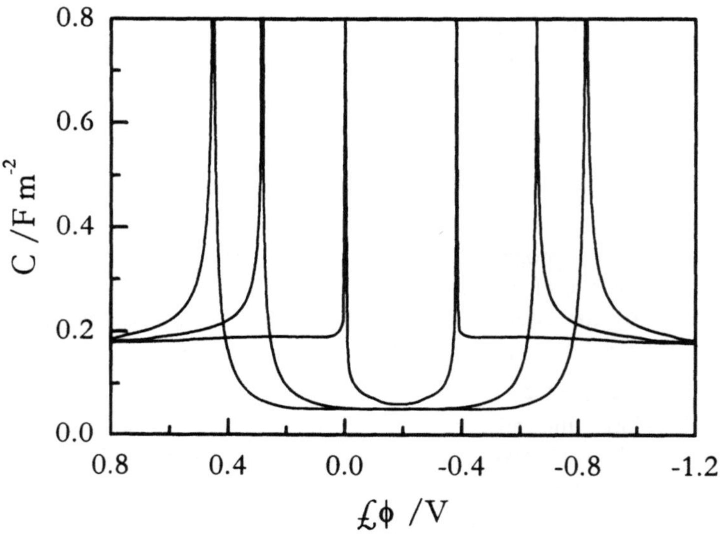

Figure 13. Plots of C *vs.* $\Delta\varphi$ due to adsorption of a neutral solute on an electrode surface when the adsorbed layer undergoes a phase separation transition. Curves were obtained by numerical differentiation of σ^M *vs.*. $\Delta\varphi$ data calculated from Eqs. (16) and (23) using the following parameters: $m = 1$, $\Delta\phi' = \Delta\phi$, $b_A - b_I = 2$ V^{-1}, $P_S = -1\times10^{-10}$ C m, $\varepsilon_A = 3$, $\varepsilon_S = 12$, $l = 0.6$ nm, $M/A = 3\times10^{18}$ molecules/m^2, $A^{AS} = 2.1$, and $\ln \beta_A x = 0, 1, 2$.

On this issue Guidelli *et al.*[36,63] expressed the view that phase transitions take place when the shape of the solute molecules hinders H-bond formation between water (solvent) molecules. In this case the water molecules are squeezed out of the adsorbed layer, leaving behind a compact film of solute molecules. This view seems to be verified by the three-dimensional lattice model, which in the presence of non-polar trimeric solute molecules does predict the occurrence of a phase transition. However, due to an inappropriate statistical mechanical approach based on the use of the grand ensemble Ξ instead of the generalized ensemble Δ, it is not possible to know whether this model predicts correctly or not the properties of the phase transitions.[67]

By analogy to phase transitions in liquid mixtures, the transitions we discuss here are first-order transitions leading to two immiscible surface solutions. One of the main features of these transitions is that the capacitance peaks become needle-like or disappear creating a capacitance pit. However, there are experimental systems where the creation of a capacitance pit is not necessarily associated with the disappearance of both peaks.[120-122] This feature may be predicted by the molecular models as described in the subsection below.

(ii) Transitions Leading to Surface Precipitation

The formation of a film of pure solute on the electrode surface may arise from an aggregation process when the aggregation number r tends to infinity.[25] In order to examine the properties of this process, we first obtain from Eqs. (43) the following equation

$$\ln\frac{\theta_{A_r}}{(\theta_1)^r} = \ln(gA_A) + (r-1)\ln(K_iA_A) \tag{46}$$

Now suppose that at a certain value of $\Delta\phi$ we have $\ln(K_iA_A) > 0$. In this case when $r \to \infty$, then $\theta_1 \to 0$ and $\theta_{A_r} \to 1$. It is easy to show that $\ln(K_iA_A)$ is a positive number when $\Delta\phi$ lies in the region $(\Delta\phi_{c_1}, \Delta\phi_{c_2})$, where

$$\Delta\phi_{c_{1,2}} = \frac{(b_r - b_1) \pm \sqrt{(b_r - b_1)^2 - 4(c_r - c_1)\ln(\beta_r\beta_A x)}}{2(c_r - c_1)} \tag{47}$$

Outside this region we have $\ln(K_iA_A) < 0$, that is, $(r-1)\ln(K_iA_A) \to -\infty$, which yields $\theta_{A_r} \to 0$.

Figure 14 depicts the shape of the capacitance plots when such a surface transformation takes place. It is seen that the most notable differentiation from the transitions leading to two immiscible surface solutions is not the orthogonal shape of the pits, which may also be predicted by the previous type of transitions at great A^{AS} values, but the possibility of existence of capacitance peaks outside the transition region, a feature which has been detected experimentally.[120-122]

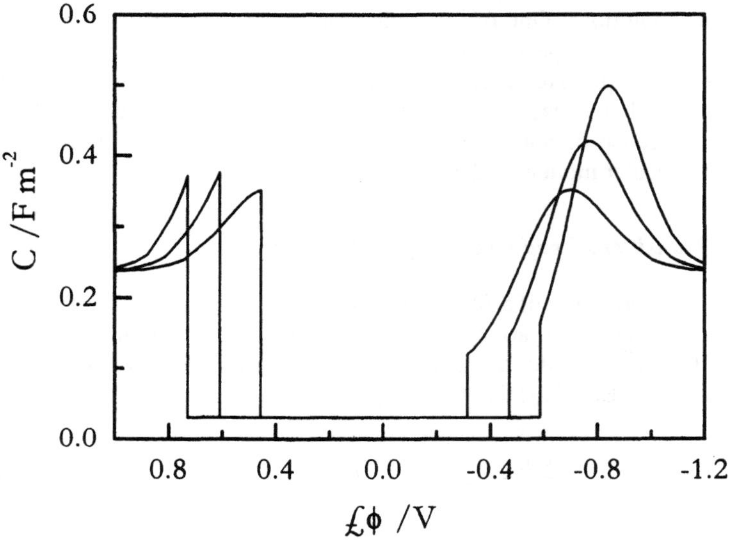

Figure 14. Plots of C *vs.* $\Delta\varphi$ due to adsorption of a neutral solute on an electrode surface when surface precipitate of pure solute is formed at the adsorbed layer. Curves were obtained by numerical differentiation of σ^M *vs.*. $\Delta\varphi$ data calculated from Eqs. (16) and (23) using Eq. (47) and the following parameters: $b_A - b_l = 2$ V^{-1}, $P_S = -1\times10^{-10}$ C m, $\varepsilon_A = 3$, $\varepsilon_S = 16$, $b_{Ar} - b_l = -1$ V^{-1}, $c_{Ar} - c_l = -7$ V^{-2}, $l = 0.6$ nm, $M/A = 3\times10^{18}$ molecules/m^2, $A^{AS} = 0$, ln $\beta_r = 0$, and ln $\beta_A x = 1, 2, 3$.

(iii) The Lattice-Gas Approach

An alternative treatment of the first-order phase transitions has been suggested by Retter[123,124] who has introduced the *lattice-gas approach.* The most striking feature of this approach is that it disregards completely the role of the solvent molecules in the properties of the adsorbed layer. Thus, the adsorbed layer behaves as a two-dimensional lattice gas where the solute molecules are distributed over the sites of a regular lattice.

The surface pressure Π of the *gaseous* state of the adsorbed layer is described by that of a two-dimensional lattice gas and, therefore, it may be written as[125-129]

$$\Pi / RT = \sum_i L_i(x) z^i \tag{48}$$

where $L_i(x)$ are tabulated polynomials, $x = \exp(w/2RT)$, w is the lateral interaction energy between two adsorbed species and $z = q \exp(\mu/kT)$. Here, q is the partition function of a single adsorbed molecule and μ its chemical potential. Retter approximates z by βx, where $\beta = \exp(-\Delta G_{ads}/RT)$ and calculates β from $\beta = \beta_{max}\exp\{-b(E - E_{max})^2\}$.

Now the surface coverage can be calculated from $\theta = z(\partial\Pi / \partial z)/RT$, which in fact gives the surface coverage at the gaseous state of the adsorbed layer. The surface coverage at the *liquid* (condensed) state may be obtained by means of the symmetry relation, $\theta(z_{gas}) + \theta(z_{liquid}) = 1$ with $z_{gas}z_{liquid} = x^{2c}$.[130] Finally, the differential capacity is calculated by means of the empirical relationship, $\theta = (C_2 - C)/(C_2 - C_l)$, where C_2 is the capacity of the gaseous state and C_l is the saturation capacity.

The model has been used to describe the cathodic transition step observed in the C *vs.* E plots of 5-iodocytosine adsorption on a Hg electrode with good results.[124] However, it is not clear whether these results are circumstantial or not, since there is not a systematic study of the model that will involve tests of the whole pit region as well as of the stable (gaseous) region. It seems that the problems with the lattice-gas approach are that

a) it disregards completely the role of the solvent, and
b) the very accurate treatment of the distribution of the adsorbed solute molecules over the lattice sites is, in fact, cancelled by the crude approximations of $\beta = \beta_{max}\exp\{-b(E - E_{max})^2\}$ and $\theta = (C_2 - C)/(C_2 - C_l)$.

Note that the lattice-gas approach might be an acceptable treatment for adsorption on solid electrodes, see for example [131-133].

VI. POLARIZATION CATASTROPHE AND OTHER ARTIFACTS

Although the study of the transition region poses no difficulties, it was for two decades a source of artifacts, big misunderstandings, and controversial views. Even respected electrochemists did not avoid the pitfalls of the transition region. The reason for all the problems seems to have been the use of the electrode charge density as the independent electrical variable but, in fact, it was the insufficient knowledge of the transition properties at charged interfaces. These properties under equilibrium conditions have been completely clarified by the combination of classical thermodynamics with experiment and model predictions,[13,14,16,119,134-137] and the most important of them are shown schematically in Figure 15.

The first indication that something went wrong with the molecular modeling of charged interfaces was detected by Cooper and Harrison[138] in 1975 and it was the prediction of infinite or negative capacities by all molecular models. This unrealistic situation was called *polarization*

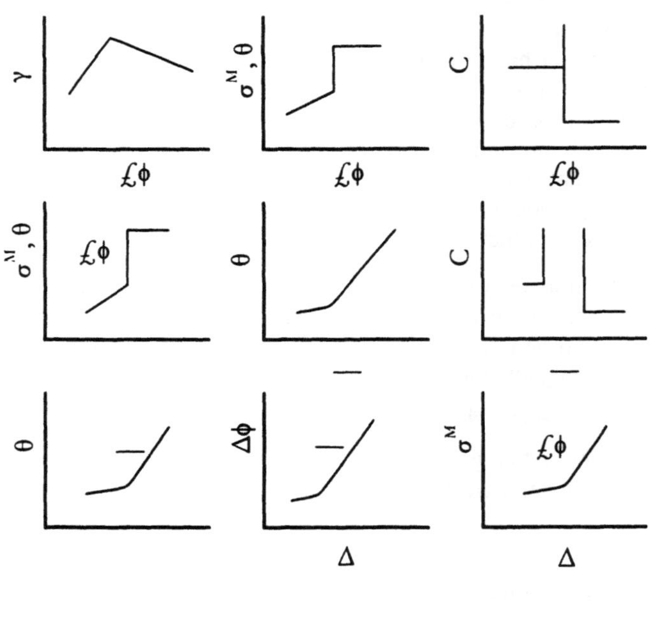

Figure 15. Diagram illustrating the equilibrium transition properties of charged interfaces. The physical quantity indicated inside each plot is kept constant.

catastrophe and it triggered many discussions and disputes[139-143] that pushed some electrochemists to think the impossible: to examine the opportunity to record experimentally negative capacities![143]

However, the solution to the polarization catastrophe enigma was surprisingly simple.[16,81] If we plot σ^M *vs..* $\Delta\phi$, we observe characteristic loops, as in the van der Waals equation, which correspond to a first-order phase transition (Figure 16). Therefore, the polarization catastrophe is nothing but the clear sign of a phase transition. The negative capacities come from the portion of the σ^M *vs..* $\Delta\phi$ curves that exhibits a negative slope and they can be eliminated if we make use of either the equal-areas theorem[83] or the generalized ensemble Δ to calculate the properties of the transition region.[16,67,82] Note that the equal-areas theorem should be applied with care, because it is not valid in all cases.[16,83]

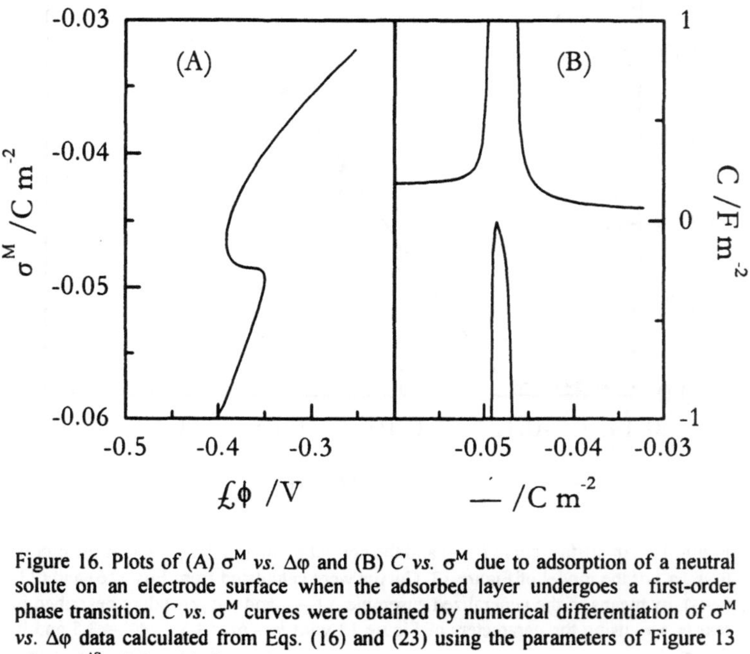

Figure 16. Plots of (A) σ^M *vs.* $\Delta\varphi$ and (B) C *vs.* σ^M due to adsorption of a neutral solute on an electrode surface when the adsorbed layer undergoes a first-order phase transition. C *vs.* σ^M curves were obtained by numerical differentiation of σ^M *vs.* $\Delta\varphi$ data calculated from Eqs. (16) and (23) using the parameters of Figure 13 when $A^{AS} = 2.2$ and $\ln \beta_A x = 0$.

It is surprising that the clarification of such a simple phenomenon lasted almost two decades and the reason was the use of σ^M instead of $\Delta\phi$ as the independent electrical variable. This observation has led me to the hyperbole that σ^M should be an inappropriate independent variable.[81] However, although the two electrical variables are not equivalent in expressing the transition properties, as clearly shown in Figure 15, we have no serious arguments that either σ^M or $\Delta\phi$ is an inappropriate variable. It is now well known that the use of σ^M as independent variable tends to mask a phase transition. But if we have in mind the properties depicted in Figure 15 and handle the equilibrium properties of the transition region with care, then we can easily avoid any artifact. However, although there are no appropriate or inappropriate electrical variables, appropriate and inappropriate statistical mechanical treatments of the transition region do exist.[67]

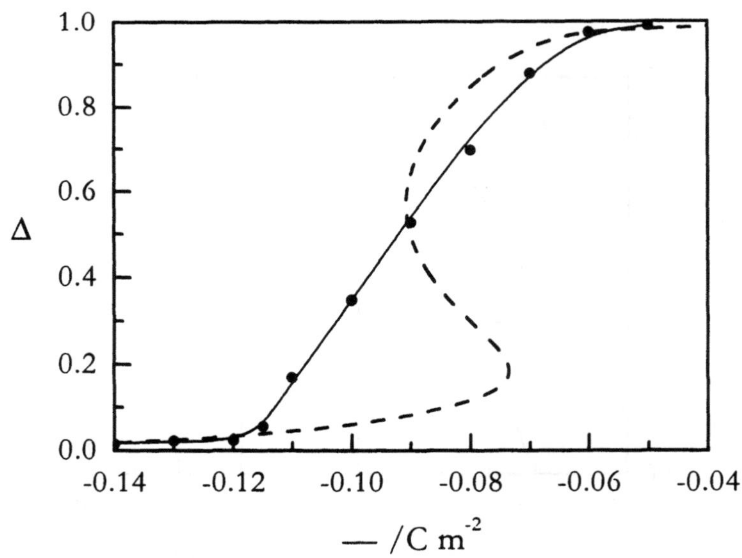

Figure 17. An artifact curve (- - -) and the correct prediction (—, • • •) at the plot of θ *vs.*. σ^M of the model of Figure 13. Curves were calculated from Eqs. (16) and (23) (- - -), by means of the generalized ensemble Δ (—), and the procedure described in text (• • •) using the parameters of Figure 13 with $b_A - b_I = 3$ V^{-1}, $A^{AS} = 3.5$ and ln $\beta_A x = -3$.

In order to clarify this point, let us consider the plots of θ *vs.*. σ^M at a constant ln $\beta_A x$ value predicted by the simple model of Figure 13 when b_A - b_I = 3 V^{-1}, A^{AS} = 3.5, and ln $\beta_A x$ = -3. This isotherm may be easily calculated from Eqs. (16) and (23) and it is shown by the broken curve in Figure 17. Note that we can be easily misled by the transition loop and replace it by a vertical step using the equal-areas theorem. Moreover, we may attempt to study the critical properties by means of the following equations

$$\partial\sigma^M / \partial\theta = 0 \tag{49a}$$

$$\partial^2\sigma^M / \partial\theta^2 = 0 \tag{49b}$$

which determine the point of inflection at this plot. Such an approach was, in fact, suggested in [68,73]. However, an inspection of Figure 15 shows that such a transition does not exist! Not only is the transition loop in Figure 17 an artifact but its replacement by a vertical step would also be an artifact, because Eqs. (49) determine the critical point of a non-existing phase transition.

Another artifact of the same type is depicted in the plots of θ *vs.* ln $\beta_A x$ at constant σ^M values, shown in Figure 18 by the broken lines. These isotherms were calculated from Eqs. (16) and (23) and the first of them, curve (1), is incompatible with the corresponding properties of Figure 15.

It is seen that inappropriate statistical mechanical treatments do exist and lead to artifacts. If we use the grand canonical ensemble Ξ to calculate the above plots, we readily find that the artifacts still exist![16,67] Therefore, the question is what the proper statistical mechanical treatment for this case is. The answer has been given in [16,67,82] and it is the use of the generalized ensemble Δ. The solid lines in Figures 17 and 18 were calculated using this ensemble. Alternatively, we may use the equations of the canonical ensemble, *i.e.*, the equations derived in the present chapter, as follows.[144]

The dependence of the interfacial tension γ upon $\Delta\phi$ at constant ln $\beta_A x$ can be calculated by means of the adsorption isotherm, Eq. (23), and the equation of state that arises from the equality $\mu_1^{ads} = \mu_S^b$:

$$\gamma / RT\Gamma_{\max} + const. = \ln(1-\theta) + A^{AS}\theta^2 + b_S\Delta\phi - c_S\Delta\phi^2 \tag{50}$$

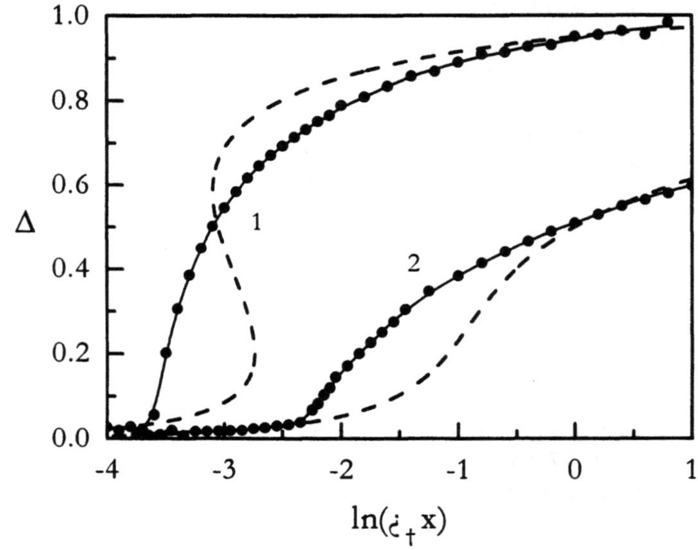

Figure 18. Artifact curves (- - -) and the correct predictions (—, • • •) at the plot of θ *vs.*. ln $\beta_A x$ at constant $\sigma^M = 0$ (1) and 0.05 C m^{-2} (2) of the model of Figure 13. Curves were calculated from Eqs. (16) and (23) (- - -), by means of the generalized ensemble Δ (—), and the procedure described in text (• • •) using the parameters of Figure 13 with $b_A - b_I = 3$ V^{-1} and $A^{AS} = 3.5$.

Therefore, at each pair (γ, $\Delta\phi$) at a certain ln $\beta_A x$ value we can calculate the quantity $\gamma + \sigma^M\Delta\phi$ and plot it as a function of $\Delta\phi$. Note that the y values of this plot contain an arbitrary constant, which nevertheless does not affect the results. In the transition region this plot exhibits two characteristic loops, which can be easily eliminated. The maximum in this plot, after the elimination of the transition loops, gives the value of Parson's function $\xi = \gamma + \sigma^M\Delta\phi$ at a given σ^M.[145] Therefore, the surface coverage θ at a certain value of σ^M is obtained by numerical differentiation of the $\xi / RT\Gamma_{max}$ values with respect to ln $\beta_A x$. These values of θ are denoted by (•) in Figures 17 and 18 and they coincide with the results of the generalized ensemble Δ.

It is seen that the polarization catastrophe is in fact an "innocent" artifact, because it is now easily detected and amended. In contrast, there are other artifacts appearing in the adsorption isotherms when σ^M is used as the independent electrical variable which are difficult to be identified. Note especially the artifact curve (2) in Figure 18, which does not exhibit any sign of phase transition. However, even these artifacts can be amended either by the use of the generalized ensemble Δ or the procedure suggested above. At any rate we observe once more that the electrode charge density σ^M, when it is used as an independent electrical variable, is a *difficult* variable, which should be handled with much care.

VII. THE ROLE OF THE METAL ELECTRODE – THE CASE OF SOLID ELECTRODES

In the models discussed in the previous sections the metal electrode has a constant contribution to the equilibrium constant(s) of the adsorption process(es). However, the role of the metal is more complex, especially in the case of solid electrodes. Up to now two effects have been analyzed in some detail: *electron spillover* from the metal into the electrolyte solution, and the *heterogeneity* of the electrode surface.

The electron spillover creates an electric field directed constantly from the metal towards the solution.[64,146,147] This electric field is superimposed on the field $\Delta\phi'/l$ and so it affects the adsorption properties, despite the fact that its contribution is restricted to distances shorter than 1 – 2 Å from the metal surface (Figure 19). Various variations of the *jellium model* have been proposed to treat the electron spillover but almost all the studies in this area concern the properties of charged interfaces in the absence of solute molecules.[146,147] In contrast, there are only some circumstantial data about the effect of the electron spillover on the adsorption of neutral organic molecules,[64] which show that it does not affect significantly the adsorption isotherms (Figure 20). However, it seems to explain qualitatively some of the differences observed when passing from mercury to gallium electrodes,[148,149] indicating that there is need of more studies in order to obtain a better understanding of this effect.

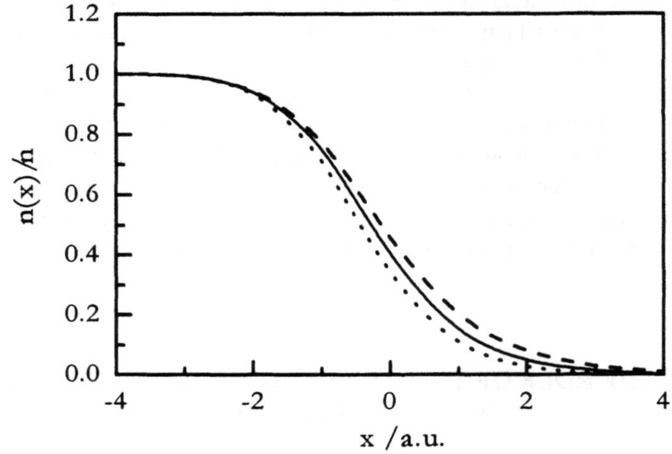

Figure 19. Electronic density profile for uncharged jellium (curve in the middle), jellium with a small negative (upper curve) and a small positive (lower curve) excess charge.[147] The distance x is measured from the metal surface in a.u. (1 a.u. = 0.529 Å).

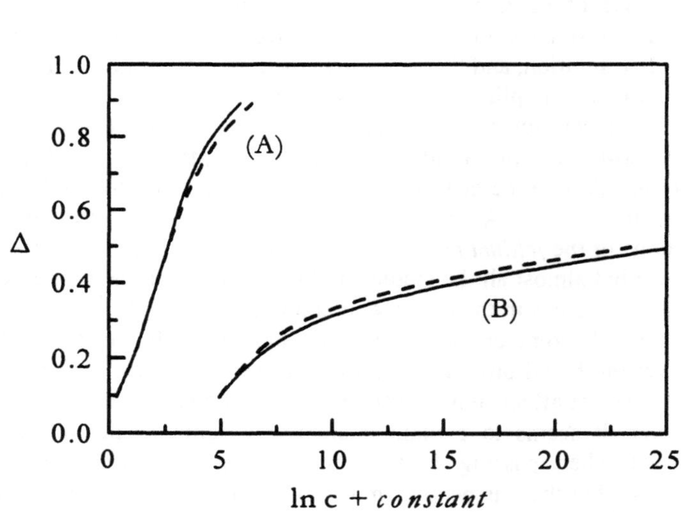

Figure 20. Effect of electron spillover (- - -) on the adsorption isotherms predicted by Guidelli's *et al.* three-dimensional model for a polar monomeric solute pointing towards the metal (A) and away from the metal (B) at $\sigma^M = -0.12\ C\ m^{-2}$.[64]

The heterogeneity of the adsorbing surface is an important factor determining the properties of adsorption on solid polycrystalline electrodes.[7,79,82,150] In this case the equilibrium properties depend upon the topography of the lattice sites with different adsorption energy. If the topography is known, the extension of the molecular models discussed in the previous sections is straightforward. For example, if the adsorption is described by a simple replacement process of pre-adsorbed solvent molecules existing in one state only (1) by the solute molecules (A) and the electric field is approximately given by $\Delta\phi/l$, then the partial adsorption isotherm, *i.e.*, the isotherm on adsorption sites of equal energy may be expressed as[79,82]

$$\theta_k/(1-\theta_k) = K_k \exp(U_k/kT) \tag{51}$$

where $K_k = \beta_A x \exp\{-(b_A - b_1)\Delta\phi + (c_A - c_1)\Delta\phi^2 - g_k\}$, θ_k is the partial surface coverage of A, $U_k = U_{kA} - U_{kl}$, U_{kA} and U_{kl} are the adsorption energies of a solute and a solvent molecule on sites of type k, respectively, and g_k, depending on the site topography, is given by $g_k = (1 - 2\theta)A^{AS}$ on *random surfaces* or $g_k = (1 - 2\theta_k)A^{AS}$ on *patchwise surfaces*. At a random surface the sites of different adsorption energy are randomly distributed over the lattice, whereas at a patchwise surface the sites of equal energy are present in groups. The total adsorption isotherm may be calculated from

$$\theta = \sum \theta_k M_k / M \tag{52}$$

where M_k is the number of adsorption sites with energy U_k and M is the total number of adsorption sites.

When there is a continuous variation of the adsorption energies on a random surface and this variation can be approximated by the uniform distribution, *i.e.*, $\chi(U) = 1/2U_0$ if $U_m - U_0 \le U \le U_m + U_0$ and $\chi(U) = 0$ outside this region, then Eq. (52) yields[79]

$$\theta = \frac{1}{2\lambda} \ln \frac{1 + Ke^{\lambda}}{1 + Ke^{-\lambda}} \tag{53}$$

where $K = K_k \exp(U_m/kT)$ and $\lambda = U_0/kT$, U_0 being half of the width of the adsorption energies distribution. Thus $\lambda = 0$ corresponds to homogeneous surfaces, whereas λ increases with increasing heterogeneity. Equation (53) is a generalized isotherm of Temkin,[151] which in the region of surface coverages where $Ke^{\lambda} >> 1$ and $Ke^{-\lambda} << 1$ results in the following Temkin's type isotherm

$$\theta = \frac{1}{f}\{\ln(\beta_A^* x) - (b_A - b_1)\Delta\phi + (c_A - c_1)\Delta\phi^2\} \tag{54}$$

where $f = 2(\lambda - A^{AS})$ and $\ln \beta_A^* = \ln \beta_A + U_m/kT + \lambda - A^{AS}$.

Figure 21 shows model predictions based on Eq. (53) when λ ranges from 0 to 5. It is seen that in general the heterogeneity effect does not seriously disturb the adsorption properties for values of λ smaller than 2 – 3. It is also interesting to observe that the model predicts Frumkin's type adsorption isotherms even at high λ values (Figure 21B). This may explain the validity of the Frumkin isotherm in a number of studies using solid polycrystalline electrodes of high hydrogen overpotential.[7]

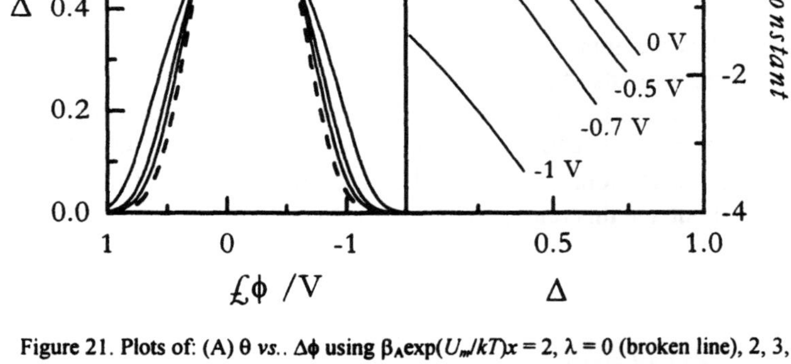

Figure 21. Plots of: (A) θ *vs.*. Δϕ using $\beta_A \exp(U_m/kT)x = 2$, λ = 0 (broken line), 2, 3, 5, and (B) $F = \ln\{\theta/x(1 - \theta)\}$ *vs.* θ using λ = 5 and Δϕ values indicated by each curve. Curves were calculated from Eq. (53) using $b_A - b_l = 2$ V^{-1}, $c_A - c_l = -6$ V^{-2} and $A^{AS} = 0$.

Adsorption on solid electrodes of high hydrogen overpotential such as Bi, Pb, Tl, and Cd, approximates to that on the Hg electrode. In contrast, the adsorption of organic compounds on the metals of the Pt group is irreversible due to chemisorption.[7] However, even in this case the adsorption may be treated using conventional techniques, usually based on the lattice gas approach.[7,131] Here, we briefly present the latest and interesting work carried out by Vayenas' group on the *electrochemical promotion.*[152-154]

The basic experimental set up for electrochemical promotion consists of a solid electrolyte, a catalyst-electrode deposited on one side of the solid electrolyte, and an inert counter electrode deposited on the other side of the solid electrolyte. On application of a potential drop between the catalyst and the counter electrode, ionic species are supplied to the catalyst and an effective double layer is established between the catalyst and the surrounding gas phase.

When a gas A_j is adsorbed on the catalyst-electrode having an effective double layer field $\boldsymbol{E}$, a partial charge transfer process, $A_j \leftrightarrow A_j^{\lambda_j +} - \lambda_j e^-$, takes place. This process creates an adsorbed dipole and if d is the distance between the centers of the positive and negative charges in this dipole, then its dipole moment is equal to $P_j = -\lambda_j e^0 d/2$. In addition, the field strength $\mathbf{E}$ may be calculated from $\boldsymbol{E} = (\Delta\Phi/e^0 l)\mathbf{n}$, where $\mathbf{n}$ is the unit vector normal to the catalyst surface, l is the double layer thickness and $\Delta\Phi$ is the work function difference between that of the actual surface and that of the surface at the potential of zero charge. Therefore, the contribution of the field-dipole interactions to the chemical potential is $\mathbf{P}_j \cdot \mathbf{E}$. In addition, if we take into account that the adsorbed layer has a regular structure where each adsorption site may be vacant or occupied by a gas molecule, we obtain the following expression for the electrochemical potential of A_j

$$\bar{\mu}_j(ads) = \mu_j^0(ads) + kT\ln\{\theta_j/(1-\theta_t)\} - \lambda_j \Delta\Phi \cos\omega d/2\ell \qquad (55)$$

where θ_t is the total surface coverage and ω is the angle between the adsorbed dipole and the field $\mathbf{E}$. Equation (55) can be straightforwardly applied to derive the adsorption isotherms that describe the adsorption step of the electrochemical promotion.

For example, if on the catalyst surface takes place the reaction D + A $\rightarrow$ products between an electron donor D ($\lambda_D > 0$) —like CO, CH_4, and C_2H_4— and an electron acceptor A ($\lambda_A<0$) —like O_2— then the catalytic rate may be written as

$$r = k_R \theta_D \theta_A \tag{56}$$

where the surface coverages θ_D, θ_A can be derived from the equilibrium equations $\mu_D(g) = \bar{\mu}_D(ads)$ and $\mu_A(g) = \bar{\mu}_A(ads)$. We readily obtain the Langmuirian type adsorption isotherms

$$\theta_j = \frac{k_j p_j \exp(\lambda_j L)}{1 + k_D p_D \exp(\lambda_D L) + k_A p_A \exp(\lambda_A L)}, \quad j = \text{D or A} \tag{57}$$

where p_j is the partial pressure of j, $k_j = \exp\{\mu_j^0(g) - \mu_j^0(ads)\}/kT$ and $L = \Delta\Phi\cos\omega d/2lkT$. Equations (56) and (57) provide an excellent semi-quantitative fit to the experimentally observed kinetics.[152-154]

VIII. COMPUTER SIMULATIONS

Molecular and macroscopic models can be effectively tested by Monte-Carlo (MC) or molecular dynamics (MD) computer simulations. These techniques are a valuable source of data for the clarification of the properties of a system and the evaluation of a theory. For this reason they have been used extensively in studies at charged interfaces.[131-133,146, 155-157] However, the majority of this work concerns the properties of solvent in the presence or absence of ions at a charged interface.

Computer simulations for charged interfaces in the presence of organic adsorption are very scarce. There is a MC study on the adsorption of urea onto Pt(100) surfaces based on the lattice-gas approach[131] and a MC study that examines the validity of the mean field approximation when it is used for the modeling of the adsorption of organic compounds on electrodes.[157] The latter study uses non-polarizable adsorbed molecules and shows that the random mixing (mean field) approximation gives results that are in reasonable agreement with those of the MC method over a wide range of molecular parameters, provided that the adsorption process is described by Eq. (2) (Figure 22 and [157]).

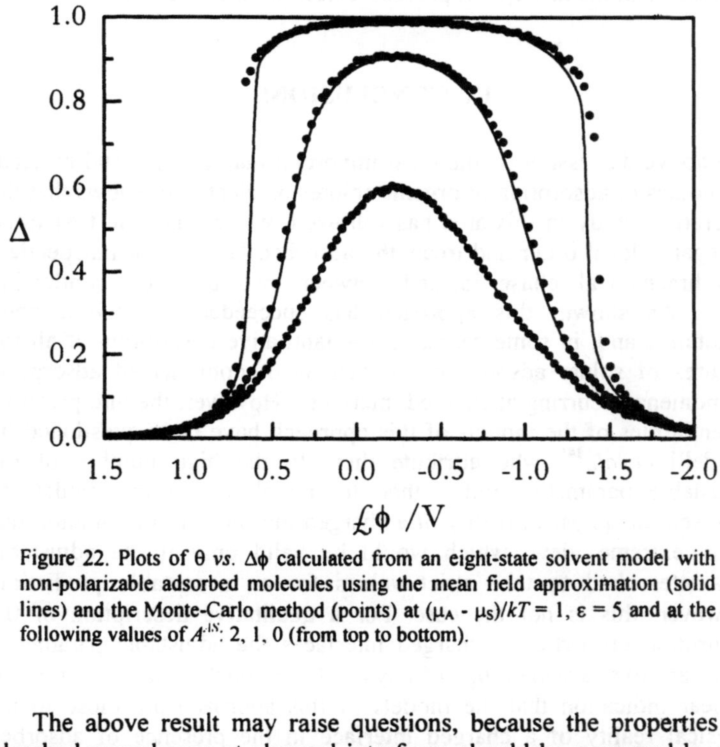

Figure 22. Plots of θ *vs.* Δϕ calculated from an eight-state solvent model with non-polarizable adsorbed molecules using the mean field approximation (solid lines) and the Monte-Carlo method (points) at $(\mu_A - \mu_S)/kT = 1$, $\varepsilon = 5$ and at the following values of A^{AS}: 2, 1, 0 (from top to bottom).

The above result may raise questions, because the properties of adsorbed monolayers at charged interfaces should be governed by the long-range coulombic interactions. However, it can be easily explained if we take into account that, when we adopt an adsorption mechanism like that represented by Eq. (2), in fact, we model the adsorbed layer as a mixture of adsorbate A molecules and solvent clusters S_A with dimensions equivalent to A. That is, the adsorbed layer consists of species with dimensions greater than 0.25 nm^2 and therefore the distance of the closest approach between two adsorbed dipoles cannot fall below 0.5 – 0.6 nm. Thus when we model the adsorbed layer as a mixture of adsorbate A molecules and solvent clusters S_A, the coulombic interactions stop to play the dominant role regarding the properties of this layer. This result is independent of whether we have polarizable or non-polarizable adsorbed molecules and, in fact, verifies the use of the mean field approximation in the models of the LSB approach, provided that the adsorption process is described by Eq. (2).

IX. CONCLUSIONS

The above discussion on the most important macroscopic and molecular models of adsorption of organic solutes on electrodes shows that the theoretical study in this area has followed two trends. The first trend disregards local order and treats the adsorbed layer as an independent two-dimensional phase located between two perfectly conducting walls. As shown, this approach has succeeded ro give a good qualitative and, in some cases, semi-quantitative description of all the features of solute adsorption, as well as of complicated adsorption phenomena occurring at charged interfaces. However, the interpretative potentialities of the models of this approach have been questioned by Guidelli *et al.*,[146] who attribute them to the high number of the adjustable parameters rather than to the ability of the models to represent the physical reality of a charged interface. In my opinion this is an extreme view, which would be valid only if the adjustable parameters had to take strict values for an acceptable prediction. However, this is not the case. For a qualitative description of the adsorption properties at charged interfaces the adjustable parameters may vary over a wide range of physically reasonable values and this is a clear indication that the models of this approach are close to the physical reality of a charged interface in the presence of adsorbed solute molecules. The fact that these models disregard local order affects the calculation of the effective field acting on each adsorbed molecule. However, this effect as well as all the effects that are not taken into account are unlikely to play the central role in the properties of the adsorbed layer and for this reason they may affect the potentialities of these models for a quantitative but not for an acceptable qualitative description of the adsorption on electrodes.

Due to its simplicity, this trend being developed for more than 20 years has reached its limits and therefore significant improvements in this area are not expected from now on. In contrast, we expect an essential development in the second trend. The second trend realizes the complexity of the charged interface and attempts to mimic it as close as possible. However, the consideration of local order and multistate solute and solvent molecules, as well as the treatment of the whole interface as a single entity lead to models with an extremely high mathematical complexity. It is interesting to point out that only one group, that of Professor Guidelli, has developed and continues to use this approach.[9,62,63] The trend towards increasing mathematical complexity should be continued and enriched, possibly with integral equation methods. The unexplainable lack of computer simulations, MC, MD and Car-Parinello type simulations,[146,158] for charged interfaces in the presence of solute molecules is a possible reason for the lack of studies following this approach in the last decade.

In any case there is still a long way until the development of a truly satisfactory molecular theory capable of predicting *a priori* and quantitatively the adsorption features of any solute. Until that occurs, the two trends mentioned above, together with computer simulations, will co-exist for different scopes: The first trend for analyzing experimental data, and for applications to complicated adsorption phenomena as well as to interfacial phenomena affected by adsorption. The second trend along with computer simulations for a better understanding of the molecular nature of the adsorption on electrodes.

REFERENCES

[1]G. Gouy, *Ann. Chim. Phys.* **29** (1903) 145; **8** (1906) 291; **9** (1906) 75.

[2]A. N. Frumkin, *Z. Phys. Chem.* **116** (1925) 466.

[3]A. N. Frumkin, *Z. Phys.* **35** (1926) 792.

[4]J. A. V. Butler, *Proc. Roy. Soc. Ser. A* **122** (1929) 399.

[5]J. O'M. Bockris, M. A. V. Devanathan, and K. Müller, *Proc. Roy. Soc. Ser. A*, **274** (1963) 55.

[6]J. O'M. Bockris, E.Gileadi, and K. Müller *Electrochim. Acta*, **12** (1967) 1301.

[7]B. B. Damaskin, O. A. Petrii, and V. V. Batrakov *Adsorption of Organic Compounds on Electrodes*, Plenum Press, New York, 1971.

[8]S. Trasatti, *J. Electroanal. Chem.* **53** (1974) 335.

[9]G. Aloisi and R. Guidelli, *J. Chem. Phys.* **95** (1991) 3679.

[10]P. Nikitas and A. Pappa-Louisi, *Electrochim. Acta* **38** (1993) 1573.

[11]A. Pappa-Louisi, P. Nikitas, and Ph. Andonoglou, *Electrochim. Acta* **38** (1993) 1585.

[12]R. de Levie, *Chem. Rev.* **88** (1988) 599.

[13]Cl. Buess-Herman, *J. Electroanal. Chem.* **186** (1985) 27; 41

[14]P. Nikitas, *J. Electroanal. Chem.* **300** (1991) 607.

[15]P. Nikitas, *J. Electroanal. Chem.* **308** (1991) 63.

[16]P. Nikitas, A. Anastopoulos, and G. Papanastasiou, *J. Electroanal. Chem.* **317** (1991) 43.

[17]K. Baikericar and R. Hansen, *J. Colloid Interface Sci.* **105** (1985) 143.

[18]P. Nikitas, *J. Electroanal. Chem.* **451** (1998) 249.

[19]P. Somasundaran, T. Healy, D. W. Fuerstenau, *J. Phys. Chem.* **68** (1964) 3562.

[20]P. Somasundaran and D. W. Fuerstenau, *J. Phys. Chem*, **70** (1966) 70; 90.
[21]N. Papadopoulos, S. Sotiropoulos, and P. Nikitas, *J. Colloid Interface Sci.* **151** (1992) 523.
[22]N. Papadopoulos, S. Sotiropoulos, and P. Nikitas, *J. Electroanal. Chem.* **324** (1992) 375.
[23]S. Sotiropoulos, P. Nikitas, and N. Papadopoulos, *J. Electroanal. Chem.* **356** (1993) 201.
[24]S. Sotiropoulos, P. Nikitas, and N. Papadopoulos, *J. Electroanal. Chem.* **356** (1993) 225.
[25]P. Nikitas and S. Andoniou, *J. Electroanal. Chem.* **375** (1994) 339.
[26]D. Bizzotto and J. Lipkowski, *J. Electroanal. Chem.* **409** (1996) 33.
[27]I. Burgess, C. A. Jeffrey, X. Cai, G. Szymanski, Z. Galus, and J. Lipkowski, *Langmuir* **15** (1999) 2607.
[28]R. H. Fowler and E. Guggenheim, *Statistical Thermodynamics*, Cambridge University Press, Cambridge, 1939.
[29]R. Guidelli, *J. Electroanal. Chem.* **110** (1980) 205
[30]P. Nikitas, *J. Chem. Soc., Faraday Trans.* **1**, 81 (1985) 1767.
[31]P. Nikitas, *J. Electroanal. Chem.* **263** (1988) 147.
[32]P. Nikitas and A. Pappa-Louisi, *J. Phys. Chem.* **94** (1990) 361.
[33]P. Nikitas, *Electrochim. Acta* **38** (1993) 2663.
[34]S. Trasatti, *Electrochim. Acta* **37** (1992) 2137.
[35]M. Rueda, I. Navarro, F. Prieto, and G. Ramirez, *J. Electroanal. Chem.* **379** (1994) 467.
[36]R. Guidelli and M. L. Foresti, *J. Electroanal. Chem.* **197** (1986) 103.
[37]P. Nikitas, *J. Electroanal. Chem.* **375** (1994) 319.
[38]P. Nikitas, *Electrochim. Acta* **39** (1994) 865.
[39]P. Nikitas, *Electrochim. Acta* **41** (1996) 2159.
[40]A. N. Frumkin, B. B. Damaskin, and A. A. Survila, *J. Electroanal. Chem.* **16** (1968) 493.
[41]B. B. Damaskin, A. N. Frumkin, and A. Chizhov, *J. Electroanal. Chem.* **28** (1970) 93.
[42]D. Schuhmann, *Electrochim. Acta* **32** (1987) 1331.
[43]D. Schuhmann, *J. Electroanal. Chem.* **252** (1988) 1.
[44]B.B. Damaskin and V.E. Kazarinov, in: *Comprehensive Treatise of Electrochemistry*, Vol. 1, Ed. by J. O'M. Bockris, B. Conway, E. Yeager, Plenum Press, New York, 1980, pp. 353-395.
[45]J. O'M. Bockris and A. K. N. Reddy, *Modern Electrochemistry*, Plenum Press, New York, 1970.
[46]D. Schuhmann, *J. Electroanal. Chem.* **201** (1986) 1.
[47]D. Schuhmann and P. Vanel, *Current Topics in Electrochem.* **2** (1993) 417.
[48]H. Nakadomari, D. M. Mohilner, and P. R. Mohilner, *J. Phys. Chem.* **80** (1976) 1761.
[49]D. M. Mohilner, H. Nakadomari, and P. R. Mohilner, *J. Phys. Chem.* **81** (1977) 244.
[50]D. M. Mohilner and M. Karolzcak, *J. Phys. Chem.* **86** (1982) 2838.
[51]M. Karolzcak and D. M. Mohilner, *J. Phys. Chem.* **86** (1982) 2840, 2845.
[52]M. Karolzcak, *J. Electroanal. Chem.* **402** (1996) 37.
[53]R. E. Boehm and D. E. Martire, *J. Phys. Chem.* **84** (1980) 3620.
[54]P. Nikitas and A. Pappa-Louisi, *Can. J. Chem.* **64** (1986) 328.
[55]P. Nikitas, *J. Chem. Soc., Faraday Trans.* 1, **80** (1984) 3315.
[56]P. Nikitas, A. Pappa-Louisis, *J. Electroanal. Chem.* **385** (1995) 257.
[57]P. Nikitas, *J. Phys. Chem.* **98** (1994) 6577.
[58]P. Nikitas, *Russian J. Electrochem.* **31** (1995) 747.
[59]R. Guidelli, *J. Electroanal. Chem.* **123** (1981) 59.

[60]R. Guidelli, *J. Electroanal. Chem.* **197** (1986) 77.
[61]M. Carla, G. Aloisi, M. L. Foresti, and R. Guidelli, *J. Electroanal. Chem.* **197** (1986) 123.
[62]R. Guidelli and G. Aloisi, *J. Electroanal. Chem.* **329** (1992) 39.
[63]R. Guidelli and G. Aloisi, *J. Electroanal. Chem.* **373** (1994) 107.
[64]R. Guidelli and G. Aloisi, in: *Electrified Interfaces in Physics, Chemistry and Biology, NATO ASI Series*, Vol. 355, Ed. by R. Guidelli, Kluwer Academic Publishers, London, 1992, pp. 337-367.
[65]S. Levine, G. M. Bell, and A. L. Smith, *J. Phys. Chem.* **73** (1969) 3534.
[66]W. L. Jorgensen, J. Chandrasekhar, J. D. Madura, R. W. Impey, and M. L. Klein, *J. Chem. Phys.* **79** (1983) 926.
[67]P. Nikitas, *J. Electroanal. Chem.* **408** (1996) 1.
[68]M. V. Sangaranarayanan and S. K. Rangarajan, *J. Electroanal. Chem.* **176** (1984) 1.
[69]M. V. Sangaranarayanan and S. K. Rangarajan, *J. Electroanal. Chem.* **176** (1984) 29.
[70]M. V. Sangaranarayanan and S. K. Rangarajan, *J. Electroanal. Chem.* **176** (1984) 45.
[71]M. V. Sangaranarayanan and S. K. Rangarajan, *J. Electroanal. Chem.* **176** (1984) 65.
[72]M. V. Sangaranarayanan and S. K. Rangarajan, *J. Electroanal. Chem.* **176** (1984) 99.
[73]M. V. Sangaranarayanan and S. K. Rangarajan, *J. Electroanal. Chem.* **176** (1984) 119.
[74]C. A. Basha and M. V. Sangaranarayanan, *J. Electroanal. Chem.* **291** (1990) 261.
[75]K. Pushpalatha and M. V. Sangaranarayanan, *J. Electroanal. Chem.* **425** (1997) 39.
[76]P. Nikitas, *J. Chem. Soc. Faraday Trans. 1* **82** (1986) 977.
[77]P. Nikitas, *J. Phys. Chem.* **91** (1987) 101.
[78]P. Nikitas, *Electrochim. Acta* **32** (1987) 205.
[79]P. Nikitas, *Electrochim. Acta* **33** (1988) 647.
[80]P. Nikitas, A. Pappa-Louisi, *Electrochim. Acta* **34** (1989) 1869.
[81]P. Nikitas, *J. Electroanal. Chem.* **306** (1991) 13.
[82]P. Nikitas, *Langmuir* **9** (1993) 2737.
[83]P. Nikitas, *Electrochim. Acta* **38** (1993) 1441.
[84]P. Nikitas, *J. Electroanal. Chem.* **425** (1997) 97.
[85]P. J. Flory, *J. Chem. Phys.* **10** (1942) 51.
[86]E. Guggenheim, *Mixtures*, Oxford University Press, Oxford, 1952.
[87]T. L. Hill, *An Introduction to Statistical Thermodynamics*, Addison-Wesley, London, 1960.
[88]P. Nikitas and S. Sotiropoulos, *J. Electroanal. Chem.* **309** (1991) 1.
[89]P. Nikitas, *Can. J. Chem.* **64** (1986) 1286.
[90]M. L. Foresti, M. R. Moncelli, G. Aloisi, and M. Carla, *J. Electroanal. Chem.* **255** (1988) 267.
[91]J. Dabkowski, M. Dabkowska, R. Pruszlowska-Drachal, Z. Koczorowski, and S. Trasatti, *Electrochim. Acta* **42** (1997) 2899.
[92]J. Ehrlich, T. Ehrlich, A. Janes, and E. Lust, *Electrochim. Acta* **45** (1999) 935.
[93]E. Lust, A. Janes, K. Lust, and J. Ehrlich, *Electrochim. Acta* **44** (1999) 4707.
[94]P. Nikitas, *Electrochim. Acta* **32** (1987) 1279.
[95]S. Trasatti, *J. Electroanal. Chem.* **28** (1970) 257.
[96]R. Parsons and R. Peat, *J. Electroanal. Chem.* 122 (1981) 299.
[97]M. Goledzinowski, J. Dojlido, and J. Lipkowski, *J. Electroanal. Chem.* **185** (1985) 131.
[98]R. S. Deinhammer, E.Y. Ting, and M. D. Porter, *J. Electroanal. Chem.* **362** (1993) 295.
[99]R. S. Deinhammer, M.D. Porter, and K. Shimazu, *J. Electroanal. Chem.* **387** (1995) 35.
[100]R. S. Deinhammer, E. Y. Ting, and M. D. Porter, *Anal. Chem.* **67** (1995) 237.
[101]E. Y. Ting and M. D. Porter, *Anal. Chem.* **69** (1997) 675.
[102]E. Y. Ting and M. D. Porter, *Anal. Chem.* **70** (1998) 94.

[103]E. Y. Ting and M. D. Porter, *J. Electroanal. Chem.* **443** (1998) 180.
[104]E. Y. Ting and M. D. Porter, *J. Chromatogr. A* **793** (1998) 204.
[105]J. A. Harnish and M.D. Porter, *Analyst* **126** (2001) 1841.
[106]B. L. Karger, L. R. Snyder, and Cs. Horvath, *An Introduction to Separation Science*, Wiley, New York, 1973.
[107]R. Tijssen, H.A.H. Billiet, and P. J. Schoenmakers, *J. Chromatogr.* **122** (1976) 185.
[108]P. Nikitas, *J. Electroanal. Chem.* **484** (2000) 137.
[109]P. Nikitas, *J. Phys. Chem.* **100** (1996) 15247.
[110]T. L. Hill, *J. Chem. Phys.* **14** (1946) 263.
[111]M. J. Dole, *J. Chem. Phys.* **16** (1948) 25.
[112]D. Schuhmann, *Progr. Colloid Polym. Sci.* **79** (1989) 338.
[113]P. Mukerjee, *J. Phys. Chem.* **76** (1972) 565.
[114]P. Nikitas, *J. Electroanal. Chem.* **348** (1993) 59.
[115]P. Nikitas, A. Pappa-Louisi, and S. Antoniou, *J. Electroanal. Chem.* **367** (1994) 239.
[116]B.B. Damaskin and S. L. Dyatkina, *Elektrokhimiya* **2** (1966) 981.
[117]W. Lorenz, *Z. Elektrochem.* **62** (1958) 192.
[118]S. Sathyanarayana, *J. Electroanal. Chem.* **10** (1965) 56.
[119]P. Nikitas, *Electrochim. Acta* **36** (1991) 447.
[120]C. Buess-Herman, *Prog. Surf. Sci.* **46** (1994) 335.
[121]Y.M. Temerk, P. Valenta, and H.W. Nurnberg, *J. Electroanal. Chem.* **100** (1979) 77.
[122]Th. Wandlowski, *J. Electroanal. Chem.* **293** (1990) 219.
[123]U. Retter, *J. Electroanal. Chem.* **236** (1987) 21.
[124]U. Retter, *J. Electroanal. Chem.* **349** (1993) 41.
[125]M. Sykes, D. Gaunt, J. Essam, and D. Hunter, *J. Math. Phys.* **14** (1973) 1060.
[126]M. Sykes, D. Gaunt, S. Mattingly, J. Essam, and C. Elliot, *J. Math. Phys.* **14** (1973) 1066.
[127]M. Sykes, D. Gaunt, J. Martin, S. Mattingly, and J. Essam, *J. Math. Phys.* **14** (1973) 1071.
[128]M. Sykes, J. Essam, and D. Gaunt, *J. Math. Phys.* **6** (1983) 283.
[129]C. Domb, in: *Phase Transitions and Critical Phenomena*, Vol. 3, Ch. 6,Ed. by C. Domb and M. Green, Academic Press, New York, 1974.
[130]T. Hill, *Statistical Mechanics*, McGraw-Hill, New York, 1956.
[131]P. A. Rikvold, M. Gamboa-Aldeco, J. Zhang, M. Han, Q. Wang, H. L. Richards, and A. Wieckowski, *Surf. Sci.* **335** (1995) 389.
[132]M. T. M. Koper, *Electrochim. Acta* **44** (1998) 1207.
[133]M. T. M. Koper, *J. Electroanal. Chem.* **450** (1998) 189.
[134]Cl. Buess-Herman, S. Bare, M. Poelman, and M. Van krieken, in: *Interfacial Electrochemistry*, Ch. 24, Ed. by A. Wieckowski, Marcel Dekker Inc., New York, 1999.
[135]Cl. Buess-Herman, C. Franck, and L. Gierst, *Electrochim. Acta* **31** (1986) 965.
[136]P. Nikitas, *J. Electroanal. Chem.* **316** (1991) 23.
[137]P. Nikitas, *J. Electroanal. Chem.* **446** (1998) 165.
[138]I. L. Cooper and J. A. Harrison, *J. Electroanal. Chem.* **66** (1975) 85.
[139]R. Parsons, *J. Electroanal. Chem.* **109** (1980) 369.
[140]S. L. Marshall and B. E. Conway, *J. Chem. Phys.* **81** (1984) 923.
[141]Z. Borkowska and J. Stafiej, *J. Electroanal. Chem.* **182** (1985) 253.
[142]Z. B. Kim, A. A. Kornyshev, and M. B. Partenskii, *J. Electroanal. Chem.* **265** (1989) 1.
[143]A. A. Kornyshev, *Electrochim. Acta* **34** (1989) 1829.
[144]P. Nikitas, *J. Electroanal. Chem.* **426** (1997) 167.
[145]J. Lawrence, R. Parsons, and R. Payne, *J. Electroanal. Chem.* **16** (1968) 193.

[146]R. Guidelli and W. Schmickler, *Electrochim. Acta* **45** (2000) 2317 and references cited therein.
[147]W. Schmickler, in: *Electrified Interfaces in Physics, Chemistry and Biology, NATO ASI Series,* Vol 355, Ed. by R. Guidelli, Kluwer Academic Publishers, London, 1992, pp. 369-425.
[148]G. Pezzatini, M. R. Moncelli, M. Innocenti, and R. Guidelli, *J. Electroanal. Chem***295** (1990) 275.
[149]G. Pezzatini, M. L. Foresti, M. R. Moncelli, and R. Guidelli, *J. Colloid Interface Sci.* **146** (1991) 452.
[150]J. O'M. Bockris and S. U. M. Khan, *Surface Electrochemistry, A Molecular Level Approach*, Plenum Press, New York, 1999.
[151]M. I. Temkin, *Zh. Fiz. Khim.* **25** (1941) 296.
[152]C. Vayenas and S. Brosda, *Stud. Surf. Sci. Catal.* **138** (2001) 197.
[153]C. Vayenas, C. Pliangos, S. Brosda, and D. Tsiplakides, *Chem. Ind.* **55** (2001) 551.
[154]S. Brosda and C. Vayenas, *J. Catal.* **208** (2002) 38.
[155]M. L. Berkowitz, In-Chul Yeh, and E. Spohr, in: *Interfacial Electrochemistry*, Ch. 3, Ed. by A. Wieckowski, Marcel Dekker Inc., New York, 1999.
[156]G. Brown, P. A. Rikvold, S. J. Mitchell, and M. A. Novotny, in: *Interfacial Electrochemistry*, Ch. 4, Ed. by A. Wieckowski, Marcel Dekker Inc., New York, 1999.
[157]P. Nikitas, *Electrochim. Acta* **47** (2002) 1765.
[158]R. Car and M. Parinello, *Phys. Rev. Lett.* **55** (1985) 2471.

4

Electrochemical Promotion of Catalysis

György Fóti, Ivan Bolzonella and Christos Comninellis

Unit of Electrochemical Engineering, Institute of Chemical and Biological Process Science, School of Basic Sciences, Swiss Federal Institute of Technology CH-1015 Lausanne, Switzerland

I. INTRODUCTION

Electrochemical promotion (EP) denotes electrically controlled modification of heterogeneous catalytic activity and/or selectivity.[1, 3] This recently discovered phenomenon has made a strong impact on modern electrochemistry,[4] catalysis,[5] and surface science.[6] Although it manifests itself also using aqueous electrolytes,[7] the phenomenon has mainly been investigated in gas-phase reactions over metal and metal oxide catalysts.[8, 9] In the latter case, the catalyst, which is an electron conductor, is deposited in the form of a porous thin film on a solid electrolyte support, which is an ion conductor at the temperature of the catalytic reaction. Application of an electric potential on the catalyst/support interface or, which is equivalent, passing an electric current between catalyst and support, causes a concomitant change also in the properties of the adjacent catalyst/gas interface, where the catalytic reaction takes place. This results in an alteration of the catalytic behaviour, controllable with the applied potential or current.

Modern Aspects of Electrochemistry, Number 36, edited by C. G. Vayenas, B. E. Conway, and R. E. White, Kluwer Academic / Plenum Publishers, New York, 2003

The current theory of electrochemical promotion[8, 9] attributes the effect to promoting species. These are generated electrochemically at the three-phase boundaries (tpb) between solid electrolyte, catalyst and gas phase, then spread out over the gas-exposed catalyst surface. Besides sophisticated techniques like XPS, UPS, PEEM or STM, allowing identification and inspection of particular promoting species, electrochemical promotion may be well investigated by phenomenological techniques such as catalytic rate measurements, solid state cyclic voltammetry and catalyst work function measurements, providing global parameters of the complex phenomenon.[10] Based on equilibrium considerations, the promotion due to backspillover of promoters may be interpreted in terms of catalyst overpotential.[11-13] Reliable estimate of the catalyst potential in a solid oxide electrochemical cell is not trivial due to the delicate choice of an appropriate reference electrode. This justifies detailed electrochemical characterization of the latter.

Besides its evident scientific interest, the phenomenon of electrochemical promotion may be of great importance in future industrial applications. The single-pellet type solid electrolyte cells generally used in fundamental investigations, however, are not suitable for direct industrial use. The first step in reactor development for electrochemical promotion consists of realization of bipolar cell configurations.[10] Such units should provide high promotion of the catalytic activity at reasonable cell potentials and with low current bypass. The final goal is to achieve efficient electrochemical promotion in dispersed catalyst systems by either using porous solid electrolyte supports or by manufacturing microstructured electrochemical reactors.

In order to address both fundamental and application aspects of electrochemical promotion in this Chapter, the combustion of ethylene over IrO_2 or RuO_2 catalysts and the reduction of NO with propylene over Rh catalysts —all deposited on yttria-stabilized zirconia (YSZ) solid electrolyte— have been chosen as model catalytic systems. While the literature of electrochemical promotion deals mainly with metal catalysts, our laboratory has a long experience with promotion of metal oxide catalysts, such as IrO_2[14-17] and RuO_2.[18, 19] In fact, ethylene combustion with IrO_2/YSZ film catalyst was the first catalytic system found to exhibit the phenomenon of permanent electrochemical promotion,[15] which manifests itself as a shift in the steady-state open-circuit activity due to polarization, and is attributed to a change in the average surface oxidation state of the metal. Permanent promotion was then reported also with Rh/YSZ film catalyst in the reduction of NO with hydrocarbons.[20] The first efficient bipolar cell configuration for electrochemical promotion, a necessary step towards potential industrial applications, was also realized with a metal oxide catalyst (RuO_2) and tested for ethylene combustion.[18]

In this Chapter, an overview is given of recent developments in this field with special attention to the investigation of the mechanism of electrochemical promotion and to the development of new bipolar cell configurations aiming at future industrial applications.

II. THE PHENOMENON OF ELECTROCHEMICAL PROMOTION

In this section, a phenomenological description of electrochemical promotion will be given, followed by a mechanistic interpretation and some theoretical considerations. For illustration, the simplest possible electrochemical cell of single-pellet type with an O^{2-}-conducting solid electrolyte has been chosen.

The electrochemical cell, depicted in Figure 1, is constructed on an yttria-stabilized zirconia (YSZ) pellet serving both as solid electrolyte and as support for the electrodes. YSZ is known to be an ionic (O^{2-}) conductor showing an Arrhenius-type temperature dependence of bulk ionic conductivity.[21] YSZ with a doping level of 6 to 8 mol% Y_2O_3 in ZrO_2 is used, which gives rise to full stabilization of the cubic zirconia structure and provides an O^{2-} conductivity high enough to enable use of the cell at moderate temperatures, down to about 250 °C. The porous catalyst (metal or metal oxide) film electrode is an electron conductor. The catalyst, which is the working electrode, is deposited on one side of the YSZ pellet; the catalytically inert counter electrode and reference electrode, usually both made of gold, are deposited on the opposite side. The reference electrode is only needed for fundamental studies and, in fact, it is often omitted in cell configurations developed for industrial applications.

The cell is placed in the reactive gas atmosphere. Several interfaces may be identified in such a system: electrode/electrolyte interfaces, electrode/gas interfaces, and three-phase boundaries (tpb). Due to the different mechanisms of electric conductivity in electrodes and electrolyte, polarization of the catalyst by current application involves

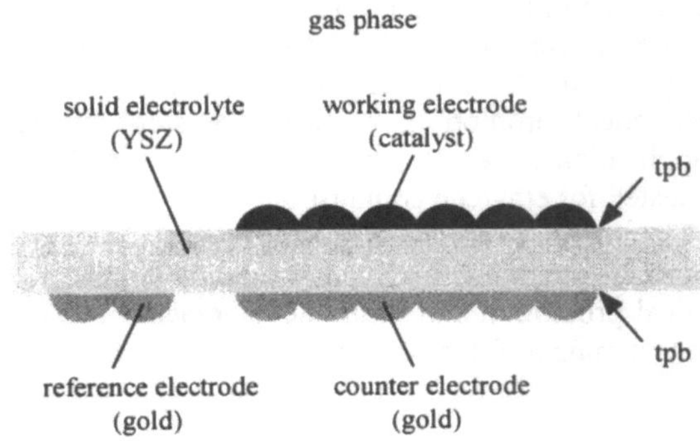

Figure 1. Schematic representation of a single-pellet cell for electrochemical promotion studies. The abbreviation tpb means three-phase boundaries where electrode, electrolyte and gas phase meet.

charge transfer reactions and a consequent charge separation, all localized, at a first glance, at the electrode/electrolyte interfaces and/or at the three-phase boundary. The resulting promotion of catalytic activity reveals, however, modification of another interface, namely the gas-exposed catalyst surface where the heterogeneous gas reactions occur. In fact, the key to understand electrochemical promotion is to elucidate the mechanism of how the effect of polarization propagates from one interface to the other.

1. Description of a Typical Electrochemical Promotion Experiment

Consider now the single-pellet configuration given in Figure 1 as the electrochemical cell and the combustion of ethylene,

$$C_2H_4\,(g) + 3\,O_2\,(g) \rightarrow 2\,CO_2\,(g) + 2\,H_2O\,(g) \qquad (1)$$

as a model reaction (see Figure 2). Under open-circuit conditions, *i.e.*, at zero electric current, the system exhibits a purely catalytic behavior.

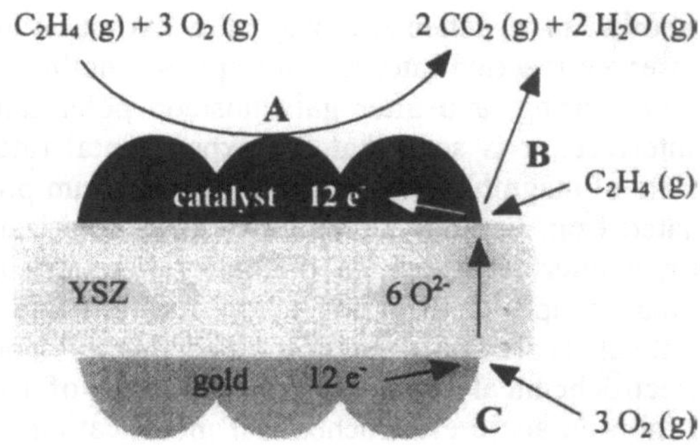

Figure 2. Scheme of ethylene combustion in a single-pellet electrochemical cell. **A**: catalytic combustion of ethylene; **B**: electrochemical oxidation of ethylene at the tpb; **C**: electrochemical reduction of oxygen.

The combustion of ethylene occurs at the gas-exposed catalyst surface with a reaction rate, r_o, expressed in terms of moles of atomic oxygen consumed per unit time (mol O s^{-1}).

Application of an anodic current between the counter and the working electrode —now the solid electrolyte is the source of O^{2-} ions and the working electrode is the collector of electrons— may result in the electrochemical oxidation of ethylene at the working electrode:

$$C_2H_4\,(g) + 6\,O^{2-} \rightarrow 2\,CO_2\,(g) + 2\,H_2O\,(g) + 12\,e^- \qquad (2)$$

Assuming a current efficiency of 100%, the maximum possible electrochemical reaction rate, r_{Farad} (mol O s^{-1}), is obtained from Faraday's law:

$$r_{Farad} = I/zF \qquad (3)$$

where I is the electric current, z is the charge number of the transported ions (for O^{2-}, $z = 2$), and F is the Faraday constant. If open-circuit catalytic reaction and Faradaic reaction were simply additive, then Eq. (3) would give the maximum expected increase in reaction rate due to polarization.

Figure 3 shows, in a schematic way, the evolution of the experimentally observed reaction rate, r, in a stepwise anodic polarization cycle, *i.e.,* before, during, and after galvanostatic polarization of the catalyst/YSZ interface. It is seen that the experimental rate increase, $r - r_0$, is by orders of magnitude higher than the maximum possible rate increase calculated from Faraday's law. Obviously, polarization of the catalyst/electrolyte interface causes a dramatic alteration in catalytic activity rather than simply contributing to the reaction rate by adding the electrochemical (Faradaic) reaction. The highly non-Faradaic character of electrochemical promotion is the origin of its currently used synonym: non-Faradaic electrochemical modification of catalytic activity (NEMCA effect).[2, 8]

The electrochemical promotion is usually quantified by three parameters, Λ, ρ, and γ. The enhancement factor, Λ, which may be

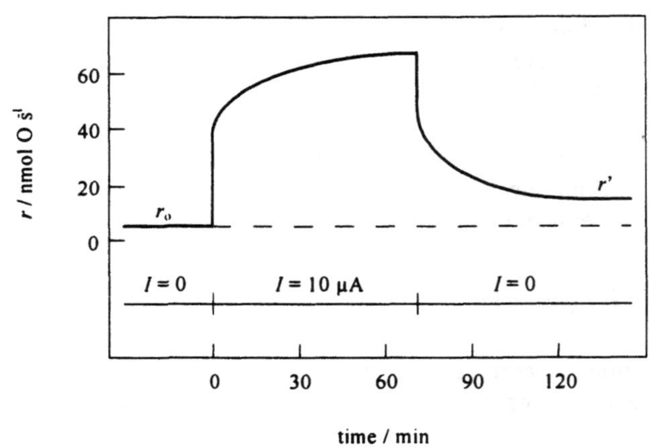

Figure 3. Schematic representation of evolution of the reaction rate due to current application of 10 μA during 70 min. Maximal electrochemical rate increase from Faraday's law: $I / 2F = 52$ pmol O s^{-1}. Catalyst: RuO_2; $p_{C_2H_4} = 100$ Pa, $p_{O_2} = 12$ kPa, flow rate: 175 mL min^{-1} STP, $T = 380$°C.

regarded as a current efficiency, is defined as the ratio of the observed rate increase to the maximum possible electrochemical rate:

$$\Lambda = (r - r_o)/(I/2F) \qquad (4)$$

For enhancement factors higher than unity the reaction is called non-Faradaic. The rate enhancement factor, ρ, is defined as the ratio of the promoted catalytic rate, *r*, to the initial open-circuit reaction rate, r_o, and it is a measure of the level of promotion:

$$\rho = r/r_o \qquad (5)$$

The open-circuit enhancement factor, γ, indicates the reversibility of the electrochemical promotion, and is defined as the ratio of two steady-state open-circuit catalytic reaction rates, one after polarization, *r*', and the other before polarization, r_o:

$$\gamma = r'/r_o \qquad (6)$$

When the factor γ is unity, the electrochemical promotion is termed reversible. Several cases of irreversible modification have been reported from our laboratory.[15, 20] This effect has also been termed permanent electrochemical promotion or permanent NEMCA effect.

2. The Mechanism of Electrochemical Promotion

Electrochemical promotion of catalysis, similarly to usual (chemical) promotion and to metal-support interactions in heterogeneous catalysis, is related to spillover-backspillover phenomena. The latter can be described as the mobility of adsorbed species from one phase on which they easily adsorb (donor) to another phase where they do not directly adsorb (acceptor). By this mechanism a seemingly inert material can acquire catalytic activity. Spillover may lead to an improvement of catalytic activity or selectivity and also to an increase in lifetime of the catalyst.

As a consequence of the occurrence of spillover in heterogeneous catalysis, the usual kinetic models can no longer be applied in a direct way. Generation of new surface sites or change in surface concentrations lead to new terms in the rate equations. A general

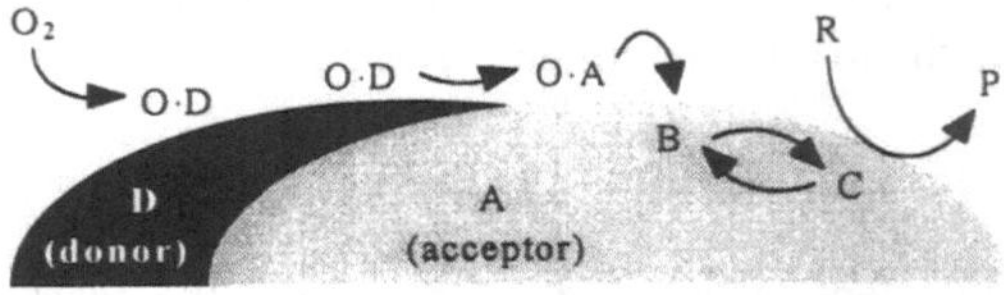

Figure 4. Schematic representation of the mechanism of oxygen spillover in a purely catalytic system. Symbols as in Eq. (7).

reaction scheme of oxygen spillover in a purely (not electro-) catalytic system[22] may be formulated as follows (see also Figure 4),

$$O_2 + 2D \rightleftarrows 2O\bullet D \tag{7a}$$

$$A + O\bullet D \rightarrow D + O\bullet A \tag{7b}$$

$$O\bullet A + B \rightarrow C \tag{7c}$$

$$R + C \rightarrow P + B \tag{7d}$$

Equation (7a) describes the dissociative adsorption and recombination of oxygen on a donor D. The transfer between the donor D and acceptor A is given by Eq. (7b). The spillover oxygen (O) is a mobile species present on the acceptor surface without being associated with a particular surface site. The mobile spillover species may interact with a particular surface site B (Eq. 7c) forming an active site C. Equation (7d) represents the deactivation of the active site C by interaction with a reactant R and the formation of product P. Kinetic models based on this mechanism are well supported by experiment.[23]

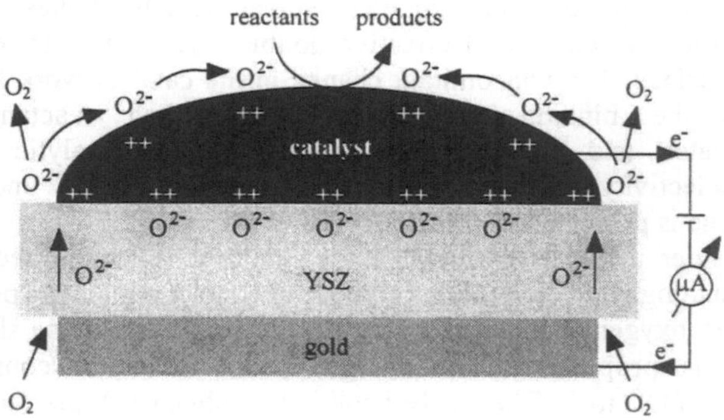

Figure 5. Schematic representation of the mechanism of electrochemical promotion under anodic current application *via* backspillover of charged promoting species (O^{2-}).

The most plausible mechanism of electrochemical promotion, illustrated in Figure 5, is based on analogous donor-acceptor interactions.[8, 9] Consider the example of anodic polarization of the catalyst. In this case, the donor is the solid electrolyte, supplying O^{2-} ions toward the catalyst (acceptor). (This migration from the support to the catalyst may be termed *backspillover* in order to distinguish it from the migration in the opposite direction, which is quite common in catalysis and is termed *spillover*.) The O^{2-} ions are released from the solid electrolyte at the three-phase boundaries. They may be consumed either along or near the tpb in electrochemical reactions (oxygen evolution and/or oxidation of adsorbed reactants) obviously obeying Faraday's law, or they may migrate over the gas-exposed catalyst surface. The existence of backspillover oxygen ions at the gas-exposed catalyst surface was demonstrated by XPS measurements.[24] No charged oxygen species were detected under open-circuit conditions but they were found in abundance after anodic polarization of the catalyst/solid electrolyte interface. Charged backspillover oxygen and neutral chemisorbed oxygen may be clearly distinguished also by temperature–programmed desorption (TPD).[25] It was shown that the oxygen adsorbed from the gas phase desorbs at lower temperature, hence is more reactive than the electrochemically generated oxygen species. During migration, the oxygen ions are accompanied by their compensating image charge in the catalyst forming surface dipoles. Spreading out of these dipoles builds up an overall neutral effective double layer at the gas-exposed catalyst surface. The concomitant change in the catalyst work function modifies the binding strength of chemisorbed reactants and intermediates, and finally causes an alteration of catalytic activity and/or selectivity. A more detailed explanation of the underlying phenomena is presented as follows.

Consider a metal (or metal oxide) catalyst (c) supported by an O^{2-}-conducting solid electrolyte (s), both in contact with a gas phase (g) containing oxygen. For oxygen species, the reaction scheme shown in Figure 6 is proposed.[12] The charge transfer reaction consists of oxidation of O^{2-} to O. The catalyst/solid electrolyte (c/s) interface is not accessible for O, therefore direct charge transfer between O^{2-}(s) and O(c) may occur only at the three-phase boundaries (tpb). Considering, however, a possible backspillover of O^{2-} from the tpb to the gas exposed catalyst surface, the charge transfer is not localised anymore at the tpb but may take place over the entire catalyst/gas (c/g) interface through O^{2-}(c) species. Under open-circuit conditions all reactions of the scheme may be in equilibrium and one may write:

$$\bar{\mu}_{O^{2-}}(s) = \bar{\mu}_{O^{2-}}(c) = 2\bar{\mu}_{e^-}(c) + \mu_O(c) = 2\bar{\mu}_{e^-}(c) + \frac{1}{2}\mu_{O_2}(g) \tag{8}$$

where μ_i is the chemical potential and $\bar{\mu}_i$ ($\equiv \mu_i + z_i F\phi$) is the electrochemical potential of species i, z_i is its charge number, and ϕ is the inner (Galvani) potential. Obviously, the driving force for O^{2-} backspillover from the solid electrolyte to the catalytically active

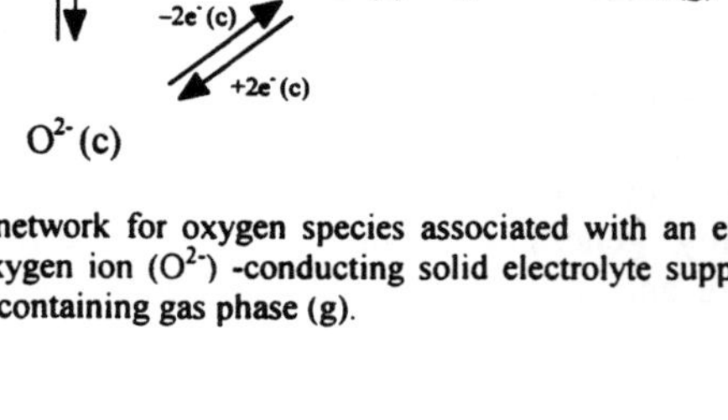

Figure 6. Reaction network for oxygen species associated with an electron conductor catalyst (c) on an oxygen ion (O^{2-}) -conducting solid electrolyte support (s) in contact with an oxygen (O_2) containing gas phase (g).

surface (c/g) is equal to $\bar{\mu}_{O^{2-}}(s) - \bar{\mu}_{O^{2-}}(c)$. This difference vanishes at equilibrium (see the left hand side of Eq. 8) when an effective double layer at the c/g interface has been established, by analogy to the case of an emersed electrode in aqueous electrochemistry.

In an electrochemical promotion experiment a potential step is applied to the electrochemical cell. In the ideal case of a perfect reference electrode having invariant potential, the applied change in the ohmic drop free potential difference between the catalyst (working electrode) and the reference electrode, V_{WR}, is equal to the change in the inner (Galvani) potential of the catalyst, ϕ:

$$\Delta V_{WR} = \Delta\phi \tag{9}$$

This difference is measurable. Changing the potential of the catalyst modifies its Fermi level, E_F, or in other terms, the electrochemical potential of the electrons in the catalyst, $\bar{\ }_{e}$ (= E_F). This latter is defined as the difference between the zero energy state of the electrons (taken at ground state at infinite distance from the solid) and the energy of a conduction electron in the bulk of the catalyst. It is common practice to count this energy difference in two conceptually different ways. One of them is common in electrochemistry, the other is common in surface science.

The electrochemical way considers the electrochemical potential of electrons as the sum of their chemical potential in the solid, μ_e, and the contribution due to the electrostatic (inner) potential:

$$\bar{\ }_{e} = \ _{e} + (-e\phi) \tag{10}$$

Under polarization, the chemical potential of electrons in the bulk solid remains invariant, and one may write:

$$\bar{\mu}_e = -e\Delta\phi = (-e\Delta V_{WR}) \tag{11}$$

Hence, the electrochemical approach formulated in terms of bulk properties, £$\bar{\ }_{e}$, μ_e, and ϕ, allows to directly relate the change in electrochemical potential of electrons to the applied potential.

Since heterogeneous catalysis is a surface phenomenon, it is preferable to express the change in electrochemical potential, $\Delta\bar{\mu}_e$, in terms of the surface science approach:

$$\bar{\mu}_e = -\Phi + (-e\psi) \tag{12}$$

where Φ is the work function of the solid surface, and ψ is the outer (Volta) potential. This approach has the clear advantage that both Φ and ψ are not only surface properties but also measurable quantities. The work function, Φ, is defined as the minimum energy required to extract an electron from the bulk of the solid (*i.e.* from the Fermi level) and to eject it with zero kinetic energy just outside the solid surface where the image force interactions are negligible (typically 0.1 to 1 μm). The Volta potential at this point determines the energy, $e\psi$, required to bring the electron from there to its zero energy level (to infinite distance from the solid). When the surface is electrically neutral (zero Volta potential), the work function alone measures the electrochemical potential of the electrons in the bulk solid. Two quantities contribute to the work function:

$$\Phi = -\mu_e + e\chi \tag{13}$$

the chemical potential, μ_e, of the electron in the solid, and the surface potential, χ, which accounts for the potential barrier the electron must overcome in order to cross the surface of the solid. Since the chemical potential, μ_e, is an invariant bulk property, any change in the work function due to potential application equals the change in surface potential. Population of the surface with dipoles *via* O^{2-} backspillover, *i.e.*, formation of an electric double layer, will increase the surface potential, and the concomitant increase in work function will affect the binding strength of chemisorbed reactants. Consequently, the adsorption energy of electron acceptors (*e.g.* atomic O) will be weakened, while that of electron donors (*e.g.* C_2H_4) will be strengthened. This results in a variation of the adsorption equilibrium constants, hence also in the coverage of the adsorbed species.[9,26,27] Then, a consequent modification in catalytic reaction rate is well to be expected.

The catalyst work function is a directly measurable quantity, by using either the Kelvin probe technique or the electron cut-off energy technique. It was found experimentally that,[3] for steady-state electrochemical promotion, the following relationship holds over a wide range of conditions.

$$e\Delta V_{WR} = \Delta\Phi \tag{14}$$

It therefore follows that solid electrolyte cells may be used as work function probes. Comparing this experimental relationship with Eqs. (11) and (12), one may conclude that the Volta potential at the catalyst surface remains invariant provided that the electric double layer has been built up at the catalyst/gas interface. Based on electrodynamic arguments it was also concluded,[9] that in an overall neutral electrochemical cell this invariant Volta potential equals zero, since the double layer formed *via* ion backspillover is overall neutral.

In summary, electrochemical promotion may be regarded as catalysis in an electric double layer controlled by current or potential application.[9] The necessary condition to achieve electrochemical promotion is the formation of a double layer at the catalyst/gas interface by the mechanism of ion backspillover from the solid electrolyte support. This may occur over a wide range of conditions. The most important experimental parameters, which favor electrochemical promotion are the following: long tpb (porous catalyst), moderate diffusion length (thin catalyst film), and adequate temperature range, *e.g.*, 250 – 500°C for YSZ (a temperature not too low to have enough ion mobility, and not too high to avoid high reactivity of backspillover ions).

3. Promotional Transients

Catalytic rate transients due to electrochemical promotion may be of great importance for a better understanding of the phenomenon. The time dependence of reaction rate following a galvanostatic step, as depicted in Figure 3, depends on the formation rate of the promoting species and on the rate of their migration to the gas-exposed catalyst surface, but may also depend on their reactivity when they are consumed in a chemical reaction (*sacrificial* promoter[9]).

A simple model for the accumulation of O^{2-} promoters was developed recently[28] on the basis of the following assumptions:

1. The formation rate of the promoter is proportional to the fraction of available surface sites at the tpb;

2. The migration of the promoter over the gas-exposed catalyst surface is rapid in comparison to its average lifetime at that surface;
3. The consumption rate of the promoter is of first order in promoter coverage.

The first assumption implies that the current efficiency for promoter formation is a decreasing function of time during polarization. The competing reaction is oxygen evolution at the tpb. The second assumption implies that the concentration of the promoter at the catalyst/gas interface is uniform and equal to that at the tpb. With thin catalyst films this is a reasonable assumption,[8,9] while for high catalyst loadings a finite surface diffusivity of the promoter has to be considered.[29] The third assumption may be best satisfied close to open-circuit conditions, i.e. at low promoter coverage.

With the above assumptions, the accumulation rate of promoters at the gas-exposed catalyst surface may be written as

$$\frac{d\theta}{dt} = \frac{I}{2FN}(1-\theta) - k\theta \tag{15}$$

where θ is the surface coverage of the promoter species at the surface, N is the total number of active sites, k is the pseudo-first order rate constant of promoter consumption, t is time, and F is the Faraday constant. The factor $1-\theta$ equals the instantaneous current efficiency for promoter formation, (ICE), *i.e.,* the fraction of current utilized for producing promoting species. The competiting reaction is oxygen evolution at the tpb:

$$(ICE) = 1 - \theta \tag{16}$$

Obviously, the current efficiency is unity at zero coverage and zero at full coverage. Solution of Eq. (15) gives the following expression of θ as a function of time of polarization:

$$\theta = \theta_0 + \frac{I/2F}{(I/2F)+kN}\left\{1 - \exp\left[-\frac{(I/2F)+kN}{N}t\right]\right\} \tag{17}$$

where θ_0 is the coverage at $t = 0$. Under typical EP conditions, the coverage of promoter under open-circuit conditions, θ_0, is negligible. Note that with the same assumption and at the limiting condition of very slow promoter consumption ($I/2F >> kN$), Eq. (17) simplifies to the expression proposed by Vayenas,[8]

$$\theta = 1 - \exp\left(-\frac{I}{2FN}t\right) \tag{18}$$

Equation (17) shows that, in the general case, the surface coverage of the promoter is an increasing function of time of polarization and that both, the steady-state coverage and the rate of accumulation, are increasing functions of the applied current.

One may assume, to a first approximation, a direct proportionality between increase in catalytic reaction rate and surface coverage of promoting species at the catalyst/gas interface:

$$\Delta r = A \cdot \theta \tag{19}$$

where the proportionality factor, A, is assumed to be independent of current, *i.e.*, $A \neq A(I)$. With this notation, the following equations are obtained:

- polarization: $$\Delta r = (\Delta r)_{st}\left[1 - \exp(-t/\tau_{pol})\right] \tag{20a}$$

  with $$\frac{1}{\tau_{pol}} = \frac{I}{2FN} + k \tag{20b}$$

- relaxation: $$\Delta r = (\Delta r)_{t=0}\exp(-t/\tau_{relax}) \tag{21a}$$

  with $$\frac{1}{\tau_{relax}} = k \tag{21b}$$

- steady-state: $$(\Delta r)_{st} = A\theta_{st} \tag{22a}$$

  with $$\theta_{st} = \frac{I/2F}{(I/2F) + kN} \tag{22b}$$

where $(\Delta r)_{st}$ is the steady-state catalytic rate increase, $(\Delta r)_{t=0}$ is the rate increase at the moment of current interruption, τ_{pol} is the time constant of catalytic rate increase during current application, τ_{relax} is the time constant of catalytic rate decrease after current interruption, and θ_{st} is the steady-state surface coverage of the promoter. The model predicts: (i) an exponential increase in reaction rate during polarization where the inverse of τ_{pol} is a linear function of the applied current (see Eq. 20); (ii) an exponential decrease in reaction rate during relaxation with a τ_{relax} independent of the current of polarization (see Eq. 21); and (iii) a steady-state reaction rate that is an increasing function of the applied current and approaches an asymptote (see Eq. 22).

A better interpretation of the proportionality factor, A, between catalytic rate increase and surface promoter coverage is possible by considering Eq. (16) and (22) to express A as a function of the applied current:

$$A = \frac{(\Delta r)_{st}}{\theta_{st}} = kN\frac{(\Delta r)_{st}}{(1-\theta_{st})(I/2F)} = kN\frac{(\Delta r)_{st}}{(ICE)_{st}(I/2F)} \tag{23}$$

Equation (23) allows for the definition of a dimensionless quantity, Λ_o, as

$$\Lambda_0 \equiv \frac{A}{kN} = \frac{(\Delta r)_{st}}{(ICE)_{st}(I/2F)} \tag{24}$$

This parameter can be compared with the Faradaic efficiency, Λ, defined in Eq. (4) and widely used in the electrochemical promotion literature. For steady-state conditions, Λ is written as:

$$\Lambda_{st} \equiv \frac{(\Delta r)_{st}}{I/2F} \tag{25}$$

One observes that the Faradaic efficiency refers to the total applied current, regardless of what fraction of current is lost for direct oxygen evolution. Consequently, Λ is a function of time during a galvanostatic transient and its steady-state value, Λ_{st}, is a function of the applied current. On the contrary, Λ_o, as proposed in the present model, refers to the current utilized for the production of promoters. It is constant during all transient experiments and is also independent of the polarizing current. Thus, it is a property of the given catalytic system. Obviously, the relation $\Lambda_o \geq \Lambda_{st}(I)$ holds. The explicit relationship between Λ_{st} and Λ_o as a function of current may be given in a suitable linearized form as

$$\frac{1}{\Lambda_{st}} = \frac{1}{\Lambda_0} + \frac{I/2F}{\Lambda_0 kN} \tag{26}$$

showing clearly that Λ_o is the maximum limiting value of the steady-state Faradaic efficiency, Λ_{st}, when the current tends to zero.

III. FUNDAMENTAL STUDIES OF ELECTROCHEMICAL PROMOTION

In this section, some recent fundamental investigations will be reviewed. Three catalytic model systems and the simplest possible electrochemical cell of the single-pellet type are chosen for illustration. Among the main techniques of investigation, special attention will be given to measurement of catalytic rate transients, cyclic voltammetry, and measurement of catalyst work function.

1. Catalytic Model Systems

We will focus on the investigation of the phenomenon of electrochemical promotion by using YSZ as the solid electrolyte. Two types of heterogeneous catalytic gas reactions will be discussed. One of them is the catalytic combustion of ethylene over RuO_2 or IrO_2 catalysts and the other is the reduction of NO by propylene in presence of oxygen over Rh catalysts.

The catalytic oxidation of ethylene on metal oxides such as RuO_2 and IrO_2 is generally highly non-selective[14] resulting in the complete oxidation of ethylene to CO_2 and H_2O (see Eq. 1). In such a case, the catalytic performance can be uniquely characterized by a single reaction rate, *e.g.*, by that of CO_2 production, r_{CO_2} (mol s^{-1}). The catalytic combustion of ethylene is a well-studied model reaction. In fact, ethylene is often used as a model compound for unsaturated hydrocarbon residues in automotive exhaust gases.

Concerning the other model system, metallic rhodium is known to be an efficient catalyst for the reduction of NO.[30, 31] The propylene-NO-oxygen system may serve as a model for the removal of nitrogen oxides (NO_x) and hydrocarbons (HC) from automotive exhaust gases aiming to transform, under lean fuel conditions, both the reducing (HC) and the oxidizing (NO_x) pollutants into harmless products such as N_2, CO_2 and H_2O:

$$18\,NO(g) + 2\ C_3H_6(g) \rightarrow 9\,N_2(g) + 6\,CO_2(g) + 6\,H_2O(g) \qquad (27)$$

The reduction of NO with propylene is, however, not selective. In addition to the desired end-product (N_2), a partially reduced by-product (N_2O) is also formed:

$$18\,NO(g) + C_3H_6(g) \rightarrow 9\,N_2O(g) + 3\,CO_2(g) + 3\,H_2O(g) \qquad (28)$$

Although N_2O is not considered as a harmful pollutant, its complete removal is strongly recommended in view of more severe future legislation. A selectivity factor, S_{N_2}, defined as the fraction of the total NO consumption converted to N_2 can be formulated in terms of production rates as:

$$S_{N_2} = r_{N_2} / (r_{N_2} + r_{N_2O}) \qquad (29)$$

Another kinetic aspect of this system arises from the presence of excess O_2 in the exhaust gases of lean-burn engines. It may prevent efficient reduction of NO_x by consuming HC *via* direct oxidation:

$$2\,C_3H_6(g) + 9\,O_2(g) \rightarrow 6\,CO_2(g) + 6\,H_2O(g) \qquad (30)$$

The aim is to induce changes in the reaction rates and/or in the selectivity of these reactions via potential or current application. The mechanism of these catalytic reactions is fairly well understood from the vast literature of heterogeneous catalysis.[32-35] The main elementary step is the dissociative chemisorption of NO which produces chemisorbed N and O. The desired product N_2 is formed by reaction of two adsorbed N atoms. The undesired byproduct N_2O comes from the reaction of coadsorbed N and NO. Further mechanistic aspects of these catalytic reactions are out of the scope of this review. As a general rule, the polarization of the catalyst does not modify the kinetic reaction scheme but affects the rate constants.[36]

2. Experimental Aspects

Since electrochemical promotion of heterogeneous catalysis is a relatively new field, it is useful to review here the main experimental aspects of the preparation and use of single-pellet electrochemical cells, such as the one shown in Figure 1. Usually, this type of cell is just suspended inside a temperature-controlled vessel. For other cell configurations and reactor designs, see Section IV.

(i) Cell Preparation

The single-pellet electrochemical cells were prepared on 1 mm thick 6 or 8 mol% yttria-stabilized zirconia pellets (ZrO_2-FSZ, Metoxit AG). They were sandblasted and then sonicated in 2-propanol before deposition of the electrodes.

The working electrodes were usually deposited by thermal decomposition[37] of suitable precursors ($RuCl_3$, H_2IrCl_6 or $RhCl_3$). The method consists of application of a dilute solution of the precursor in 2-propanol onto the YSZ surface followed by mild evaporation of the solvent and by a heat treatment at 500-550°C for 10 min. The procedure is repeated until reaching the desired catalyst loading. After having applied the last layer, the sample is kept at 500-550°C for 2 hours. The catalyst loading is calculated based on of the applied amount (*i.e.*, by neglecting any loss of metal due to eventual formation of volatile compounds) and of the geometrical area (usually around 0.5 cm^2) of the coated surface. For the deposition of rhodium catalyst an alternative technique, that of rhodium sputtering, is also used. This process, consisting of cathodic deposition of metallic rhodium onto the solid electrolyte support, was performed in the radio-frequency (RF) mode of plasma generation with a working power of 100 W on the target, in an argon atmosphere of 0.5 Pa and at ambient temperature. Under these conditions the deposition rate was 5 Å s^{-1}. The film thickness is measured *via* calibration on smooth silicon samples processed simultaneously. The porosity of the coating is efficiently enhanced when lowering the deposition rate by performing sputtering in the direct-current (DC) mode with a discharge of 700 V and at much higher argon pressures up to 20 Pa. In all cases, chemical and morphological properties of the resulting catalyst films have been determined by various methods of characterization, such as thermogravimetric analysis

(TGA), X-ray photoelectron spectroscopy (XPS), scanning electron microscopy (SEM), and X-ray diffraction (XRD).

The counter and reference electrode consist of porous gold deposited onto the YSZ by thermal decomposition of a low-temperature gold paste (Gwent Electronic Materials Ltd., C70219R4) at 550°C. The counter electrode was deposited exactly opposite to the working electrode (catalyst), as seen in Figure 1, in order to minimize the ohmic drop and to ensure a symmetrical current distribution through the solid electrolyte. The electric connections were made with gold wire (Ø = 0.127 mm, Müller Feindraht AG) directly inserted into the gold layer but stuck onto the catalyst surface with a tiny dot of the same gold paste.

(ii) Equipment for kinetic studies

Figure 7 shows the experimental set-up used in our laboratory for the measurements of electrochemical promotion. Reactants were Carbagas certified standards of C_2H_4 (99.95%), C_3H_6 (99.4%), NO (99.90%), and O_2 (99.95%) supplied as 1% (C_2H_4, C_3H_6, NO) or 20% (O_2) mixtures in He (99.996%). The balance was helium. The gas flow was controlled by mass control systems (Bronkhorst, F/201C and E-5514-FA) in the range between 7.5 and 500 mL min^{-1} STP with a precision of ±0.5 mL min^{-1} STP. Reactants and products were continuously monitored using online infrared (NDIR) analysers (Siemens, Ultramat 5E-2R for C_2H_4, C_3H_6 and CO_2, Ultramat 6 for NO and N_2O) and a quadrupole mass spectrometer (Balzers, QMG 421). Analysis of O_2 and N_2 was made using an offline gas chromatograph (Shimadzu GC-8A, detector TCD, column Alltech 5S CTR packed with Chromosorb 107 and molecular sieve 5A). Electrochemical measurements were carried out by applying constant currents between the working and the counter electrode (galvanostatic operation), or by applying constant potentials between the working and the reference electrode (potentiostatic operation) with a scanning galvano-potentiostat (EG&G, Princeton Applied Research, Model 362). For online acquisition of current, potential, and outlet gas concentration data the LabTech Note Book system was used.

The reactor used with single-pellet cells is seen in Figure 8. It consists of a glass tube, closed with a stainless steel cap and operated at atmospheric pressure. The gas composition in the reactor was assumed to be uniform and identical to that measured at the outlet (ideal CST reactor). The electrochemical cell was suspended inside the reactor

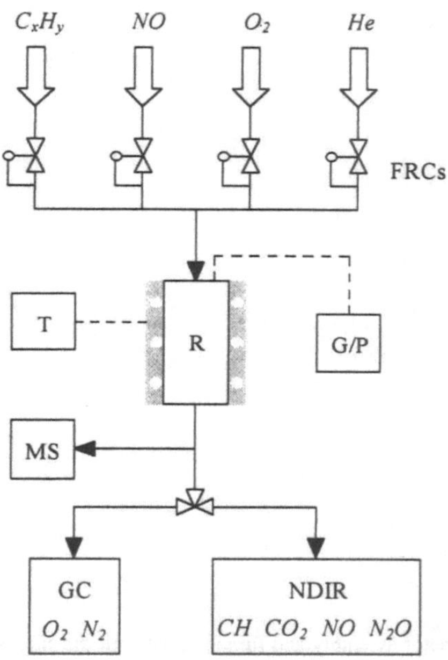

Figure 7. Schematic representation of the experimental set-up. FRCs: flow rate controllers; R: reactor; T: temperature control; G/P: galvano-potentiostat; MS: mass spectrometer; GC: gas chromatograph; NDIR: infrared analysers.

with the three gold wires serving as connections to the electrodes. Note that in such a single-compartment (single chamber) reactor, the reference gold electrode is exposed to the varying gas composition, hence must be regarded as a pseudo-reference. The temperature of the catalyst was measured with a K-type (NiCr-Ni) thermocouple placed close to the catalyst surface. The entire reactor assembly was placed in a furnace (Horst, XVA271) equipped with a heat control system (Horst, HT30).

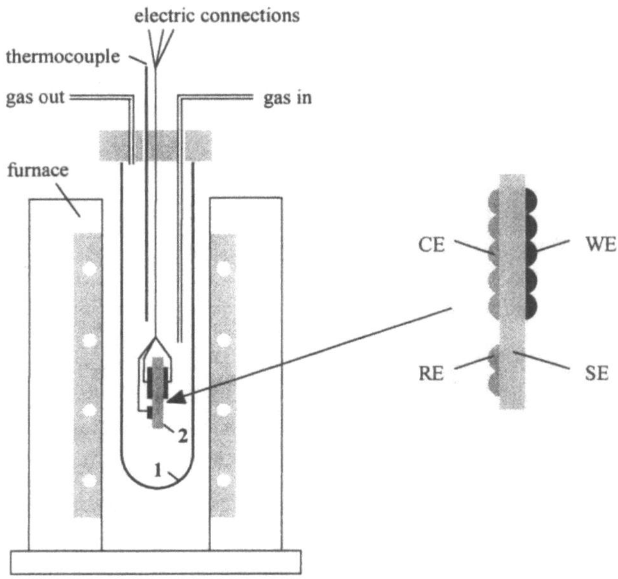

Figure 8. Reactor assembly and furnace used with the single-pellet configuration. **1**: glass tube; volume 50-100 mL. **2**: electrochemical cell. WE: working electrode; CE: counter electrode; RE: reference electrode; SE: solid electrolyte. Reprinted from *J. Electroanal. Chem.*, G. Fóti, V. Stanković, I. Bolzonella, and Ch. Comninellis, Transient Behavior of Electrochemical Promotion of as-Phase Catalytic Reactions, ©(2002) in press, with permission from Elsevier Science.

3. Electrochemical Characterization of the Single-Pellet Cell

In order to characterize the gold reference electrode and to estimate the potential distribution in the single-pellet cell, an electrochemical cell identical to that shown in Figure 1, but having all three electrodes made of gold, was used.[28] A good reference electrode for the purposes of electrochemical promotion experiments must be catalytically inert in all experimental conditions, and it must have an invariant potential defined by a (preferably reversible) redox couple. Nevertheless, a quasi-reversible behavior is tolerable, since electrochemical promotion experiments utilize low current densities, typically less than 100 μA cm^{-2}, and only a tiny fraction of this may pass through the reference electrode.

The reference electrode for EP experiments is usually prepared by deposition of a continuous porous gold film onto the solid electrolyte. In the case of YSZ (an O^{2-}-conducting solid electrolyte), the potential of this reference electrode is determined by the redox reaction

$$\tfrac{1}{2}O_2(gas) + 2e^- \leftrightarrows O^{2-}(YSZ) \qquad (31)$$

The reference electrode was characterized in the temperature range between 375 and 500°C. First, the catalytic inertness of gold electrodes was verified. No detectable catalytic activity in ethylene combustion was observed at any applied feed composition, neither at open-circuit nor at closed-circuit conditions. Then, current-potential curves were determined by varying the concentration of oxygen, both in the absence and in the presence of ethylene in the reactive gas mixture. No significant deviation due to changes in gas composition was observed at any temperature. This confirms that ethylene does not interfere with the potential determining redox reaction. A current-potential curve obtained in the low overpotential region at 375°C is shown in Figure 9. It is seen that the charge transfer resistance is relatively high and the corresponding exchange current density is low (0.14 μA cm^{-2}), indicating that the behavior of the reference electrode is not ideal. However, taking into account that the current applied during typical NEMCA experiments is small (less than 100 μA), and assuming that the fraction of current passing through the reference is lower than 0.1%, the uncertainty in the potential of the reference electrode is smaller than 20mV. Also the activation energy of the charge transfer reaction at the gold/YSZ interface was calculated from current-potential curves measured between 375°C and 500°C. A value of 165 kJ mol^{-1} was obtained, in good agreement with literature data.[38]

In order to obtain the true ohmic drop free potential of the catalyst (working electrode), the potential difference measured with respect to the reference, V_{WR}, must be corrected from the ohmic drop in the solid electrolyte. This calculation requires knowledge of the potential distribution in the solid electrolyte, which may be obtained either with the primary or with a secondary current distribution. The approximation of primary current distribution is justified when most of the resistance is located in the solid electrolyte and the charge transfer resistances are negligible. In the present case, a charge transfer resistance of 370 kΩ was obtained from the slope of the current-potential curve (see Figure 9), while the calculated resistance of the solid electrolyte between the working and counter electrodes was 1.8 kΩ, using the value of 9.2 kΩ cm[21] for the bulk resistivity of YSZ at 375°C. This results in a Wagner number of about 200, indicating that a secondary current distribution would be more appropriate to simulate the cell behavior. Nevertheless, it is preferred to calculate the potential distribution with primary current distribution in order to avoid uncertainties due to erroneous estimation of the Wagner number. Knowing that the primary current distribution results in the least

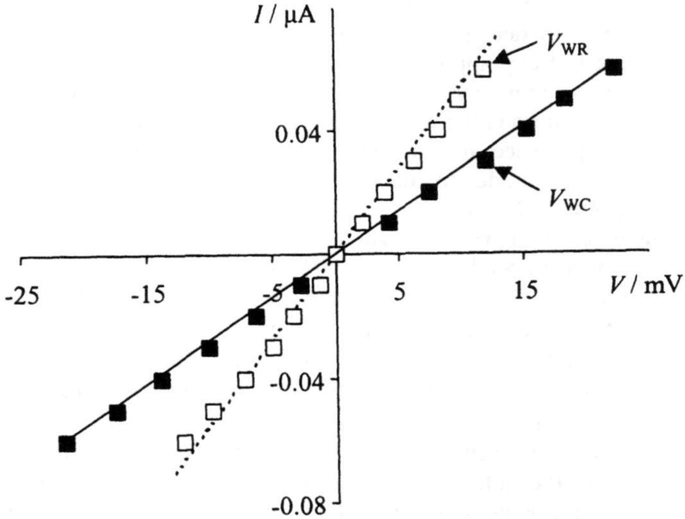

Figure 9. Electrochemical characterization of the three-gold-electrode cell in the region of polarization resistance; gas composition: 10% O_2 in He; $T = 375°C$. Relation between total current, I, passed through the cell, and working electrode potential with respect to counter and reference electrode, V_{WC} (full symbols) and V_{WR} (open symbols), respectively. Solid line shows the fitted current-potential curve, $I = 0.0027 \cdot V_{WC}$. Dotted line represents the function I *vs* $\frac{1}{2}V_{WC}$. Reprinted from *J. Electroanal. Chem.*, G. Fóti, V. Stanković, I. Bolzonella, and Ch. Comninellis, Transient Behavior of Electrochemical Promotion of as-Phase Catalytic Reactions, ©(2002) in press, with permission from Elsevier Science.

uniform potential distribution in the proximity of the counter electrode where the reference is located, this is the worst possible case for correct estimation of the ohmic drop.

Figure 10 shows the calculated potential distribution in the electrochemical cell assuming primary current distribution and solving the Laplace equation with the Femlab2.0 software. The boundary conditions are the following:

- $\phi_{we} = 1$ for the interface between working electrode and solid electrolyte,
- $\phi_{ce} = -1$ for the interface between counter electrode and solid electrolyte,
- $\frac{\partial \phi}{\partial \vec{n}} = 0$ for the other boundaries,

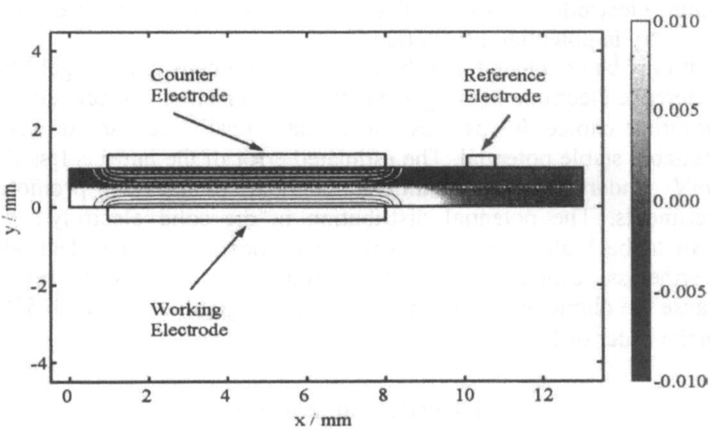

Figure 10. Calculated equipotential lines in a pellet section for primary current distribution. The counter electrode potential is set to -1 V and the working electrode potential to +1 V. The equipotential lines are drawn at steps of 0.2 V. Gray levels show the potential distribution between -0.01 V and +0.01 V, see right scale. Reprinted from *J. Electroanal. Chem.*, G. Fóti, V. Stanković, I. Bolzonella, and Ch. Comninellis, Transient Behavior of Electrochemical Promotion of as-Phase Catalytic Reactions, ©(2002) in press, with permission from Elsevier Science.

where ϕ is the inner potential of the electrolyte and $\vec{n}$ is the normal vector. It is seen that in an electrochemical cell where working and counter electrodes are symmetrically deposited on the solid electrolyte of 1 mm thickness and the reference electrode is separated from the counter electrode by a distance of 2 mm, the reference electrode is found at a quasi-uniform potential of the same value as it would have if located in the middle of the electrolyte. The error is less than 1%. Assuming a secondary current distribution with a Wagner number of 200, the same error would be estimated as less than 0.1%. The predicted symmetrical potential distribution in the solid electrolyte was verified by comparing potential differences between working and counter (V_{WC}) and between working and reference (V_{WR}) electrodes measured at identical closed-circuit conditions (see Figure 9). In a perfectly symmetrical cell configuration, V_{WC} should be twice as high as V_{WR}. The measurements confirm this prediction with an error of 6% independently of the temperature up to 500°C. This error may be attributed to a slight asymmetry in the positions of the working and counter electrodes, knowing that such deviations gives rise to an asymmetry in potential distribution.[39]

It may be concluded that the use of a continuous, porous gold film as reference electrode in the given solid oxide electrochemical cell is an appropriate choice. It was shown to be catalytically inert and to have a reasonably stable potential. The estimated error of the latter is less than 20 mV under typical conditions of electro-chemical promotion experiments. The potential distribution in the solid electrolyte was shown to be highly sensitive to the exact position of the electrodes. Nevertheless, estimation of the catalyst potential remains reliable because the ohmic drop correction is almost negligible, *e.g.*, at 375°C it is in the order of 1 mV μA^{-1}.

4. Cyclic Voltammetry

The effect of polarizing the catalyst/electrolyte interface in a solid electrolyte electrochemical cell may be well investigated by the technique of high-temperature cyclic voltammetry. In a simple case, the voltammogram would be related only to the reversible charge transfer reaction between O^{2-} (s) and O (c). In presence of electroactive gaseous species, other reactions must also be considered. Such a case was reported with porous Pt/YSZ interfaces in presence of oxygen[40] where an asymmetry of the voltammogram, showing higher cathodic than anodic peak currents, was observed. This suggests formation of O (c) species not only by electro-chemical oxidation of oxygen ions, but also *via* dissociative adsorption of gaseous oxygen at free sites during the anodic scan.

In our laboratory, a different approach was used to investigate electrochemical promotion by means of cyclic voltammetry. It consists of measuring the voltammetric charge. For this purpose, cyclic voltammetry is performed in a narrow potential range around the open-circuit potential by varying the potential scan rate. The voltammetric charge is determined by integrating the area under the voltammogram. Due to the different accessibility of different interfaces where electrochemical reactions may occur, the voltammetric charge, $q = q(\nu)$, depends on the scan rate, ν. Extrapolation to zero and infinite scan rates allows for division of the total voltammetric charge, q_{tot}, to a promptly accessible (called outer, q_{out}) and a slowly accessible (called inner, q_{in}) charge,

$$q_{tot} = q_{out} + q_{in} \tag{32}$$

using the following formulae,[41]

$$\lim_{\nu^{-1/2} \to 0} q(\nu) = q_{out} \tag{33a}$$

$$\lim_{\nu^{1/2} \to 0} [q(\nu)]^{-1} = (q_{tot})^{-1} \tag{33b}$$

The voltammetric charge of IrO_2/YSZ interfaces was determined in ethylene/oxygen gas mixtures at $T = 380$ °C.[16] Recorded over a range of ±25 mV around the equilibrium potential with scan rates ranging from 0.5 to 500 mV s^{-1}, fairly symmetrical cyclo-voltammograms exhibiting no peaks were obtained, as shown in Figure 11. The voltammetric charge was attributed to formation of an oxygen capacitor according to the reaction:

$$IrO_2 + \delta O^- \rightleftarrows IrO_{2+\delta} + \delta\, e^- \tag{34}$$

An increase in voltammetric charge with increasing thickness of the catalyst film was observed, contributed by the inner charge, q_{in}, alone.

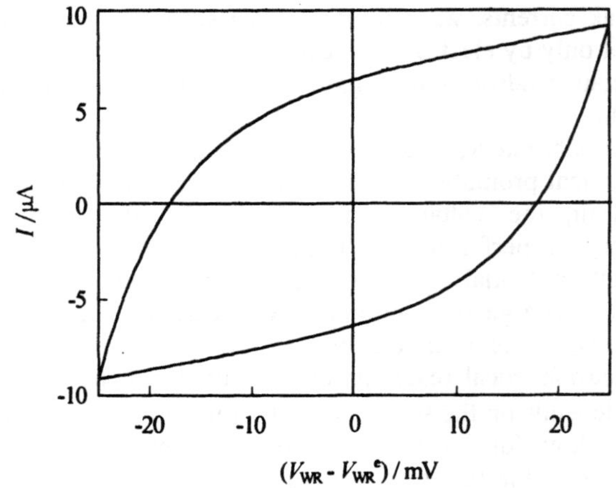

Figure 11. Typical cyclo-voltammogram of an IrO_2 catalyst in a narrow potential range around the equilibrium (V_{WR}^{e}) potential. Scan rate: 20 mV s^{-1}, p_{O_2} = 20 kPa, T = 380°C.

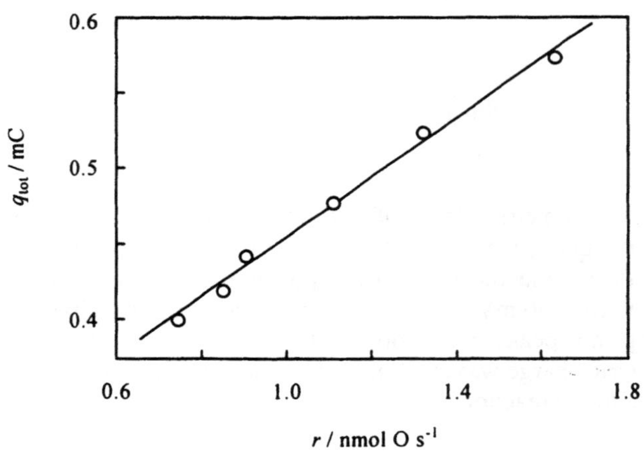

Figure 12. Relation between the total voltammetric charge and rate of ethylene combustion over an IrO_2 catalyst. $p_{C_2H_4}$ = 120 Pa, p_{O_2} = 180 Pa, T = 380°C.

The outer charge, q_{out}, was rather independent of electrode thickness. This behavior suggests that, once the effective double layer at the catalyst / gas interface has been formed, the entire catalyst surface —and not merely the IrO_2/YSZ interface— can be charged. As seen in Figure 12, a linear correlation was observed between the total voltammetric charge and the reaction rate of ethylene combustion during relaxation of the reaction rate after current interruption. This shows clearly that the electrochemical promotion of IrO_2 is closely related to the capacitance of the gas-exposed catalyst surface.

5. Fast-Galvanostatic Transients

The galvanostatic-transient behavior of the catalyst potential and that of the reaction rate due to anodic current steps, was investigated as a function of the applied current.[28] The reaction was the combustion of ethylene at 375°C over IrO_2/YSZ catalyst at highly oxidative feed conditions ($C_2H_4 : O_2 = 1 : 100$). Transients under current application (termed *polarization*) and after current interruption (*relaxation*) were recorded. In order to achieve well-established steady-states, long polarization of at least 100 min and relaxation of at least 200 min were applied.

Figure 13 shows typical transient cycles due to such stepwise changes of the applied current. The change in the ohmic-drop- corrected catalyst potential, ($V_{WR} - IR$), between its open-circuit and promoted steady-state values was very fast, if not instantaneous; it remained essentially invariant, and this behavior was valid for both, polarization and relaxation steps. This means that in the given system, there would be no significant difference between galvanostatic and potentiostatic operation. The catalytic reaction rate exhibited a quite different transient behavior, being not only much slower than the corresponding potential transient, but also suggesting a complex structure. In fact, rather than to follow a simple exponential evolution usually found for metal catalysts,[8, 9] rate transients on the IrO_2 catalyst were composed of a fast initial step completed in a few minutes, and a much slower final step characterized by a time constant in the order of about 20-30 min, in good agreement with earlier observations.[16] The subsequent steps of a galvanostatic cycle may be identified in Figure 13 as follows:

(a) fast initial step during polarization,
(b) slow final step during polarization,

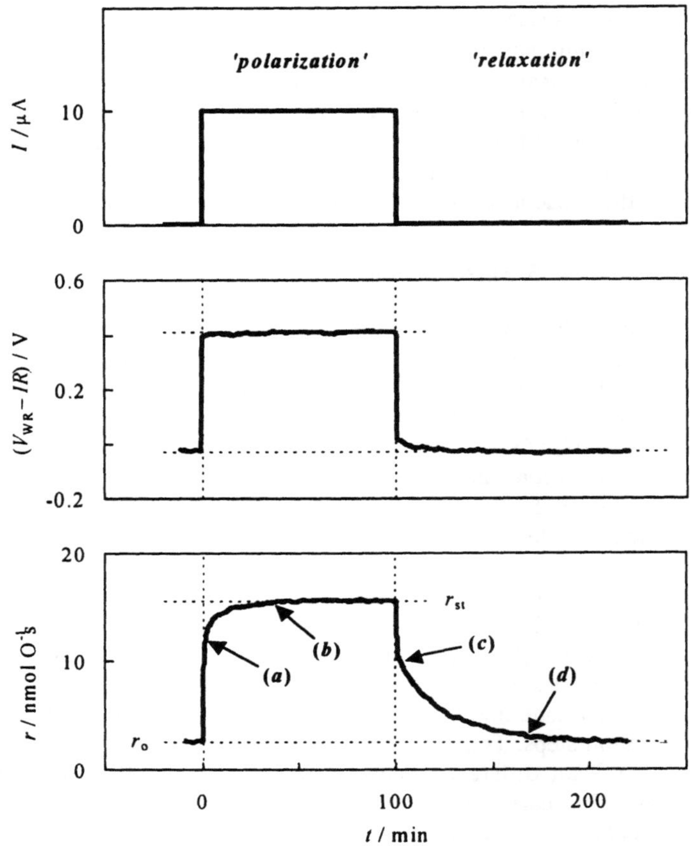

Figure 13. Typical polarization and relaxation transients of catalyst potential, $(V_{WR} - IR)$, and ethylene combustion reaction rate, r, on IrO_2/YSZ catalyst. Galvanostatic anodic polarization with $I = 10$ μA during 100 min. (*a*), (*b*), (*c*) and (*d*) designate the subsequent transient steps. Catalyst loading: 77 μg IrO_2. Feed composition at $p_{tot} = 100$ kPa: $p_{C_2H_4} = 12.5$ Pa, $p_{O_2} = 1.25$ kPa, balance: helium. Flow rate: 200 mL min^{-1} STP ; $T = 375$°C. Reprinted from *J. Electroanal. Chem.*, G. Fóti, V. Stanković, I. Bolzonella, and Ch. Comninellis, Transient Behavior of Electrochemical Promotion of Gas-Phase Catalytic Reactions, ©(2002), in press, with permission from Elsevier Science.

(c) fast initial step during relaxation, and
(d) slow final step during relaxation.

The reaction rate transient was faster under polarization than during relaxation, and experiments made evident that the kinetics of polarization and relaxation depended differently upon the applied current. On one hand, the higher the current, the faster the increase in reaction rate under polarization. On the other hand, the decrease in reaction rate during relaxation was rather independent of the current applied in the preceding polarization.

The variation of the reaction rate between the two extremes, *i.e.*, the open-circuit reaction rate, r_o, and the steady-state promoted reaction rate, r_{st}, was fully reversible. A possible irreversible contribution to the promotion effect (*permanent NEMCA*) was avoided by limiting the polarizing current to 10 μA. The current dependence of the steady-state promoted catalytic reaction rate was also investigated. The total increase in reaction rate, $r_{st} - r_o$, was a continuously increasing function of the applied current as shown in Figure 14.

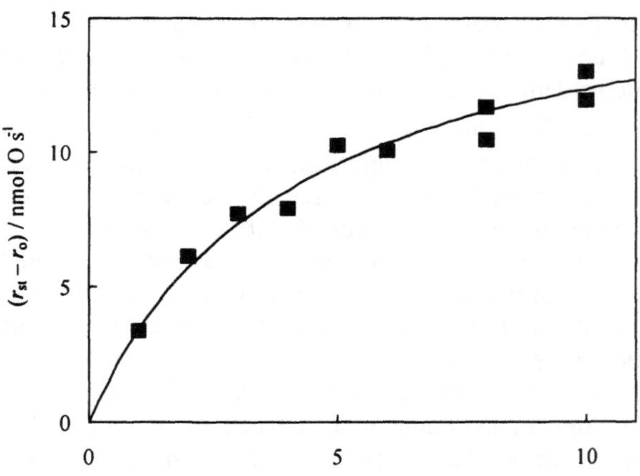

Figure 14. Plot of the total steady-state reaction rate increase, $r_{st} - r_o$, as a function of the anodic polarizing current, I. Reprinted from *J. Electroanal. Chem.*, G. Fóti, V. Stanković, I. Bolzonella, and Ch. Comninellis, Transient Behavior of Electrochemical Promotion of Gas-Phase Catalytic Reactions, ©(2002), in press, with permission from Elsevier Science.

The complexity of the reaction rate transients, which consist of one fast and one slow stage, is in agreement with the cyclic voltammetric evidence about the existence of differently accessible regions for surface charging.[16] The first rapid step (*a*) is believed to be due to accumulation of promoting species over the gas-exposed catalyst surface by the mechanism of backspillover, while the second step (*b*) is due to current-assisted chemical surface modification. Since no correlation between potential transients and reaction rate transients was manifested, a dynamic approach is justified and the applied current —rather than the catalyst overpotential— may be an appropriate parameter to describe the transient behavior of ethylene combustion rate at electrochemically promoted IrO_2/YSZ film catalysts. For the interpretation of the fast transient steps (*a*) and (*c*), a dynamic model of electrochemical promotion has been developed,[28] as presented in detail in Section II.3.

The model was applied to the present experiments as follows. After having subtracted the contribution of the slow steps (*b*) and (*d*), approximated —for lack of an adequate model— by simple exponentials based on the well-established limiting values at infinite time, the remaining fast transient components, steps (*a*) and (*c*), were fitted with Eqs. (20) and (21), respectively. Figure 15 shows the thus obtained time constants. The relaxation time constants were found to be fairly independent of the applied current giving an average value of $\bar{\tau}_{relax}$ = 3.1 ± 0.2 min (or k = 5.4×10^{-3} s^{-1}), while the polarization time constants decreased significantly with increasing current. In agreement with model prediction (see Eq. 20), the inverse of the time constant of polarization is a linear function of the applied current as shown in Figure 16. Such a behavior was already reported for metal electrodes[8, 42] but it is the first time that this observation is made for an oxide and is modeled. An extrapolated value of 3.4 min of the polarization time constant at zero current (*i.e.*, $k = 4.8 \times 10^{-3}$ s^{-1}) was obtained, in accord with the average of the relaxation time constants. From the slope of the same plot, a value of $N = 3.2 \times 10^{15}$ atom was calculated for the number of active sites in a complete monolayer. This is in a reasonable agreement with the number of available active sites obtained by surface titration (4.0×10^{15} atom).

The current dependence of the steady-state enhancement of catalytic reaction rate achieved in the fast step (*a*) was evaluated by using Eq. (26). Figure 17 shows that the inverse of the steady-state Faradaic efficiency is a linear function of the applied current as

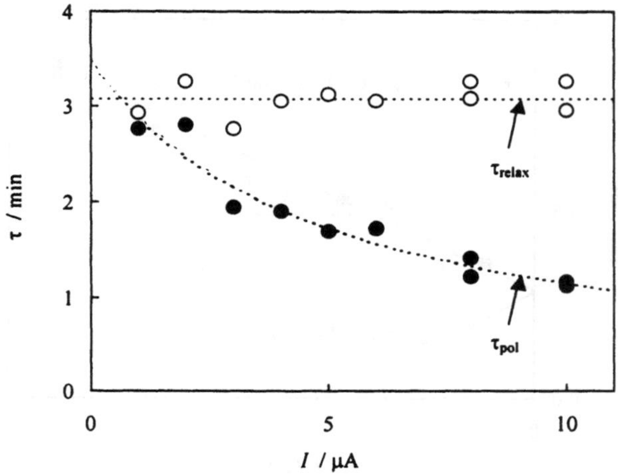

Figure 15. Polarization and relaxation time constants in the fast steps as a function of the applied current, *I*. Full symbols: τ_{pol} in step (*a*); open symbols: τ_{relax} in step (*c*). Reprinted from *J. Electroanal. Chem.*, G. Fóti, V. Stanković, I. Bolzonella, and Ch. Comninellis, Transient Behavior of Electrochemical Promotion of Gas-Phase Catalytic Reactions, ©(2002), in press, with permission from Elsevier Science.

predicted by the model. A limiting Faradaic efficiency of $\Lambda_o = 430$ at zero current was found, indicating a strong non-Faradaic behavior of this catalytic system. From the slope, a value of kN = 16 pmol O s^{-1} was calculated, which is to be compared with the product of k and N obtained from the transients (27 pmol O s^{-1}). This is a satisfactory level of agreement, given the approximations in the model.

It was concluded that a dynamic approach considering the applied current as key parameter is well adapted for the interpretation of galvanostatic electrochemical promotion of ethylene combustion over IrO_2 catalysts. Both, the transient and the steady-state behavior of the system, were satisfactorily described by the proposed model, which assumes free surface site dependent formation, rapid spreading-out (backspillover), and first order rate consumption of the O^{2-} promoter.

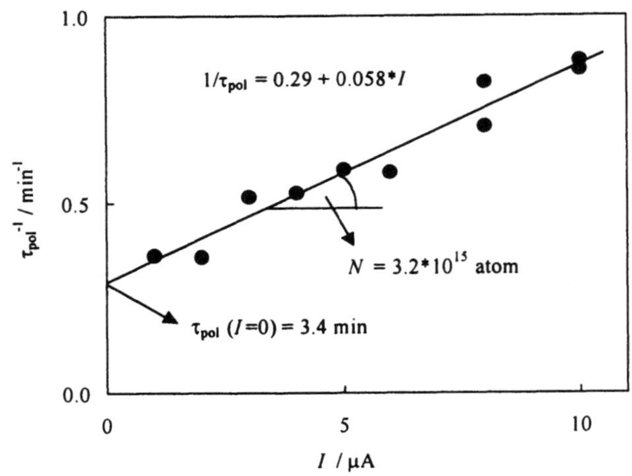

Figure 16. Plot of the inverse of the polarization time constant, τ_{pol}^{-1}, of fast step (*a*) as a function of the applied current, *I*. Reprinted from *J. Electroanal. Chem.*, G. Fóti, V. Stanković, I. Bolzonella, and Ch. Comninellis, Transient Behavior of Electrochemical Promotion of Gas-Phase Catalytic Reactions, ©(2002), in press, with permission from Elsevier Science.

6. Permanent Electrochemical Promotion

As was seen in the previous section, galvanostatic transients may have a complex structure indicating a multi-step mechanism of promotion. In this section, the slow parts of complex galvanostatic transients will be discussed. In order to get a large contribution of the slow step to the total effect, application of high polarizing currents is preferable.

Figure 18 depicts a galvanostatic transient experiment, showing the rate response to an applied current step (+300 μA for 120 min) during ethylene oxidation on an IrO_2 catalyst.[15] The evolution of the catalyst film potential with respect to the reference electrode, V_{WR}, is also given. Initially the circuit is open (I = 0) and the unpromoted catalytic rate is very slow. At t = 0 a constant current (I = +300 μA) is applied between

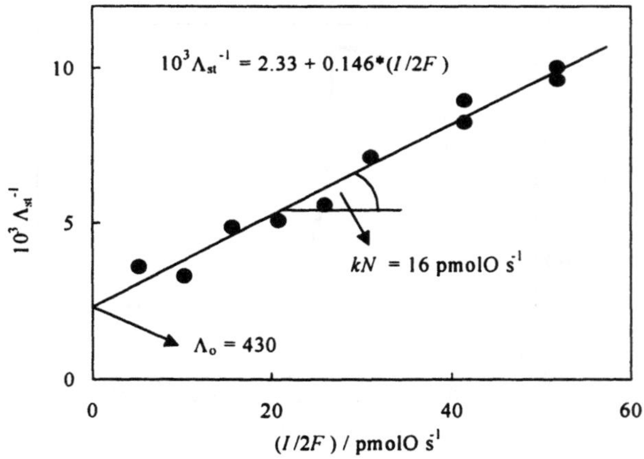

Figure 17. Plot of the inverse of the steady-state Faradaic efficiency, Λ_{st}^{-1}, of fast step (*a*) as a function of the polarizing current in terms of the rate of electrochemically supplied oxygen, $I/2F$. Reprinted from *J. Electroanal. Chem.*, G. Fóti, V. Stanković, I. Bolzonella, and Ch. Comninellis, Transient Behavior of Electrochemical Promotion of Gas-Phase Catalytic Reactions, ©(2002), in press, with permission from Elsevier Science.

the catalyst (working electrode) and the counter electrode. This current corresponds to an O^{2-} supply to the catalyst through the solid electrolyte at a rate of $I/2F = 1.55\times10^{-9}$ mol O/s. The reaction rate increases rapidly at the beginning and, within the first few minutes of current application, it reaches a value, which is about eight times higher than the initial rate. After this rapid increase, the rate increases slowly to a steady-state value, which is 13 times higher than the initial open-circuit rate. The final increase in catalytic rate is strongly non-Faradaic, *i.e,.* it is two orders of magnitude higher than the rate $I/2F$ of O^{2-} supply to the IrO_2 catalyst. Upon current interruption, the rate decreases rapidly at the beginning followed by a slow decrease to a new steady-state value. The remaining value is about three times higher than the initial rate, which was measured before current application. Consequently, the open-circuit enhancement factor, γ, defined in Eq. (6), is equal to 3. Thus, the steady-state open-circuit catalytic activity of the IrO_2 catalyst for ethylene combustion has been enhanced by the electrochemical

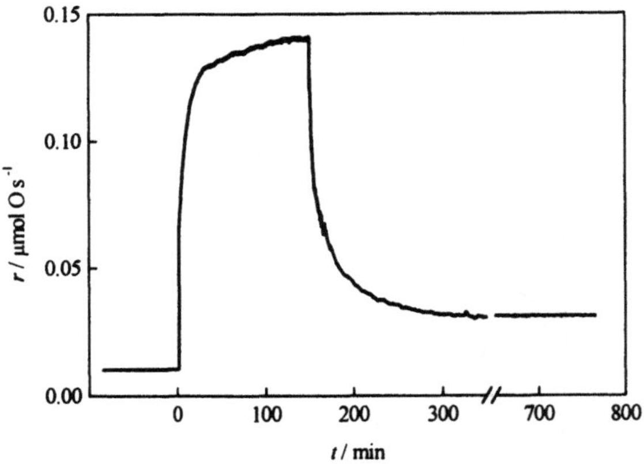

Figure 18. Permanent EP: polarization and relaxation transient of the rate of ethylene combustion on IrO_2/YSZ catalyst due to galvanostatic polarization with +300 µA. Conditions: $p_{C_2H_4}$ = 0.14 kPa, p_{O_2} = 17 kPa, and T = 380°C.

activity acquired by the IrO_2 catalyst persists after current interruption ($\gamma > 1$). It is apparent that the slow step is responsible for the permanent effect.

The proposed model of two-stage process is well supported by the cyclic voltammetric experiments presented in Section III.4. The fast reversible stage, attributed to formation of an electric double layer at the catalyst/gas interface *via* backspillover of promoters, has been discussed in detail in Section III.5. To explain the slow irreversible pretreatment. This phenomenon is called permanent electrochemical promotion or permanent NEMCA effect. The similarity between the regions of rate increase and decrease indicates that similar mechanisms are involved during current application and interruption, but the enhancement of the open-circuit rate indicates that the electrochemical promotion of the IrO_2 catalyst is not reversible. This behavior of an oxide catalyst is different from that of a metal catalyst for which the electrochemical promotion is usually reversible.[8, 9]

The extent of permanent promotion depends on the time of current application. As shown in Figure 19, the open-circuit rate enhancement

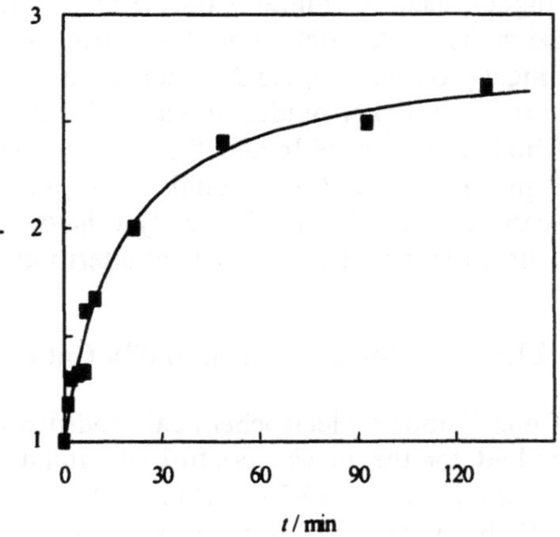

Figure 19. Influence of the duration, *t*, of the galvanostatic step on the open-circuit enhancement factor, γ, (defined by Eq. 6). Experimental conditions as in Figure 18.

factor, γ, increases with the duration of the galvanostatic step, *t*, reaching a plateau for polarization times of about 90 min. For very short current application times that correspond to the region of fast increase in rate, γ is very close to unity, *i.e.*, the effect of current is reversible. These results suggest again that electrochemical activation of IrO_2 is a two-stage process. It consists of a rapid reversible stage, where the catalyst restores its initial catalytic activity after current interruption(γ = 1), followed by a slow irreversible stage, where the increase in catalytic stage, it is believed that the surface of the catalyst is chemically modified by the formation of a higher oxide ($IrO_{2+\delta}$). The catalytic activity for ethylene combustion on $IrO_{2+\delta}$ is supposed to be higher than on IrO_2, and this causes the observed catalytic rate enhancement after current interruption. The time constant of the higher oxide formation at the IrO_2/gas interface is typically of the order of 50-100 min, which corresponds well to the experimental time constant in the slow part of the rate transient. The assumption of the formation of $IrO_{2+\delta}$ is also supported by the fact that the catalyst work function after current application is higher than its initial value, since oxygen adsorbates cause an increase in the work function of clean surfaces.[43]

Electrochemical promotion of RuO_2 catalysts has also been studied. It is well known that the higher oxides of Ru (RuO_3 and RuO_4) are less stable than the higher oxides of Ir and thus it was expected that the electrochemical promotion of RuO_2 would be reversible. Indeed, EP measurements carried out with RuO_2 catalyst have shown that the catalyst restores its initial activity after current interruption.[44]

7. Electrochemical Activation of a Catalyst

The concept of non-Faradaic electrochemical modification of catalysis may also be applied for the *in situ* control of catalytic activity. This phenomenon was studied using YSZ-interfaced Rh film catalysts for the reduction of NO by C_3H_6 in presence of excess oxygen.[20] The measurements were performed in a slightly oxidizing gas mixture (0.1 kPa C_3H_6; 0.1 kPa NO; 0.5 kPa O_2) at $T = 300$°C. At this composition, a temporary (1 hour) heating up to 400°C resulted in a strong deactivation of the catalyst due to partial oxidation of the rhodium surface; the new steady-state catalytic activity at 300°C was ~90% smaller than before heat treatment (see Figure 20). Under open-circuit conditions, no spontaneous reactivation occurred at 300°C, allowing systematic investigation of the catalytic performance at both the active (*a*) and the deactivated (*d*) state.

In both states, application of a low positive current (+5 μA) gave rise to an enhancement of N_2 and CO_2 production, the latter exceeding several hundred times the Faradaic rate. While active rhodium exhibited a fairly reversible behavior, electrochemical promotion on the deactivated catalyst was quite unusual. On one hand, no apparent steady-state has been reached in reasonable time of polarization and, on the other hand, the promotion was highly irreversible, as seen in Figure 21.

The effect obtained over the deactivated catalyst was split into two parts (see the example of N_2 production in Figure 22): a reversible electrochemical promotion (EP_r) being rather independent of the time of polarization and an irreversible electrochemical promotion (EP_i) increasing with longer polarization time. The reversible electrochemical promotion (EP_r), vanishes after current interruption, is related to the steady-state accumulation of promoting species at the gas-exposed catalyst surface. The irreversible effect (EP_i), persisting after current

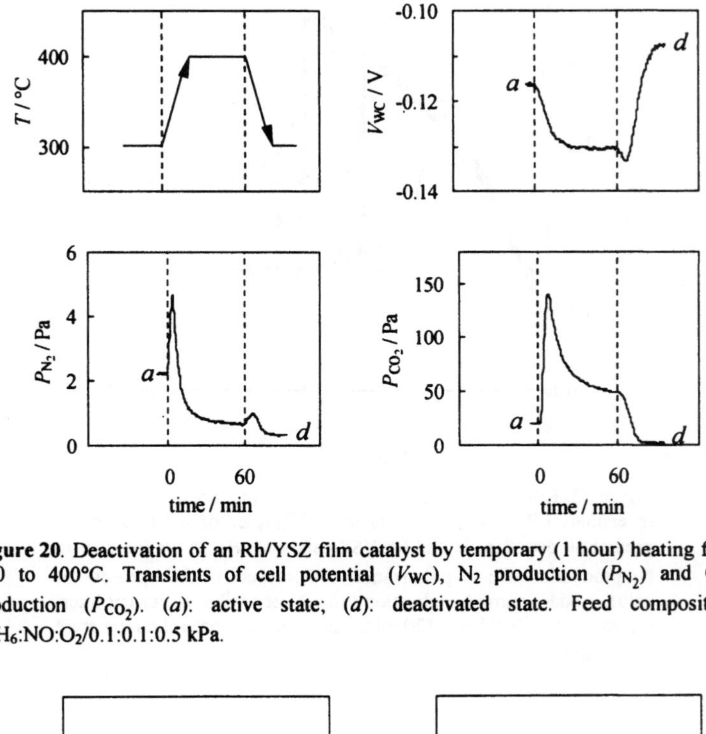

Figure 20. Deactivation of an Rh/YSZ film catalyst by temporary (1 hour) heating from 300 to 400°C. Transients of cell potential (V_{WC}), N_2 production (P_{N_2}) and CO_2 production (P_{CO_2}). (*a*): active state; (*d*): deactivated state. Feed composition: C_3H_6:NO:O_2/0.1:0.1:0.5 kPa.

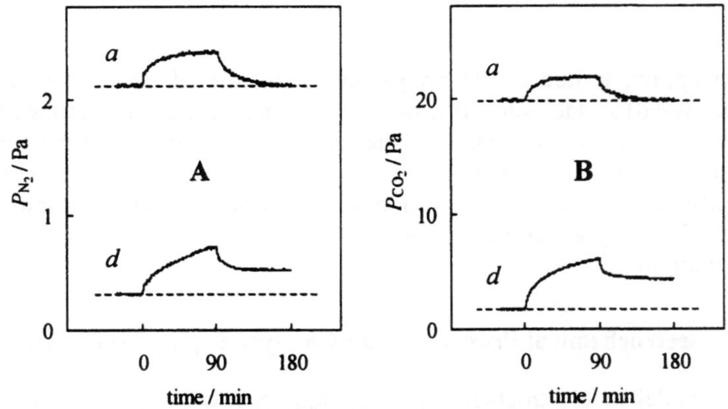

Figure 21. Electrochemical promotion of N_2 (**A**) and CO_2 (**B**) production over the active (*a*) and the deactivated (*d*) Rh/YSZ catalyst during a galvanostatic polarization cycle. Anodic current application (+5 μA) at $t = 0$ min; current interruption at $t = 90$ min. Feed composition as in Figure 20, $T = 300$°C.

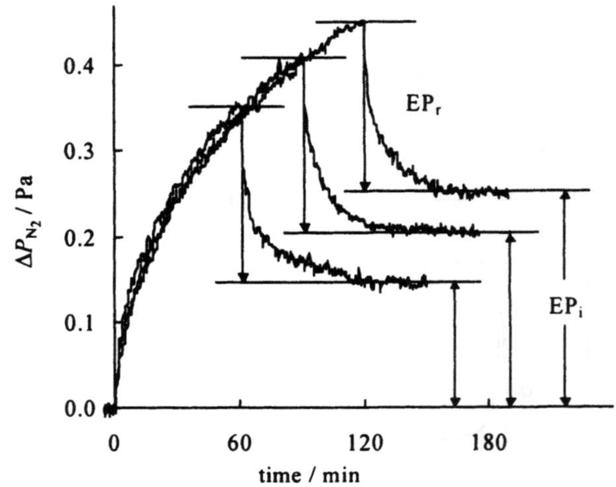

Figure 22. Effect of the duration of galvanostatic polarization on the reversible (EP_r) and the irreversible (EP_i) part of electrochemical promotion over deactivated Rh/YSZ catalyst. ΔP_{N_2} is the increase in N_2 production due to polarization with respect to the open-circuit value. Anodic current application (+5 μA) at $t = 0$ min; current interruption at $t = 60$, 90 and 120 min. Feed composition and temperature as in Figure 20.

interruption, is due to the progressive reduction of oxidized (hence, deactivated) surface sites. It is believed that positive current application increases the rate of rhodium oxide reduction *via* weakening the Rh-O bond similarly to the case of platinum oxides.[45] As a practical consequence, a current application with negligible power consumption may efficiently assist the recovery of accidentally lost catalytic activity of rhodium.

8. Electrochemical Promotion and Catalyst-Support Interactions

The model of electrochemical promotion regards the phenomenon as catalysis in presence of an electrically controlled double layer formed by spillover-backspillover mechanism at the gas-exposed catalyst surface. This shows strong analogy with catalyst-support interactions and also with classical promotion of catalysis.[9] Recent experiments in our laboratory provided further evidence that electrochemical promotion and catalyst-support interactions are functionally identical and only operationally different phenomena.[17]

In order to relate electrochemical promotion and catalyst-support interactions, it was proposed to prepare mixed $IrO_2 - TiO_2$ catalysts consisting of micro– and nanoparticles of IrO_2 (active phase) and TiO_2 (inert support) in intimate contact.[17] The method of thermal co-deposition was similar as for pure IrO_2 films (see Section III.2.*i*) using appropriate mixtures of H_2IrCl_6 and $TiCl_4$ precursors. A series of catalyst films of varying composition was prepared where the IrO_2 loading was kept constant (~130 μg/cm^2). XPS analysis confirmed that Ir and Ti are present in the form of IrO_2 and TiO_2, and that the surface composition matches within 5% with the bulk composition. SEM analyses indicated that the structure of the film becomes more and more similar to that of the YSZ support with decreasing IrO_2 content in the mixed film.

Figure 23 shows galvanostatic transients of ethylene oxidation obtained with IrO_2 - TiO_2 samples of different composition. Upon increasing the TiO_2 content, the magnitude of the rate enhancement factor, ρ, decreases, and the reaction rate tends to return very slowly towards its initial unpromoted value (permanent electrochemical promotion).

A decrease in the effect of electrochemical promotion with increasing TiO_2 content is clearly manifested in Figure 24, showing open circuit ($I = 0$) and electrochemically promoted ($I = 200$ □A) steady-state catalytic rates as a function of film composition. It is seen that

(a) pure TiO_2 is catalytically inactive,
(b) pure IrO_2 shows moderate unpromoted (open-circuit) catalytic activity,
(c) addition of TiO_2 to the IrO_2 film causes a pronounced enhancement in open-circuit catalytic rate, hence considerably limits the room left for electrochemical promotion.

In fact, the equimolar IrO_2 - TiO_2 catalyst mixture can only be marginally promoted electrochemically, while its open-circuit activity is about ten times higher than that of pure IrO_2. This is the same level of rate enhancement as is achieved with pure IrO_2 catalyst when it is

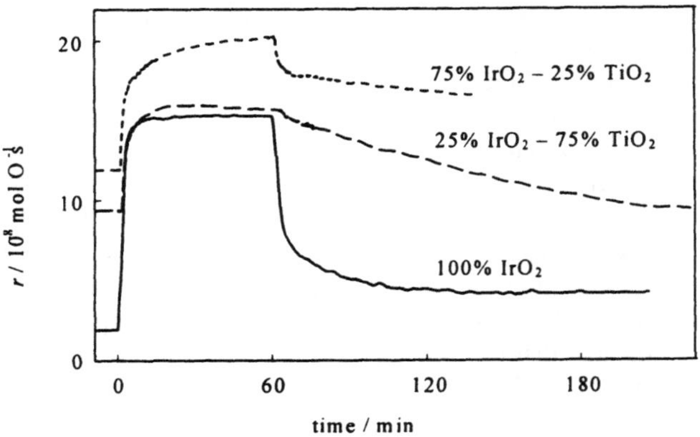

Figure 23. Galvanostatic transients of ethylene oxidation over IrO_2-TiO_2/YSZ catalyst films with IrO_2 content of 25, 75 and 100%. Anodic current application: I=200 μA during 60 min. Feed composition: $p_{C_2H_4}$ = 0.15 kPa, p_{O_2} = 20 kPa. T=380°C. Reprinted from Ref. [17] with permission from Academic Press.

promoted electrochemically. These observations strongly suggest that the same type of promoting mechanism is operative for the $IrO_2 - TiO_2$ mixed catalyst under open-circuit conditions (catalyst-support interaction) and for pure IrO_2 catalyst under closed-circuit conditions (electrochemical promotion). One may consider the $IrO_2 - TiO_2$ catalyst-support interaction as a remote control mechanism[46, 47] or a *wireless* type NEMCA configuration,[18, 48] where TiO_2 continuously provides promoting O^{2-} species to the IrO_2 catalyst surface, and gaseous O_2 continuously replenishes spent O^{2-} in the TiO_2.

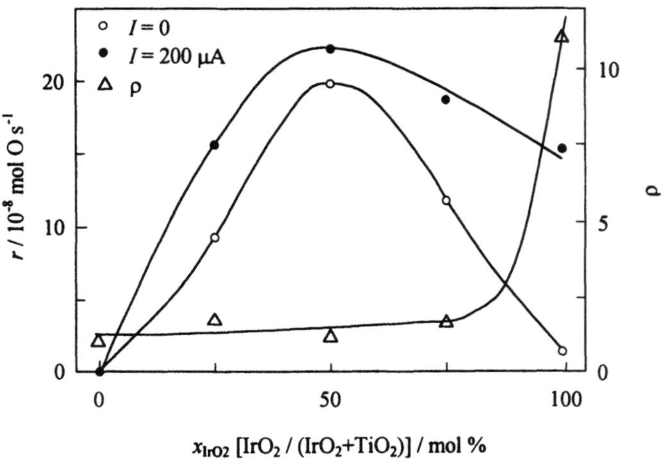

Figure 24. Effect of the composition of the IrO_2-TiO_2/YSZ catalyst films on the rate of C_2H_4 oxidation under open-circuit (open circles) and electrochemical promotion (full circles) conditions. Triangles indicate the corresponding rate enhancement ratio (ρ) value. Experimental conditions as in Figure 23. Reprinted from Ref. [17] with permission from Academic Press.

9. Work Function Measurements

The most commonly used technique, allowing work function measurements in a reacting atmosphere, is known as the *Kelvin* method or the vibrating capacitor method.[49-51] Its principle is as follows. By placing a conductor, the *Kelvin* probe, to the immediate proximity of the catalyst surface and connecting them together by an external circuit, a capacitor is formed. The contact potential difference between the two conductors, originated from their differing work functions, results in a charge separation between the opposing surfaces. Vibration of the *Kelvin* probe generates an alternative current *via* periodic variation of the capacitance. The external potential necessary to make the current vanish is the measure of the work function difference between the two conductors. By using a *Kelvin* probe made of an inert material that has

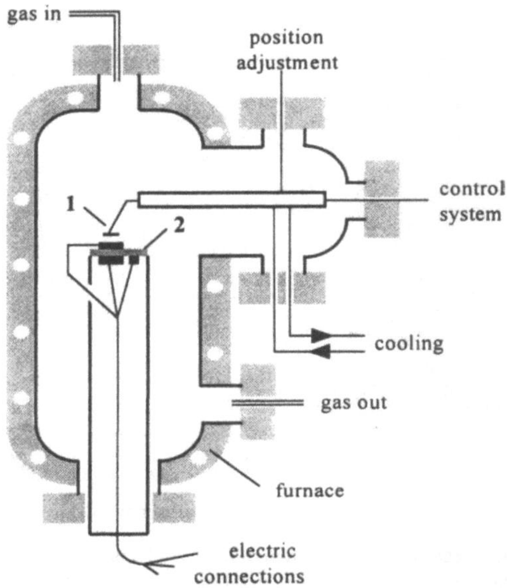

Figure 25. Reactor assembly for *in situ* work function measurements. **1**: *Kelvin* Probe S. **2**: electrochemical cell.

an invariant work function, variation of the response due to varying experimental conditions may be attributed to the change in work function of the catalyst alone.

The reactor built in our laboratory for *in situ* determination of the work function during catalytic measurements is shown in Figure 25. It was adapted to a commercially available probe, the *Kelvin* Probe S (Besocke Delta PHI GmbH). The use of this probe is limited to a maximum working temperature of 250°C, whereas catalytic measurements should usually be made in the temperature range of 300 to 400°C. A cooling system was, therefore, added in order to allow measurements with this probe at temperatures up to 400°C in the reacting gas phase. The *Kelvin* method is known to be very sensitive to the orientation of the sample relative to the probe. The experimental setup presented in Figure 25 allowed both, the adjustment of the distance between sample and probe with an external screw, and the selection of different spots on the sample by simply turning the glass cylinder on the top of which the sample is attached. The system provided work function data relative to the gold grid of the *Kelvin* Probe S, assumed to be an invariant reference. Two main operating parameters, the vibration amplitude of the grid and the integration time for potential compensation were controlled by an electronic unit, the *Kelvin* Control 07 (Besocke Delta PHI GmbH).

For illustration of the usefulness of work function studies in electrochemical promotion, a dynamic approach is presented, *i.e.*, the time dependence of work function changes due to polarization steps was investigated. The model reaction was ethylene combustion over RuO_2 catalyst.[52] Such experiments should provide insight into the migration of promoting species towards the gas-exposed catalyst surface. Figure 26 shows the time dependence of the work function change ($\Delta\Phi$), the corresponding reaction rate (r), and the catalyst potential (V_{WR}) following galvanostatic steps of 50 μA.

The work function transients during anodic polarization of the catalyst were composed of two main steps: a sharp increase just after application of the current and a subsequent slow increase leading to a new steady-state value under polarization. Between these two steps, a sharp peak might also be observed in certain cases as exemplified in Figure 26A. The initial jump in work function is associated with the surface region easily accessible for promoting species. This region is supposed to be close to the three-phase boundary (tpb) where the promoting species are generated. The adsorption of charged oxygen on sites next to the tpb leads to the observed increase in work function. The final step, where the work function exhibits a slow increase, is due to migration of the promoting species from the tpb towards the entire gas-exposed catalyst surface. The work function continues to change until steady-state coverage of the promoting species is reached and, in parallel, the catalytic reaction rate reaches its new promoted steady-state value. The eventual appearance of a sharp peak preceding the final step depends sensibly on the relative position of the catalyst and the *Kelvin* probe. In principle, the latter must be placed in a region of the catalyst where the tpb is accessible for the work function measurements. If the *Kelvin* probe is too far away from the tpb, no sharp peak but only a smooth transition between initial jump and final slow progression may be observed. Verification and satisfactory interpretation of the sharp peak in work function transients requires further investigation.

After current interruption, the work function does not return immediately to its initial open-circuit value. The remaining difference to the initial steady-state open-circuit value is explained by the promoting species still remaining on the catalyst surface. This difference, of 84 meV, is in good agreement with the increase in work function due to the migration of promoting species to the gas-exposed catalyst surface during the third phase (79 meV, computed by extrapolating the work function changes occurring during the third phase back to time zero). The initial decrease in work function is sharp, but slower than the corresponding increase during the first phase of anodic polarization, and neither an extreme value in the time dependence of the work function nor a transition phase were observed. The final slow decrease in work function was also slower than the increase during the third phase under current application. The entire decrease after current interruption is related to slow disappearance of the promoting species due to their reaction with ethylene from the gas phase.

IV. CELL DEVELOPMENT FOR ELECTROCHEMICAL PROMOTION

In Section III, electrochemical promotion studies with single -pellet type electrochemical cells were presented. This type of cell, especially when equipped also with a reference electrode as shown in Figure 1, is well suited for fundamental studies of the phenomenon. For practical applications of electrochemical promotion, however, this configuration is obviously inadequate. In this chapter, development of more advanced, bipolar cell configurations will be presented. These results may be considered as the first successful steps towards achievement of electrochemical promotion with highly dispersed catalysts.

1. Bipolar Configuration for Electrochemical Promotion

The only viable way to achieve electrochemical promotion in dispersed systems involves indirect bipolar polarization of the catalyst in a suitable electrochemical cell. Electrochemical promotion in bipolar configuration is feasible. This was first demonstrated by polarizing a platinum stripe catalyst deposited between two gold feeder electrodes, each on the same side of an YSZ plate.[48] The promotion achieved in

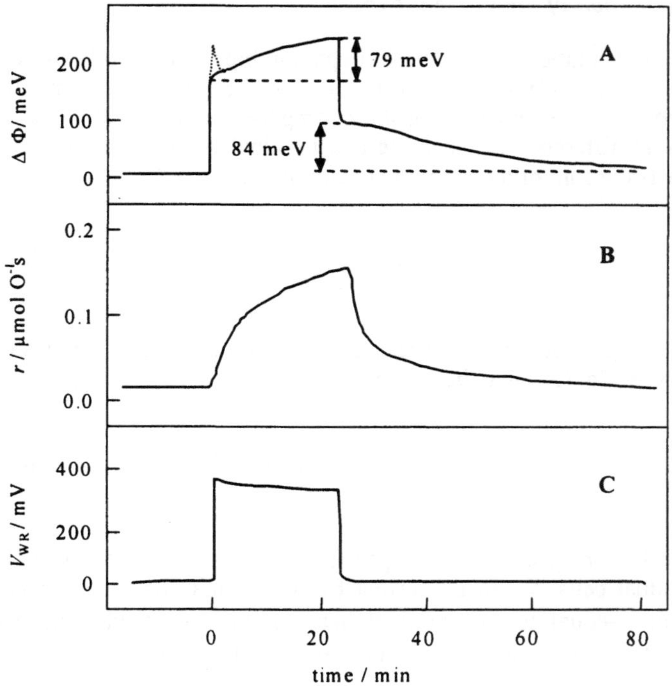

Figure 26. Time dependence of the work function change (**A**), the reaction rate (**B**) and the catalyst potential (**C**) following galvanostatic steps. Catalyst: RuO_2 (m = 0.4 mg; A = 0.5 cm^2), I = 50 μA during 23 min, $p_{C_2H_4}$ = 114 Pa, p_{O_2} = 17.7 kPa, flow rate : 175 mL min^{-1} STP, T = 380°C.

this simple configuration was much lower than in usual single-pellet cells. The poor performance was attributed to the unfavourable geometry of the bipolar cell, leaving room for an important current bypass and causing a non-uniform distribution of the catalyst work function. In fact, the main disadvantage of bipolar systems is that a fraction of the current may bypass the working electrode, hence be lost for electrochemical promotion. Limitation of the current bypass is of major importance in bipolar cell design.

(i) Estimation of the Current Bypass

For estimation of the current bypass in bipolar electrode stacks, the following model was proposed for water electrolysis in liquid electrolytes.[53] In an electrode stack composed of N individual cells, the potential difference between the two terminal feeder electrodes, V_N, is given as the sum of the individual cell voltages, V_j:

$$V_N = \sum_{j=1}^{N} V_j \tag{35}$$

At high potentials where a linear current-voltage relationship is observed, Eq. (35) may be written as:

$$V_N = \sum_{j=1}^{N} \left(V_o + R_e I_j \right) = N V_o + R_e \sum_{j=1}^{N} I_j \tag{36}$$

where V_o and R_e are reaction-specific constants, hence identical for all individual cells. When no current bypass occurs, the same amount of current —equal to that applied between the two feeder electrodes, I_{feed}— flows through each cell, and Eq. (36) simplifies to:

$$NV_1 = NV_o + NR_e I_{feed} \tag{37}$$

where V_1 is the potential drop in any single cell. Use of Eqs. (36) and (37) in Eq. (38) and defining the current bypass, Ψ, as:

$$\Psi = \frac{I_{feed} - \frac{1}{N} \sum_{j=1}^{N} I_j}{I_{feed}} \tag{38}$$

gives the expression of the current bypass in terms of voltages measured at a given current, that is,

$$\Psi = \frac{NV_1 - V_N}{NV_1 - NV_o} \tag{39}$$

The values of NV_1, V_N and NV_o can be found graphically in Figure 27. As seen, the method needs the knowledge of the current-voltage curve of the electrode stack, the current-voltage curve of a single cell, and the number of the individual cells.
The model was first verified for the case of a bipolar stack with fourteen Ni electrodes.[54] The electrochemical reaction used for the determination of current bypass was the electrolysis of water in alkaline solutions. Good agreement between experimental and estimated current bypass was observed, especially at high current densities. The model was adapted successfully for solid electrolyte cells of bipolar configuration,[10, 55] as discussed below in Section IV.2.*i*.

(ii) Equipment for Kinetic Studies

Kinetic measurements with bipolar electrochemical cells were made using the same experimental setup as with single-pellet cells described in Section III.2.*ii* and seen in Figure 7. Two types of bipolar electrochemical cells have been developed in our laboratory, a ring-shaped and a multi-channel configuration. As shown in Figure 28,

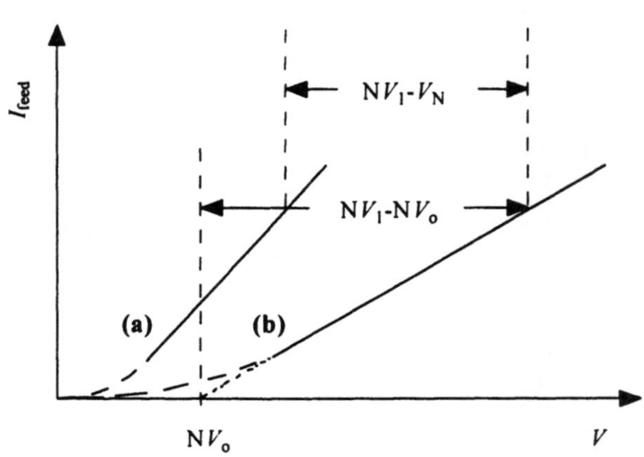

Figure 27. Graphical determination of the current bypass from current-voltage curves. I_{feed} as a function of **(a)** the cell potential of the bipolar stack with N cells (V_N) and of **(b)** N times the cell potential of a single cell (NV_1).

both configurations have a cylindrical geometry, with two gold feeder electrodes on the outside surface of the YSZ cylinder and with a catalyst film inside, *i.e.*, on the inner wall of the ring in the first case and on that of 37 parallel channels in the second case. The reactor used with these cylindrical cells was of tubular type, also seen in Figure 28. It was made of a pyrex glass tube closed with two stainless steel caps with appropriate feed-through and operated at atmospheric pressure. Between gas inlet (top) and gas outlet (bottom), the cylindrical electrochemical cell was placed in the center of the glass tube at a location where the latter changes in diameter. The gold electric connections were inserted through the bottom cap while the K-type thermocouple was introduced through the top cap. The entire reactor assembly was placed in a furnace (Horst, XVA271) equipped with a heat control system (Horst, RD4456).

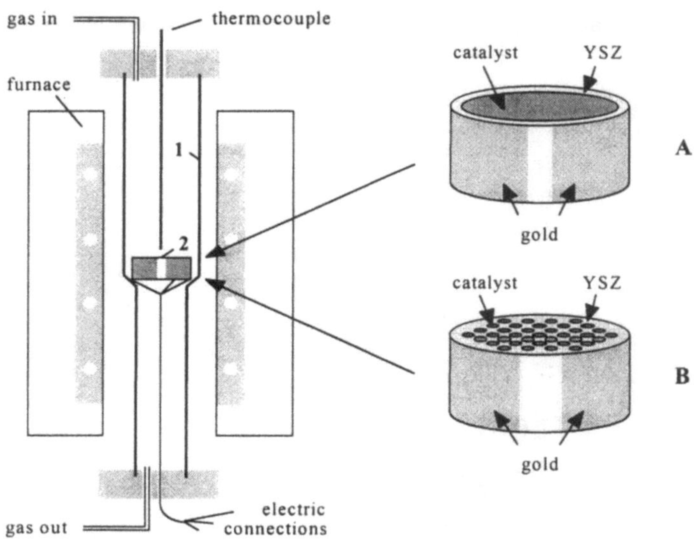

Figure 28. *Left:* tubular reactor assembly and furnace used with bipolar cell configurations. **1**: Pyrex tube (Ø=25/18 mm, length: 400 mm, volume: 150 mL); **2**: electrochemical cell. *Right:* bipolar electrochemical cells. **A**: ring-shaped cell configuration; **B**: multiple-channel cell configuration.

2. Ring-Shaped Electrochemical Cell

As a first step in the development of new cell designs, ring-shaped electrochemical cells were prepared in our laboratory. This configuration, also termed bipolar configuration of the first generation, allows for experimental determination of the current bypass. Nearly bypass-free configurations have been realized. The feasibility of electrochemical promotion in such bipolar cells has been successfully demonstrated.

(i) Cell Preparation and Characterization

The ring-shaped electrochemical cells were prepared with cylindrical rings (OD 20 mm, ID 17 mm; height 10 mm) made of 8 mol% yttria-stabilized zirconia (Technox 802, Dynamic Ceramic Ltd). The catalyst film was deposited on the internal surface of the ring, as illustrated in Figure 28A. The two gold electrodes were deposited symmetrically on the outside surface of the ring, separated from each other by two bands of the same width (3 to 15 mm). The pre-treatment procedure of the YSZ support and that of the deposition of the electrodes (either thermally or by sputtering) were the same as with single-pellet cells. For details, see Section III.2.*i*.

Indirect bipolar (IB) polarization of the catalyst film in a ring-shaped electrochemical cell is realized by using the two gold electrodes as feeder electrodes. For advanced characterization of the cell, a third electric connection may be added. This latter, connected to the catalyst film itself, permits measurements also in the direct (monopolar) polarization mode, which is useful for the determination of the current bypass.

For the estimation of current bypass in the ring-shaped electrochemi-cal cell, the model developed for electrode stacks with liquid electrolytes (see Section IV.1.*i*) was adapted.[10, 55] It is now allowed for the individual monopolar cells to be different, which is, in fact, almost inevitable due to the poor reproducibility of electrode deposition. The method requires knowledge of the steady-state current-voltage curves in three polarization modes (see Figure 29): direct cathodic (DC) polarization of the catalyst film relative to gold electrode 1, direct anodic (DA) polarization of the catalyst film with respect to gold electrode 2, and indirect bipolar (IB) polarization between the two terminal gold electrodes, gold 1 being the anode and gold 2 the cathode. Since the method is based on the additivity of the potential at a given current in the linear (ohmic) region, the current bypass is measured by the excess of the sum of cell voltages in the two monopolar polarization modes over the cell potential measured in the bipolar polarization mode. The fraction of current bypass at a given current is then calculated from the following expression:

$$\Psi = \frac{(V_{DC} + V_{DA}) - V_{IB}}{(V - V_0)_{DC} + (V - V_0)_{DA}} \tag{40}$$

where V are the cell potentials measured at the same current in the linear region, and V_0 are the overvoltages in the monopolar polarization mode extrapolated to zero current.

The method was first applied to ring-shaped RuO_2/YSZ cells.[55] It is illustrated here with the example of a Rh/YSZ cell of the same configuration.[10] The rhodium film was deposited by the thermal decomposition technique. The metal loading was 610 µg (1.1 g m^{-2}, catalyst thickness ~100 nm). The steady-state polarization curves of the cell are shown in Figure 30. Although the two gold electrodes on the outside surface of the YSZ ring were separated from each other by thin uncoated bands of only 3 mm width, the obtained current bypass

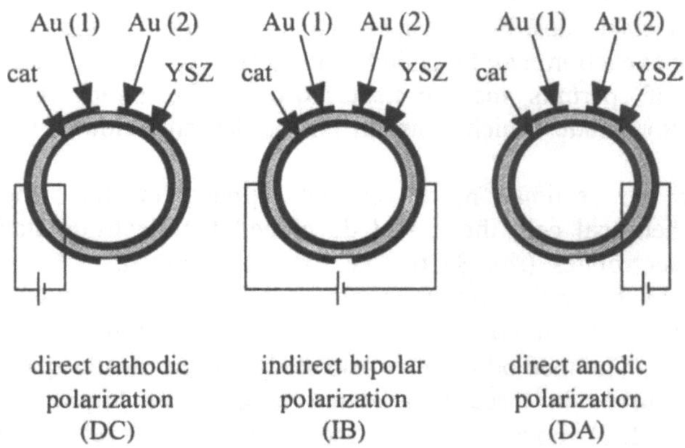

Figure 29. Schematic of various modes of polarization using a ring-shaped electrochemical cell.

was very low and only slightly increased with temperature: 2-4% at 350°C and 5-6% at 400°C. Ψ was found to be fairly insensitive to the gas composition and also to the applied current, up to 3 mA. A nearly bypass-free bipolar configuration is of great importance, as it is a necessary condition for achieving efficient electrochemical promotion in bipolar cells.

(ii) Electrochemical Promotion

The ring-shaped bipolar electrochemical cells were applied successfully to achieve electrochemical promotion in several systems. Efficient promotion of the combustion of ethylene using RuO_2/YSZ cell of this type of configuration was reported first.[55] Here the feasibility of electrochemical promotion with ring-shaped cells is illustrated with the example of the reduction of NO by propylene using a Rh/YSZ cell, of the same type presented in detail in the previous section.[10]

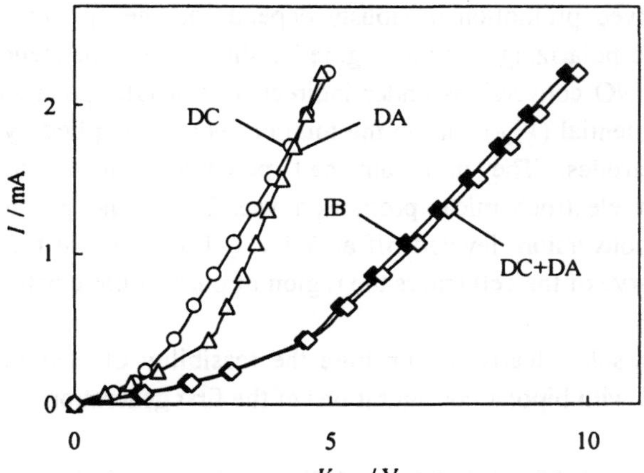

Figure 30. Current-potential curves of a ring-shaped Rh/YSZ cell at different polarization modes. O: direct cathodic (DC); Δ: direct anodic (DA); ◇: DC + DA; ◆: indirect bipolar (IB). V_{cell}: cell potential between the two feeder electrodes. Feed composition: C_3H_6:NO/75:500 Pa, $T = 375°C$.

Catalytic performance tests were performed at various feed compositions and polarization conditions in the temperature range between 350 and 400°C. Figure 31 shows typical results obtained at 375°C and in the absence of O_2 in the feed. Analogous experiments in the presence of O_2 gave similar results. The polarization of the Rh film was performed by applying constant cell potentials, i.e., 5 V in the direct polarization modes (DC or DA) and 10 V in the indirect bipolar polarization mode (IB). As shown in Figure 31, direct cathodic polarization (DC) was rather inefficient. Both, direct anodic (DA) and indirect bipolar (IB) polarizations, resulted in a very significant promotion, although the initial conversions of C_3H_6 and NO in open-circuit (OC) condition were already quite high (32% and 51%, respectively). Under indirect bipolar polarization, the steady-state conversions rose up to 48% (C_3H_6) and 72% (NO), both being about 5% less than the promotion achieved in the DA mode. Note, for the sake of completeness, that the nitrogen selectivity of the catalytic reaction, defined by Eq. (29), was rather invariant (~0.7) during potential application at any polarization mode.

The observed promotion obviously depends on the applied potential and on the polarizing current. Figure 32 shows the dependence of the C_3H_6 and NO conversions under indirect bipolar (IB) polarization, on the cell potential (V_{cell}) and on the total current (I) supplied by the two gold electrodes. The minimum cell potential required to induce significant electrochemical promotion was 2-3 V; the promotion, in terms of conversion, leveled off at 6-7 V (~1 mA) where the current-voltage curve of the cell enters the region of purely ohmic behavior (see Figure 30).

The results clearly demonstrate the feasibility of electrochemical promotion with bipolar configurations of the first generation.

3. Multiple-Channel Electrochemical Cell

As a second step towards achievement of efficient electrochemical promotion over dispersed catalysts, a multiple-channel electrochemical cell, termed bipolar configuration of the second generation, has been

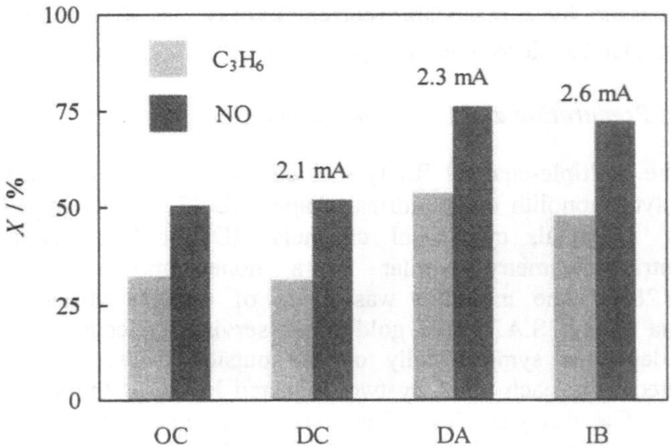

Figure 31. Conversion (X) of C_3H_6 and NO in a ring-shaped Rh/YSZ cell at open-circuit (OC) condition and at three different potentiostatic polarization modes: direct cathodic (DC; -5V), direct anodic (DA; +5V), and indirect bipolar (IB; 10V). The obtained current is also indicated. Feed composition and temperature as in Figure 30, flow rate: 200 mL min^{-1} STP.

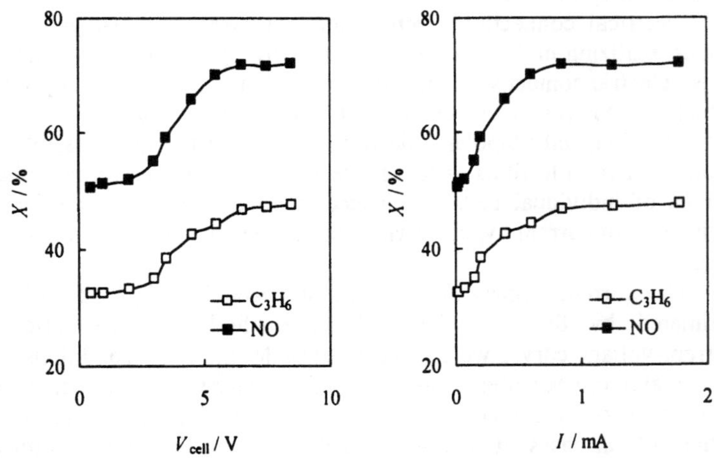

Figure 32. Effect of the cell potential (V_{cell}) and of the total current (I) on the conversion (X) of C_3H_6 and NO under indirect bipolar (IB) polarization in a ring-shaped Rh/YSZ cell. Feed composition, gas flow and temperature as in Figure 31.

constructed and investigated in our laboratory.[10, 18, 55] This cell, characterized by a reasonable current bypass, has also shown good performance for electrochemical promotion.

(i) Cell Preparation and Characterization

The multiple-channel RuO_2/YSZ cell was prepared with a solid electrolyte monolith of cylindrical shape (OD 22 mm, height 10 mm), having 37 axial, cylindrical channels (ID 2 mm) arranged in a concentric symmetry similar to a honeycomb structure (see Figure 28B). The monolith was made of 6 mol% yttria-stabilized zirconia (Maret S.A.). Two gold films, serving as feeder electrodes, were deposited symmetrically on the outside surface of the cell, separated from each other by two uncoated bands of the same width (5 mm). The RuO_2 catalyst films were deposited on the walls of the channels, nominal loading: 4 mg (1.7 g m^{-2}). The procedures of thermal deposition were the same as with the single-pellet cells described in Section III.2.*i*. The experimental setup used for cell characterization and for kinetic measurements was the same one described in Section III.2.*ii*, but supplemented with a power supply (Dr. K. Witmer Elektronik AG, WS 150-36) allowing for potential application up to 35 V.

Electrical connections were made only with the two gold electrodes, realizing in this manner a *wireless* configuration. This means no direct electric contact with the catalyst, as should be the case anyway in future electrochemical cells with dispersed catalysts. In this configuration, only bipolar polarization of the catalyst is possible, by using the two gold films as feeder electrodes. Since the current-voltage curves of individual cells are inaccessible, the method used for the estimation of current bypass in ring-shaped cells is not applicable in this case.

The current bypass in the multiple-channel RuO_2/YSZ cell was estimated by the following indirect method.[55] Two series of current-voltage curves were determined under the same conditions, one before and another after deposition of the catalyst. Measurements were made at various gas compositions and at different temperatures, the same in both cases. It was assumed that coating the support with the catalyst does not change the current distribution in the solid electrolyte, but simply opens new parallel conduction pathways. The current bypass was then calculated from the currents measured at a same cell potential, before and after deposition of the catalyst. Both, the current bypass and the cell resistance, were found to be much higher than in the ring-shaped cells. At the temperature of subsequent kinetic measurements (360°C), the ohmic cell resistance was in the order of 20 kΩ and a cell potential of 25-35 V was necessary to obtain a total current of 1-1.5 mA. The current bypass in the same current range and at the same temperature was about 34%, rather independent of both, the cell potential and the gas composition. Although the use of this cell needs a relatively high potential, successful polarization of the catalyst in the multiple-channel configuration is very promising for practical applications of electrochemical promotion.

(ii) Electrochemical Promotion

Successful electrochemical promotion using the multiple-channel RuO_2/YSZ cell was achieved with the combustion of ethylene.[18] Kinetic measurements were made at various feed compositions in the temperature range between 300 and 410°C. A maximal promotion of ethylene conversion was obtained with a feed composition of 0.2 kPa C_2H_4 and 12 kPa O_2 at 360°C. Figure 33 shows the temporal evolution of the ethylene conversion under these conditions due to potentiostatic steps of 30 V. A significant increase from 36 to 51% was observed after 135 min. The current was slowly increasing during polarization reaching a plateau at 1.5 mA.

The effect of the applied cell potential on the electrochemical promotion is shown in Figure 34. The increase in ethylene conversion becomes significant above a cell potential of 2-3 V, similarly to the behavior of the ring-shaped bipolar cell. The ethylene conversion then increases with increasing cell voltage and tends towards a plateau above 30 V. The enhancement factor, Λ, characterizing the non-Faradaic behavior of the promotion, can only be estimated, due to the lack of exact knowledge of the current distribution in the multiple-channel cell. In fact, estimation of the average effective current is difficult because, on one hand, only a part of the total current passes through a given individual cell and, on the other hand, the same current may pass through several cells, hence used several times for polarization. As a rough estimate, overall enhancement factors were calculated using the total current supplied between the two feeder electrodes. Λ values between 30 and 90 were obtained with the maximum at $V_{cell} = 5$ V. Considering the number and the arrangement of individual cells in the given bipolar configuration, such high enhancement factors may only be explained by successful non-Faradaic electrochemical modification of the catalytic activity.

Achievement of a considerable NEMCA effect in the multiple-channel cell is very promising for future applications of electrochemical promotion using dispersed catalysts.

4. Perspectives

The development of a dispersed catalyst system for electrochemical promotion should become a major practical goal in future work. The actual progress, ranging from single-pellet cells to various bipolar configurations either with a single bipolar element (ring-shaped configuration) or with multiple bipolar elements (multiple-channel configuration), may be extended to dispersed systems. A possible approach in this direction is the use of porous solid electrolytes as the support. The catalyst may be deposited in the form of a discontinuous film on the porous support, with the two feeder electrodes placed on the

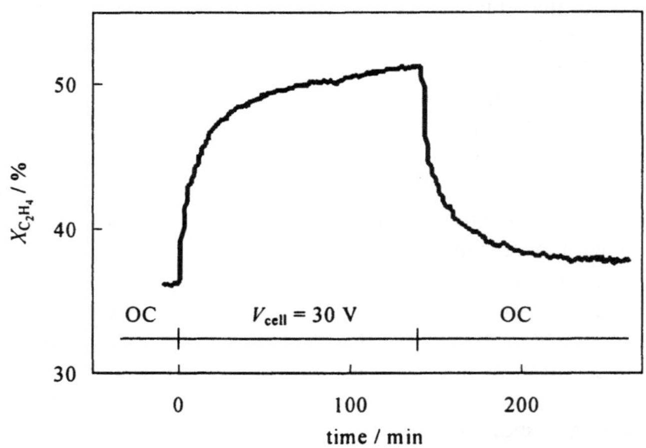

Figure 33. Temporal evolution of the ethylene conversion in the multiple-channel RuO_2/YSZ cell during a potentiostatic step of V_{cell} = 30 V. OC: open-circuit. Feed composition: C_2H_4:O_2/0.2:12 kPa, flow rate: 175 mL min^{-1} STP, T = 360°C. Reproduced from Ref. [18] by permission of the Electrochemical Society, Inc.

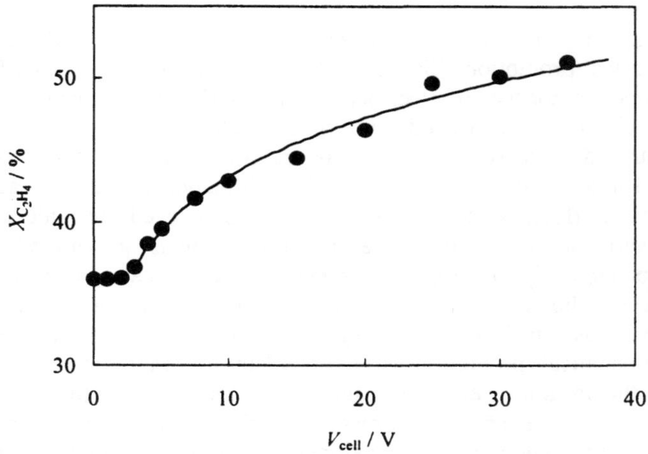

Figure 34. Influence of the applied cell potential on the ethylene conversion in the multiple-channel RuO_2/YSZ cell. Feed composition, gas flow and temperature as in Figure 33. Reproduced from Ref. [18] by permission of the Electrochemical Society, Inc.

outside surface of the solid electrolyte. Another possible approach to achieve electrochemical promotion in dispersed systems is based on the fabrication of microstructured cells. The application of thin catalyst films interfaced with solid electrolytes requires small reactor volumes, hence requires also the development of suitable microreactors. In both cases, high cell potentials are anticipated and careful optimisation is required in order to avoid excessive current bypass.

Improvement of the ring-shaped configuration may also be envisaged. Its rather simple design and its high performance make it a readily applicable system. Since the catalyst film deposited on the inside wall of the tube is polarized in an indirect way by the two feeder electrodes placed on the outside wall of the tube, the electric part of the system is well separated from the gas-feeding part making the device quite easy to handle. This type of cell has the advantage of being almost free from current bypass, but an improvement of its hydrodynamic behavior remains desirable. Further efforts in these directions are in progress in our laboratory.

V. CONCLUSIONS

In this Chapter, the progress recently made in the field of electrochemical promotion (EP) of catalytic gas reactions is reviewed. The phenomenon consists of electrochemical polarization of metal or metal oxide electrodes interfaced with solid electrolytes which result in a pronounced increase in the catalytic reaction rate. The effect is also termed non-Faradaic electrochemical modification of catalytic activity (NEMCA effect), since the rate increase may exceed the ionic current by several orders of magnitude. The promotion is not limited to the electrochemically polarized interface between catalyst and solid electrolyte, but extends to the entire catalyst surface exposed to the reactive gas. In fact, one of the major challenges in the field of electrochemical promotion is to elucidate the exact mechanism by which the promoting effect propagates from one interface to the other.

Two main aspects are addressed, the phenomenological description of the phenomenon and the development of bipolar cell configurations suitable for practical applications. The discussion was restricted to O^{2-}-conducting solid electrolytes, *i.e.*, yttria-stabilized zirconia (YSZ). This choice enables EP measurements at moderate temperatures, typically in the range of 300-400°C.

For fundamental studies, the simplest possible electrochemical cell of single-pellet type was used. The behavior of the reference electrode, prepared by deposition of a gold film, was found to be quasi-reversible, $\frac{1}{2}O_2/O^{2-}$ being the potential determining redox couple. The potential distribution in the single-pellet cell was estimated to be fairly symmetrical, and the IR drop correction was shown to be negligible for low currents typical of EP experiments.

Cyclic voltammetric experiments with IrO_2/YSZ catalysts for ethylene combustion provided a linear relation between voltammetric charge and reaction rate during relaxation after current interruption. This indicates that the promotion effect and the capacitance of the gas-exposed catalyst surface are closely related, the latter being dependent on the population of charged promoters at the catalyst surface. In good agreement with the cyclic voltammetric evidence of easily and slowly accessible parts of the active surface, the catalytic reaction rate transients often exhibit a complex behavior both under polarization and during relaxation. In order to describe the fast component of complex transients, a model was proposed, assuming free surface site dependent formation, rapid spreading and first order rate consumption of O^{2-} promoters. The model predicts exponential catalytic rate transients with current dependent polarization and current independent relaxation. It enables determination of the maximum limiting value of the Faradaic efficiency, which is a current independent property of the catalytic system. The validity of the model was confirmed for anodic galvanostatic EP of ethylene combustion over IrO_2/YSZ catalyst. Prolonged application of higher polarizing currents (>100 μA) may cause an irreversible effect (permanent promotion). The phenomenon, first reported with IrO_2/YSZ catalysts for ethylene combustion, is explained with current assisted modification in the surface oxidation state of the catalyst, similarly to the current assisted recovery of lost catalytic activity at low temperature as observed for the reduction of NO by C_3H_6 over Rh/YSZ catalyst in presence of excess O_2 at 300°C. Experiments performed with mixed IrO_2 - TiO_2/YSZ catalysts for ethylene oxidation, provided evidence for the functional identity of electrochemical promotion and catalyst-support interactions. The O^{2-} ions responsible for promotion are being supplied by the electric current in the first case, and by the TiO_2 donor in the second case. *In situ* work function measurements over RuO_2/YSZ catalysts for ethylene combustion showed a close relationship between reaction rate and catalyst work function transients, indicating migration (spreading) of charged promoting species over the gas-exposed catalyst surface during polarization and their slow removal when returning to steady-state open-circuit conditions. All these results support well the current theory of electrochemical promotion, which attributes the phenomenon to the existence of promoting species originated from the charge transfer reaction at the catalyst/electrolyte interface and to their migration onto the gas-exposed surface where the catalytic gas reaction occurs.

The single-pellet electrochemical cell used for fundamental investigations is obviously not suitable for practical applications. In view of future industrial exploitation of the phenomenon, two new cell designs were developed, a ring-shaped and a multiple-channel configuration. Both were designed to operate in the bipolar polarization mode, i.e. by supplying the polarizing potential or current via two catalytically inert feeder electrodes. Bipolar operation may create a significant current bypass, resulting in a decrease in current efficiency. Hence, it is a fundamental parameter in the development of bipolar configurations. A method for the estimation of current bypass from current-voltage curves was adapted to solid-state electrochemical systems. Using this method, the ring-shaped configuration was found to be almost free of current bypass during the electrochemical promotion of the catalytic reactions. In the multiple-channel configuration, the current bypass was estimated via an indirect method and was found to be much higher (34%). The feasibility of electrochemical promotion was demonstrated successfully with both configurations. Promotion of the reduction of NO by propylene in ring-shaped bipolar Rh/YSZ cells and that of the combustion of ethylene in both ring-shaped and multiple-channel bipolar RuO_2/YSZ cells were highly non-Faradaic, even at high open-circuit conversions. Realization of efficient bipolar cell configurations for electrochemical promotion is of great importance in view of future applications in dispersed catalytic systems.

ACKNOWLEDGEMENTS

Financial support from the *Fonds national suisse de la recherche scientifique* and from the *Office fédéral suisse de l'éducation et de la science* is gratefully acknowledged. Sincerest thanks are expressed to Professor Costas G. Vayenas, University of Patras, for fruitful discussions and to Professors Richard M. Lambert, University of Cambridge, and Francis Levy, Swiss Federal Institute of Technology in Lausanne, for the sputter deposition of rhodium.

REFERENCES

[1]M. Stoukides, and C.G. Vayenas, *J. Catal.* **70** (1981) 137.
[2] C.G. Vayenas, S. Bebelis, and S. Neophytides, *J. Phys. Chem.* **92** (1988) 5083.
[3]C.G. Vayenas, S. Bebelis, and S. Ladas, *Nature* **343** (1990) 625.
[4]J.O'M. Bockris, A.K.N. Reddy, and M. Gamboa-Aldeco, *Modern Electrochemistry*, Kluwer Academic/Plenum Publishers, New York, 2000.
[5]*Handbook of Catalysis*, Ed. by G. Ertl, H. Knötzinger, and J. Weitkamp, VCH Publishers, Weinheim, 1997.
[6]J. Pritchard, *Nature* **343** (1991) 592.
[7]S. Neophytides, D. Tsiplakides, P. Stonehart, M.M. Jaksic, and C.G. Vayenas, *J. Phys. Chem.* **100** (1996) 14803.
[8]C.G. Vayenas, M.M Jaksic, S. Bebelis, and S.G. Neophytides, in: *Modern Aspects of Electrochemistry*, Eds. J.O'M. Bockris, B.E. Conway, and R.E. White, Vol. 29, p. 57, Kluwer Academic/Plenum Publishers, New York, 1996.
[9]C.G. Vayenas, S. Bebelis, C. Pliangos, S. Brosda, and D. Tsiplakides, *Electrochemical Activation of Catalysis: Promotion, Electrochemical Promotion, and Metal-Support Interactions*, Kluwer/Plenum Press, New York, 2001.
[10]G. Fóti, S. Wodiunig, and Ch. Comninellis, in : *Current Topics in Electrochemistry*, Research Trends, Vol. 7, p.1,Trivandrum, 2000.
[11]J. Janek, M. Rohnke, B. Luerssen, and R. Imbihl, *Phys. Chem. Chem. Phys.* **2** (2000)1935.

[12]I.S. Metcalfe, *J. Catal.* **199** (2001) 247.
[13]I.S. Metcalfe, *J. Catal.* **199** (2001) 259.
[14]E. Varkaraki, J. Nicole, E. Plattner, Ch. Comninellis, and C.G. Vayenas, *J. Appl. Electrochem.* **25** (1995)978.
[15]J. Nicole, D.T. Tsiplakides, S. Wodiunig, and Ch. Comninellis, *J. Electrochem. Soc.* **144** (1997) L312.
[16]J. Nicole, and Ch. Comninellis, *Solid State Ionics* **136-137** (2000) 687.
[17]J. Nicole, D. Tsiplakides, C. Pliangos, X.E. Verykios, Ch. Comninellis, and C.G. Vayenas, *J. Catal.* **204** (2001) 23.
[18]S. Wodiunig, F. Bokeloh, J. Nicole, and Ch. Comninellis, *Electrochem. Solid State Lett.* **2** (1999) 281.
[19]S. Wodiunig, V. Patsis, and Ch. Comninellis, *Solid State Ionics* **136-137** (2000) 813.
[20]G. Fóti, O. Lavanchy, and Ch. Comninellis, *J. Appl. Electrochem.* **30**, (2000) 1223.
[21]O. Yamamoto, in: *Solid State Electrochemistry*, Ed. by P.G. Bruce, p. 292 Cambridge Univ. Press, Cambridge, 1995.
[22]T. Rebitzki, B. Delmon, and J.H. Block, *AIChE Journal* **41** (1995) 1543.
[23]F.M. Faus, B. Zhou, H. Matralis, and B. Delmon, *J. Catal.* **132** (1991) 200.
[24]S. Ladas, S. Kennou, S. Bebelis, and C.G.Vayenas, *J. Phys. Chem.* **97** (1993) 8845.
[25]S.G. Neophytides, D. Tsiplakides, and C.G. Vayenas, *J. Catal.* **178** (1998) 414.
[26]C.G. Vayenas, S. Brosda, and C. Pliangos, *J. Catal.* **203** (2001) 329.
[27]S. Brosda, and C.G. Vayenas, *J. Catal.* **208** (2002) 38.
[28]G. Fóti, V. Stanković, I. Bolzonella, and Ch. Comninellis, *J. Electroanal. Chem.* (2002) in press.
[29]C.G. Vayenas, and G.E. Pitselis, *I&EC Res.* **40** (2001) 4209.
[30]J.T. Kummer, *J. Phys. Chem.* **90** (1986) 4747.
[31]K.C. Taylor, *Cat. Rev. Sci. Eng.* **35** (1993) 457.
[32]R. Sant, D.J. Kaul, and E.E. Wolf, *AIChE Journal* **35** (1989) 267.
[33]A.P. Walker, *Catal. Today* **26** (1995) 107.
[34]M. Iwamoto, *Catal. Today* **29** (1996) 29.
[35]A. Fritz, and V. Pitchon, *Appl. Catal. B: Environmental* **13** (1997) 1.
[36]S. Bebelis, and C.G. Vayenas, *J. Catal.* **118** (1989) 125.
[37]Ch. Comninellis, and G.P. Vercesi, *J. Appl. Electrochem.* **21** (1991) 335.
[38]D.Y. Wang, and A.S. Nowick, *J. Electrochem. Soc.* **128** (1981) 55.
[39]S.H. Chan, X.J. Chen, and K.A. Khor, *J. Appl. Electrochem.* **31** (2001) 1163.
[40]C.G. Vayenas, A. Ioannides, and S. Bebelis, *J. Catal.* **129**(1991) 67.
[41]S. Ardizzone, G. Fregonara, and S. Trasatti, *Electrochim. Acta* **35** (1990) 263.
[42]M. Stoukides, and C.G. Vayenas, in: *ACS Symp. Series*, Ed. by A.T. Bell, and L.L. Hegedüs, **178** (1982) 181.
[43]J. Hölzl, and F.K.-Schulte, in: *Solid Surface Physics*, , Springer, Berlin, **85** (1979) 1.
[44]S. Wodiunig, and Ch. Comninellis, *J. Eur. Ceram. Soc.* **19** (1999) 931.
[45]S.G. Neophytides, and C.G. Vayenas, *J. Phys. Chem.* **99** (1995) 17063.
[46]B. Delmon, and G.F. Froment, *Catal. Rev.-Sci. Eng.* **38** (1996) 69.
[47]*Spillover and Migration of Surface Species on Catalysts*, Ed. by C. Li, and Q. Xiu, Vol. 112, Elsevier, Amsterdam, 1997.
[48]M. Marwood, and C.G. Vayenas, *J. Catal.* **168** (1997) 538.
[49]J.C. Rivière, in: *Solid State Surface Science*, Ed. by M. Green, Vol. 1, p.179, Marcel Dekker, New York, 1969.
[50]K. Besocke, and S. Berger, *Rev. Sci. Instrum.* **47** (1976) 840 .
[51]N.A. Surplice, and R.J. D'Arcy, *J. Phys. E.* **3** (1970) 477.

[52]S. Wodiunig, Ch. Comninellis, and C. Mousty, in: *Energy and Electrochemical Processing for a Cleaner Environment*, Ed. by E.J. Rudd and C.W. Walton, The Electrochemical Society Proceedings Series, PV 97-28, p. 147, Pennington, NJ, 1997.
[53]Ch. Comninellis, in: *Electrochemical Engineering and Energy*, Ed. by F. Lapicque., p. 177, Plenum Press, New York, 1995.
[54]Ch. Comninellis, E. Plattner, and P. Bolomey, *J. Appl. Electrochem.* **21** (1991) 415.
[55]S. Wodiunig, F. Bokeloh, and Ch. Comninellis, *Electrochim. Acta*, **46** (2000) 357.

5

Mechanisms of Lithium Transport through Transition Metal Oxides and Carbonaceous Materials

Heon-Cheol Shin and Su-Il Pyun

Department of Materials Science and Engineering, Korea Advanced Institute of Science and Technology, 373-1, Guseong-dong, Yuseong-gu, Daejeon, 305-701, Republic of Korea

I. INTRODUCTION

The kinetics of lithium transport through transition metal oxides and carbonaceous materials has been studied extensively due to its great importance for high power output rechargeable batteries. In most cases, the kinetic analysis and the determination of the chemical diffusivity of lithium in the electrode has been performed assuming the *diffusion control* concept.[1-15] This assumes that the diffusion of lithium in the electrode is very slow, while other reactions, including interfacial charge transfer, are too fast to affect the kinetics of lithium transport; thereby lithium diffusion in the electrode is assumed to be the rate-controlling process of lithium intercalation.

However, various kinds of anomalous behavior of lithium transport have been reported in many current transients (CTs) and voltammetric curves obtained with a number of transition metal oxides and carbonaceous materials. In spite of the fact that many researchers

Modern Aspects of Electrochemistry, Number 36, edited by C. G. Vayenas, B. E. Conway, and R. E. White, Kluwer Academic / Plenum Publishers, New York, 2003

have tried to interprete these atypical phenomena (a) under the above *diffusion control* concept, combined with the effects of irregular geometry,[16-18] electric field,[9,19] intercalation-induced stress,[20-22] growth of a new phase in the electrode,[19-26] *etc.*, and (b) from other viewpoints than the *diffusion control* concept, such as the charge transfer limited mechanism[27-35] and the trapping mechanism,[36-39] most of the abnormalities in transient curves have not been clearly understood.

Recently, the analysis of the anomalous current response is beginning to enter a new phase with the help of the concept based upon a *cell-impedance controlled* constraint, which has been systematically studied by Pyun *et al.*[40-45]

Pyun *et al.* started[40-43] their exploration of the anomalous current response with the CTs obtained from $Li_{1-\delta}CoO_2$ which is the cathode material of almost all commercially available rechargeable lithium batteries today. They reported that the CTs obtained from $Li_{1-\delta}CoO_2$ composite[40] and thin film[41-43] electrodes hardly exhibit a typical trend of *diffusion controlled* lithium transport, *i. e.,*, Cottrell behavior. Furthermore, they have found that the current-potential relation obeys Ohm's law during the CT experiments. They thus suggested that lithium transport at the interface of the electrode and the electrolyte is mainly limited by the internal cell resistance, and not by lithium diffusion in the bulk electrode. This concept is called the *cell-impedance controlled* lithium transport.

They have extended the kinetic study of lithium intercalation to such transition metal oxides as $Li_{1-\delta}NiO_2$,[44,45] $Li_{1-\delta}Mn_2O_4$,[46-48] $Li_{1+\delta}[Ti_{5/3}Li_{1/3}]O_4$,[44] V_2O_5,[18,44] and carbonaceous materials.[49] In these works, they have reported that the theoretical CTs, based upon the *cell-impedance control* concept, matched quantitatively those experimentally measured: all the anomalous features of the experimental transients were readily interpreted in the *cell-impedance controlled* transients with such simplified parameters as electrochemical active area, dimensionality of the diffusion path, cell impedance, *etc.*

Before the works by Pyun *et al.*, several other research groups[50-58] have discussed independently their CTs in terms of internal cell resistance. They focused on the determination of the diffusion coefficient of lithium in such electrochromic materials as WO_3, $Mo_{0.5}V_{0.5}O_{2.75}$, and $LiCoO_2$. Nagai *et al.*[50,54] tried to estimate the diffusion coefficient of lithium inside the WO_3 film electrode during the bleaching and coloring processes by means of fitting CTs theoretically calculated using simplified electrode potential curves to their experimental CTs. Similarly, Li *et al.*[56-58] derived the current-time relation assuming an one-dimensional (linear) electrode potential curve to estimate the diffusion coefficient of lithium from rough comparison of the calculated CTs at various values of parameters with the experimental CTs. Finally, Honder *et al.*[51-53,55] calculated the values of diffusion coefficient and total cell resistance from a combination of the approximate equations in the short time and long time ranges.

However, these works were performed without any clear evidence that the internal cell resistance governs lithium transport during the whole lithium intercalation/deintercalation, including the duration of phase transition which almost all the electrode materials experience during battery operation. Thus, their works could not generally be applied to calculate the diffusion coefficient and to explain the abnormal behavior of CTs in intercalation compounds. In addition, this approach oversimplified the electrode potential curve and at the same time overlooked the strong variation of cell resistance with lithium content, so that one could hardly examine, on the basis of these models, the quantitative dependence of lithium transport through the electrode on the variation in electrode potential and in internal cell resistance with varying lithium content in the electrode.

This Chapter discusses lithium transport through transition metal oxides and carbonaceous material (graphite) during CT experiments. The structure of this review is as follows: in Section II, the conventional and modified *diffusion control* models for explaining the CTs are briefly summarized. Typical experimental CTs from transition metal oxides and carbonaceous material (graphite) are presented and then several anomalous behaviors in these curves are pointed out in Section III. In Section IV, the physical aspects of the CTs are discussed in terms of total cell resistance. Finally, Section V is devoted to the theoretical consideration of the *cell-impedance controlled* lithium transport, and to the comparison of experimental curves with theoretical ones.

II. BIRD'S EYE VIEW OF THE MODELS FOR CURRENT TRANSIENTS IN LITHIUM INTERCALATION SYSTEMS : DIFFUSION CONTROLLED LITHIUM TRANSPORT

The most frequent method for the analysis of CTs from intercalation electrodes is based on the grounds that lithium diffusion in the electrode is the rate-determining process in lithium intercalation/deintercalation.[1-3] This involves the following: the system is so kinetically fast that the equilibrium concentration of lithium is quickly reached at the interface between electrode and electrolyte at the moment of potential stepping in CT experiments. The instantaneous depletion and accumulation of the lithium concentration at the interface caused by the chemical diffusion away from and to the interface, and to and away from the bulk electrode, respectively, is completely compensated by the supply and release away from and into the electrolyte, respectively. This condition is referred to as *real potentiostatic constraint* at the interface between the electrode and the electrolyte.

If one assumes that the electrode material is homogeneous, the concentration gradient of lithium through the electrode is the only factor that drives lithium transport, lithium enters/leaves the planar electrode only at the electrode/electrolyte interface, and finally lithium cannot penetrate into the back of the electrode, *i. e.,,* impermeable (impenetrable) constraint, then the electric current can be expressed by Eq. (1) in the initial stage of diffusion, and by Eq. (2) in the later stage.[2]

$$I(t) = FA_{ea}\left(c^1 - c^0\right)\left(\frac{\tilde{D}_{Li}}{\pi t}\right)^{1/2} \quad \text{(Cottrell relation)}, \qquad t \ll \frac{L^2}{\tilde{D}_{Li}} \qquad (1)$$

$$I(t) = \frac{2FA_{ea}\left(c^1 - c^0\right)\tilde{D}_{Li}}{L}\exp\left(\frac{\pi^2 \tilde{D}_{Li}}{4L^2}t\right), \qquad t \gg \frac{L^2}{\tilde{D}_{Li}} \qquad (2)$$

here t is the lithium injection/extraction time; F is the Faraday constant; A_{ea} is the electrochemical active area of the electrode; $\tilde{D}_{Li}$ is the chemical diffusivity of lithium; L is the thickness of the electrode, and c^l and c^0 represent the final and initial equilibrium concentration of lithium in the electrode, respectively.

Using Eqs. (1) and (2), a number of CTs have been analyzed. The slopes of $I(t)$ *vs.* $t^{-1/2}$ (or the values of $I(t)\cdot t^{1/2}$) and ln $I(t)$ *vs.* t curves have been determined in the initial and later stages of lithium injection/extraction, respectively, to estimate the chemical diffusivity of lithium in the electrode. However, it has been pointed out[9,31,59,60] that there is a great discrepancy between the values of the chemical diffusivity determined by the CT technique using the *diffusion control* concept and those values obtained by other electrochemical techniques such as the galvanostatic intermittent titration technique (GITT) and the electrochemical impedance spectroscopy (EIS). In addition, anomalous behaviors in CTs, which are hardly explained using the conventional *diffusion control* concept, have been reported for various transition metal oxides and carbonaceous materials.[18-26]

Several efforts have been made to explain the atypical lithium transport behavior using modified *diffusion control* models. In these models the boundary conditions, *i. e., real potentiostatic* constraint at the electrode/electrolyte interface and the impermeability constraint at the back of the electrode, are still valid, while lithium transport is strongly influenced by, *e.g.*, 1) the geometry of the electrode surface,[16-18], 2) the growth of a new phase in the electrode,[19-26] and 3) the electric field in the electrode.[9,19]

1. The geometry of the Electrode Surface

In the case of cylindrical diffusion, potential step chronoampero-metry (PSCA) suggested by Takehara *et al.*[61,62] has been frequently used to estimate the value of the diffusivity. They obtained the following flux-time relation from the solution of the diffusion equation for a cylinder using the *diffusion control* concept.

$$\frac{I(t)}{j_\infty} = 2\sum_{n=1}^{\infty} \exp\left(\beta_n^2 \tau\right) \tag{3}$$

$$j_\infty = F\,\tilde{D}_{Li}\left(c^0 - c^1\right)/\,R^\# \tag{4}$$

$$\tau = \tilde{D}_{Li}\,t\,/\,R^{\#2} \tag{5}$$

where β_n is the root of a Bessel function of the first kind of order zero ($J_0(\beta_n) = 0$), and $R^\#$ represents the radius of the cylindrical electrode. The chemical diffusivity of lithium has been estimated by curve fitting of the observed CTs to those CTs calculated according to Eq. (3).

As for electrodes with irregular shape (e.g. rough, porous, scratched, partially blocked, etc.), fractal geometry[63] has been used to analyze the CTs. Following the fractal concept, the generalized Cottrell equation is given for the self-similar fractal electrode[64-68] by the equation:

$$I(t) = \frac{FS_0^2\sqrt{\tilde{D}_{Li}^{*}}\left(c^1 - c^0\right)}{\Gamma\left(\frac{3-D_f}{2}\right)} t^{-\frac{D_f-1}{2}} \tag{6}$$

where S_o is the side length of a square electrode; D_f is the fractal dimension of the electrode/electrolyte interface; $\Gamma(x)$ is the gamma function of x, and $\tilde{D}_{Li}^{*}$ represents the fractional diffusivity defined as $K^{(4-2D_f)}S_0^{(2D_f-4)}\tilde{D}_{Li}^{(3-2D_f)}$ (K is a constant related to the fractal dimension).

Equation (6) holds in a certain limited range of time bordered by temporal cut-offs τ,[65] *i.e.*, inner cut-off and outer cut-off, which are connected to the corresponding spatial cut-offs λ by $\lambda \approx \left(2\tilde{D}_{Li}\tau\right)^{1/2}$. Beyond the temporal cut-offs, when the diffusing species hardly recognize the fractal character of the surface, the Cottrell relation in Eq. (1) for planar electrode is retrieved. With a combination of the CT technique and topographic analysis of the electrode using scanning electron microscopy (SEM), atomic force microscopy (AFM), scanning tunneling microscopy (STM), *etc.*, the effect of surface irregularity on the CT response has been analyzed.[69,70]

2. The Growth of a New Phase in the Electrode

The role of phase transformation of lithium-poor phase to lithium-rich phase in lithium transport during the CT experiments has been discussed independently by Pyun *et al.*[19,23] and Funabiki *et al.*[24-26] under the assumption that phase boundary movement is controlled by the lithium diffusion in each phase. The former authors have focused on the shape of the CTs and the onset/end of the phase boundary movement by solving numerically the modified diffusion equation, while the latter authors have determined the rate of the phase boundary movement by solving analytically the diffusion equation based upon Wagner's approach.[71]

3. The Electric Field in the Electrode

The contribution of electric field to lithium transport has been considered by a few authors. Pyun *et al.*[19] argued on the basis of the Armand's model[72] for the intercalation electrode that lithium deintercalation from the $LiCoO_2$ composite electrode was retarded by the electric field due to the formation of an electron-depleted space charge layer beneath the electrode/electrolyte interface. Nichina *et al.*[9] estimated the chemical diffusivity of lithium in the $LiCoO_2$ film electrode from the current-time relation derived from the Nernst-Planck equation[73] for combined lithium migration and diffusion within the electrode.

In spite of the many efforts briefly described above and some other attempts which are not mentioned here, the atypical trajectories of CTs from various transition metal oxides and carbonaceous materials for rechargeable lithium battery, and the great difference between the chemical diffusivities determined by the CT technique and other electrochemical techniques were not clearly understood.

III. GENERAL PERSPECTIVE ON CURRENT TRANSIENTS FROM TRANSITION METAL OXIDES AND GRAPHITE

Let us begin with a brief description of electrode potential curves. The electrode potential curves of the carbon-dispersed composite electrodes of $Li_{1-\delta}NiO_2$,[44,45,74] $Li_{1-\delta}CoO_2$,[40,75,76] $Li_{\delta}V_2O_5$,[18,44,77] $Li_{1+\delta}[Ti_{5/3}Li_{1/3}]O_4$,[44,78] and graphite[49] are obtained from the respective (intermittent) galvanostatic charge-discharge curves (Figures 1a-e). In the present review, all the electrode potential curves and CTs were measured in a 1M solution of lithium perchlorate ($LiClO_4$) in propylene carbonate (PC) with the exception of those curves from graphite, which were obtained in a 1M solution of lithium hexafluorophosphate ($LiPF_6$) in ethylene carbonate (EC) / diethyl carbonate (DEC) (50:50 vol %).

The changes in the equilibrium phase with lithium content are given in Figures 1(a)-(e). Specifically, two different phases coexist in the lithium content range where the potential plateau appears. For the details about the chemical and physical properties of the equilibrium

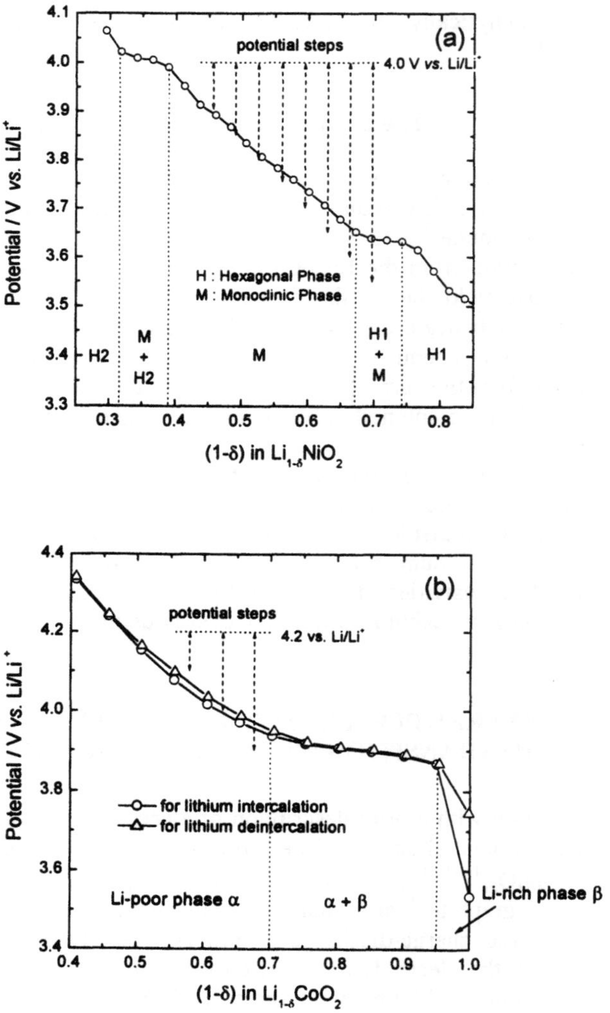

Figure 1. Electrode potential curves obtained from the (intermittent) galvanostatic charge-discharge curves of the carbon-dispersed composite electrodes of (a) $Li_{1-\delta}NiO_2$, (b) $Li_{1-\delta}CoO_2$, (c) $Li_{\delta}V_2O_5$, (d) $Li_{1+\delta}[Ti_{5/3}Li_{1/3}]O_4$, and (e) graphite. Reprinted from [40] ©(1999), [44]©(2001), and [45]©(2001), with permission from Elsevier Science.

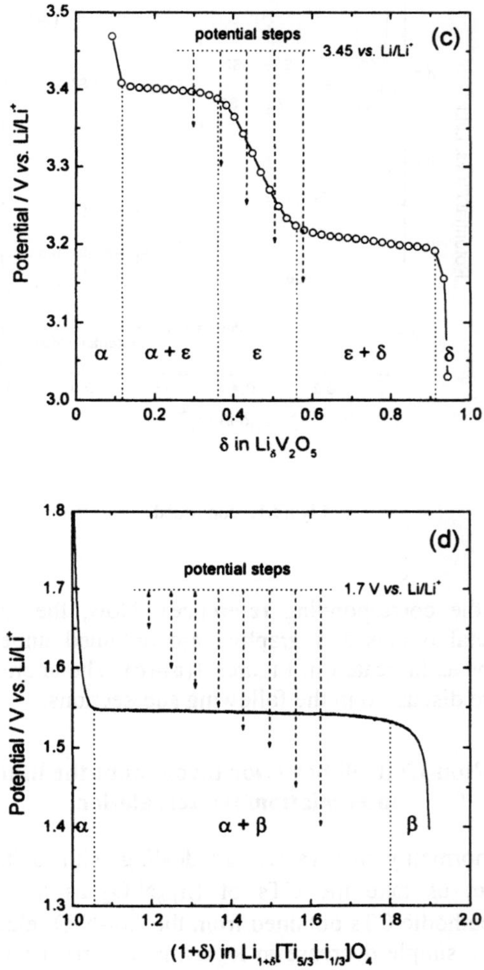

Figure 1. (continued)

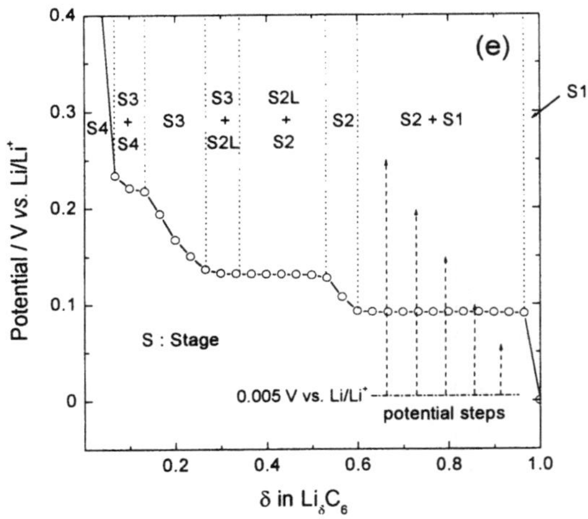

Figure 1. (continued)

phases, see the corresponding references. Now, the typical CTs of transition metal oxides and graphite are obtained under the various potential steps as indicated in Figures 1(a)-(e). The anomalous features of the CTs are discussed in the following sub-sections.

1. Non-Cottrell behavior throughout the lithium intercalation/deintercalation

The first abnormality in CTs we are dealing with is its non-Cottrell character. Let us take the CTs of $Li_{1-\delta}NiO_2$ as an example. The logarithmic cathodic CTs obtained from the $Li_{1-\delta}NiO_2$ electrode (Figure 2(a)) exhibit a simple decrease in logarithmic current with logarithmic time for potential drops above the plateau potential of 3.63 V *vs.* Li/Li^+, while those curves below 3.63 V *vs.* Li/Li^+ are characterized by an inflexion point (or quasi-current plateau). As we mentioned in Section II, the $I(t)\cdot t^{1/2}$ *vs.* ln *t* plots of Figure 2(b), which is reconstructed from the CTs of Figure 2(a), can be an effective way of analyzing the CTs based upon the *diffusion control* concept. To date, several researchers[79-87] have believed that the plateau in this type of plot, which plateau has been said to be extremely narrow, described the semi-infinite planar diffusion and represented the Cottrell behavior.

However, one is hardly able to find any region with the Cottrell character in Figure 2(b). In fact, even the flattest region (see the inset!) in Figure 2(b) shows nothing but a local maximum. Under this circumstance, it is very doubtful that the pseudo-plateau in Figure 2(b) is due to the semi-infinite diffusion of lithium through the oxide, *i. e.*,

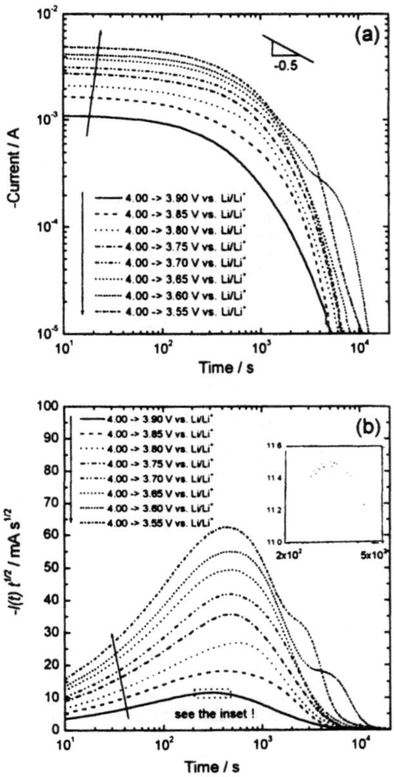

Figure 2. (a) Logarithmic cathodic current transients obtained experimentally with the $Li_{1-\delta}NiO_2$ electrode and (b) $I(t)\cdot t^{1/2}$ *vs. ln t* plots obtained from Figure 2(a). Reprinted from [45] with permission from Elsevier Science.

the physical meaning of the pseudo-plateau has been overestimated. Rather, it is more reasonable to say that the upward convex shape of the $I(t)\cdot t^{1/2}$ *vs.* ln t plots reflects simply a continuous decrease in current with time. The simplicity of the CTs implies that lithium transport runs without any critical transition in transport mechanism during the entire lithium intercalation.

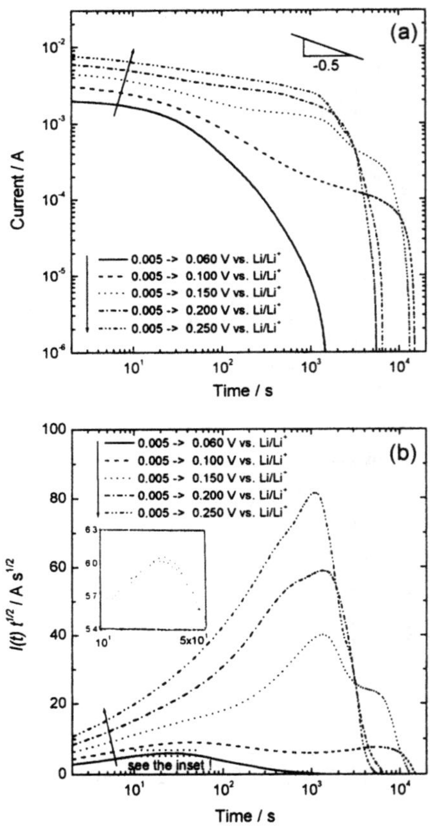

Figure 3. (a) Logarithmic anodic current transients obtained experimentally with a graphite electrode. (b) $I(t)\cdot t^{1/2}$ *vs. ln t* plots obtained from Figure 3(a).

The above is still valid for CTs of other transition metal oxides and carbonaceous material (graphite). For instance, the $I(t) \cdot t^{1/2}$ *vs.* $\ln t$ plots from graphite electrode (Figure 3b) do not show any Cottrell region, similarly to the case of $LiNiO_2$ in Figure 2(b). The appearance of shoulders and of more than one local maxima in Figures 2(b) and 3(b) will be addressed again in Section III.3.

2. Intersection of anodic and cathodic current transients

The intersection of the anodic and the corresponding cathodic CTs has been first reported by Pyun *et al.*[40,42-44] and it is definitely one of the clear evidences that the main factor controlling the interfacial lithium ion transfer, *i. e.*, the current, during potential stepping is not the lithium diffusion inside the electrodes. One finds the intersection phenomenon in the cathodic and anodic CTs (Figure 4) obtained from the $Li_{1-\delta}CoO_2$ electrode at the potential drops and jumps, respectively, as indicated in Figure 1(b): the cathodic CTs at the potential steps of 4.2 to 4.1, 4.2 to 4.0, and 4.2 to 3.9 V *vs.* Li/Li^+ intersect those anodic CTs of 4.1 to 4.2, 4.0 to 4.2, and 3.9 to 4.2 V *vs.* Li/Li^+, respectively. Considering that the chemical diffusivity of lithium is almost invariable in the electrode potential range of 3.9-4.2 V *vs.* Li/Li^+,[75] these cathodic and anodic CTs should coincide exactly with each other under the assumption of *diffusion controlled* lithium transport (Eqs. (1) and (2)). A mutual intersection of the CTs is also observed in the case of $Li_{1+\delta}[Ti_{5/3}Li_{1/3}]O_4$ electrodes (Figure 5).

In particular, it is noticeable that all the anodic CTs in Figures 4 and 5 exceed in current value the corresponding cathodic CTs in the short time range, while the cathodic curves overwhelm the corresponding anodic curves in the long time range. This tendency of the CTs can be hardly explained even by such modified *diffusion control* concepts as non- contiguous reaction sites of the electrode[65,68,88,89] (*e.g.* electrochemically active surface with a fractal dimension D_f less than 2) and a surface film covering the electrode which has been regarded as the reason for the severe suppression of the Cottrell behavior.[79,87] This is because the electrochemical active area proved to be greater than the apparent surface area,[90] *i.e.*, $D_f > 2$, and the properties of the surface film were reported to be independent of electrode potential,[30,83] *viz.* the properties of surface film is nearly invariable during the CT experiments.

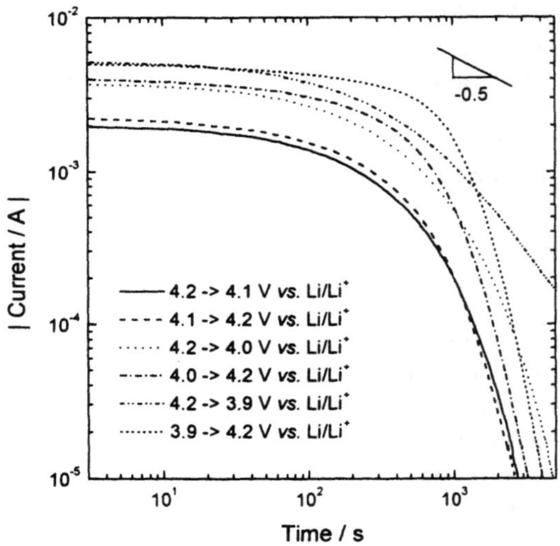

Figure 4. Logarithmic cathodic and anodic current transients obtained experimentally with a $Li_{1-\delta}CoO_2$ electrode. Reprinted from [40] with permission from Elsevier Science.

On the other hand, it is very interesting that the CTs in Figures 4 and 5 seem to reflect the concavity of the electrode potential curves within the single α phase region: the electrode potential curves (Figures 1b and c) first fall rapidly and finally slowly in the cathodic direction. In other words, the electrode potential first rises up slowly and finally rapidly in the anodic direction. The rate of change in electrode potential with lithium content in the cathodic and anodic directions equals qualitatively the rate of change of current with time in the cathodic and anodic directions, respectively.

3. (Quasi-)Current Plateau

When one chooses the potential steps to encounter the plateau potential, the CTs deviate severely from the simple shape of the CTs from single-phase electrodes. The logarithmic cathodic CTs (Figure 6a) obtained with the $Li_{1+\delta}[Ti_{5/3}Li_{1/3}]O_4$ electrode at potential drops from 1.700 V *vs.* Li/Li^+ to various lithium injection potentials below the plateau

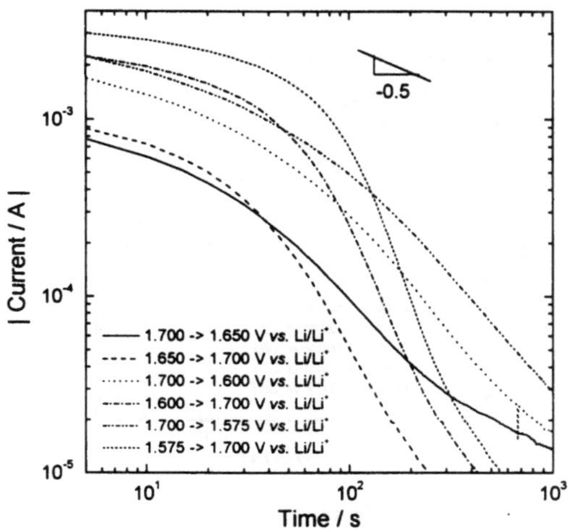

Figure 5. Logarithmic cathodic and anodic current transients obtained experimentally from the $Li_{1+\delta}[Ti_{5/3}Li_{1/3}]O_4$ electrode at the potential drop and potential jump, respectively, between applied potentials above the plateau potential. Reprinted from [44] with permission from Elsevier Science.

potential, show clearly the effect of phase transition, *i. e.,* three-stage CT behavior. In other words, the logarithmic current of Figure 6 decreases first slowly, then remains almost constant, and finally decays steeply. As far as we know, there is no literature available that gives any clear explanation for this three-stage behavior, specifically the second stage, *i. e.,* the current plateau.

The onset time (solid circle) / end time (open circle) of the current plateau are determined graphically as the times at which the tangent line of the first-/second-stage curve intersects that of the second- / third-stage curve as indicated in Figure 6(a). It is worthwhile to note that the values of cumulative charge corresponding to the onset (solid circle in Figure 6b) and to the end (open circle in Figure 6b) of the current plateau nearly equal the maximum solubility limit of lithium in lithium-poor α phase (solid triangle in Figure 6b) and the minimum solubility limit of lithium in lithium-rich β phase (open triangle in Figure 6b), respectively, irrespective of the magnitude of the potential drop. This

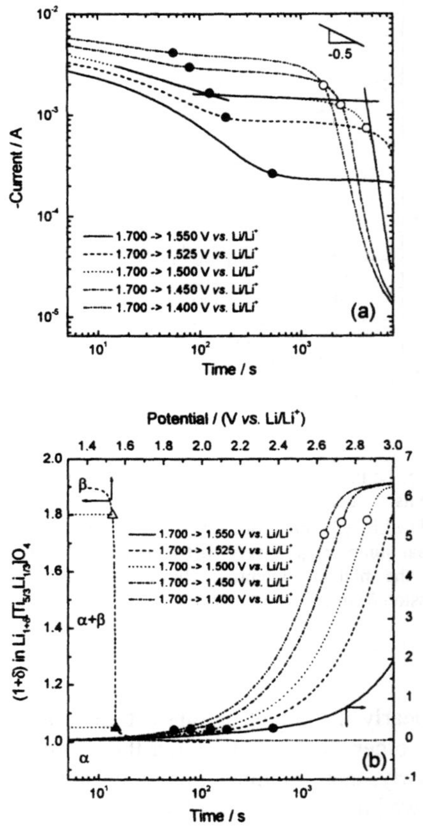

Figure 6. (a) Logarithmic cathodic current transients obtained experimentally with a $Li_{1+\delta}[Ti_{5/3}Li_{1/3}]O_4$ electrode at potential drops from 1.700 V *vs.* Li/Li^+ to various lithium injection potentials below the plateau potential. (b) *Cumulative charge vs. t* plots obtained from Figure 6(a), along with electrode potential curve of $Li_{1+\delta}[Ti_{5/3}Li_{1/3}]O_4$. Reprinted from [44] with permission from Elsevier Science.

implies that the phase transformation of α to β occurs during the *current plateau* interval in the CTs.

Keeping in mind that the *current plateau* implies a constant driving force for lithium intercalation, it is unlikely that the phase transformation of α to β is governed by *diffusion controlled* lithium transport where the rate of the phase boundary movement, *i. e.*, current, decreases significantly with time.[23-26]

Similarly to the case of $Li_{1+\delta}[Ti_{5/3}Li_{1/3}]O_4$, the (quasi-) current plateau appears in the CTs obtained with other intercalation compounds as well. For instance, the logarithmic cathodic CTs (Figure 7a) obtained with a $Li_\delta V_2O_5$ electrode show a single current plateau in the same manner as the curves obtained with $Li_{1+\delta}[Ti_{5/3}Li_{1/3}]O_4$, at the potential drops from 3.45 to 3.35, 3.45 to 3.30, and 3.45 to 3.25 V *vs.* Li/Li^+, which passed only through one potential plateau at 3.40 V *vs.* Li/Li^+. On the other hand, the CTs for the potential drops from 3.45 to 3.20 and 3.45 to 3.15 V *vs.* Li/Li^+, which involve the effect of two potential plateaus at 3.40 and 3.21 V *vs.* Li/Li^+, are characterized by double (quasi-) current plateaus.

In the case of $Li_\delta V_2O_5$, the current plateaus are not so clear as those in $Li_{1+\delta}[Ti_{5/3}Li_{1/3}]O_4$. The most plausible reason for this deviation from an ideal current plateau is the combined effect of the relatively narrow potential plateau region and the size distribution of the oxide particles, which makes the onset/end time of the phase transformation quite ambiguous.

In spite of the ambiguity in current plateau interval, it is possible to reasonably estimate the value of the plateau current as follows. Since the (quasi-) current plateau on a linear scale is roughly composed of two curves, *i. e.,* downward convex shaped curve and upward convex shaped curve in series, it is reasonable that the current values at the inflexion points of *current vs. time* curve on a linear scale, in particular the current values at the local maxima on the *d(current)/d(time) vs. time* curve, can be considered as the plateau currents. The resulting inflexion points were designated as half-filled circles in Figure 7(a). All the values of cumulative charge corresponding to the first and second inflexion points (half-filled circles in Figure 7b), fall just in the two-phase regions (between solid and open triangles in Figure 7b) of $\alpha+\varepsilon$ and $\varepsilon+\delta$, respectively. This is a strong indication that the occurrence of the (quasi-) current plateaus on CTs is caused by the phase transformation.

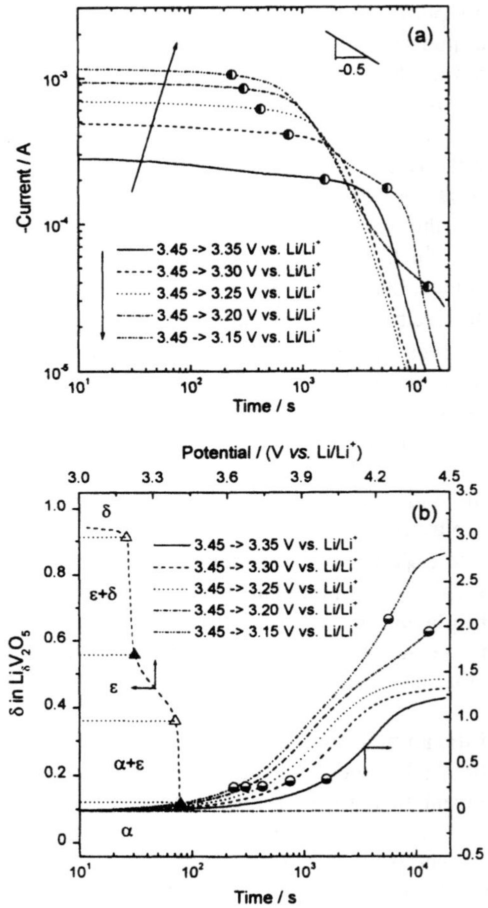

Figure 7. (a) Logarithmic cathodic current transients obtained experimentally with a $Li_\delta V_2O_5$ electrode at potential drops from 3.45 V *vs.* Li/Li^+ to various lithium injection potentials below the plateau potential. (b) *Cumulative charge vs. t* plots obtained from Figure 7(a), along with the electrode potential curve of $Li_\delta V_2O_5$. Reprinted from [18,44] with permission from Elsevier Science.

The inflexion points or (quasi-) current plateaus in Figures 2(a) and 3(a) of Section III.1 also prove to originate from the phase transition, from the comparison between the cumulative charges during the CT experiments and during the measurement of electrode potential. Furthermore, the origin of the shoulders and of the more than one local maxima in Figures 2(b) and 3(b) is, undoubtedly, the phase transition.

4. Depression of the initial current value

It is readily predicted from Eq. (1) that under the *diffusion controlled* lithium transport, the initial current level of CT should increase with rising potential step, *i. e.,* with rising difference between the initial and final equilibrium lithium contents. This is not true in the case of the anodic CTs obtained with the $Li_{1-\delta}NiO_2$ electrode. Figure 8(a) presents on a logarithmic scale the anodic CTs measured with the $Li_{1-\delta}NiO_2$ electrode by jumping the electrode potential from a value in the range 3.55 to 3.90 V *vs.* Li/Li^+ to a lithium extraction potential of 4.00 V *vs.* Li/Li^+.

All the anodic CTs hold the non-Cottrell character during the entire lithium deintercalation. Moreover, it should be pointed out that, in spite of the linear increase in the magnitude of the potential jump, the initial current level increases very slowly and eventually decreases at the large potential jump (> 0.25 V). In other words, the relationship between initial current level and potential jump takes the form of a parabola (Figure 8b), rather than a monotonically increasing curve.

Considering that the larger potential jump means the lower initial electrode potential, the parabolic relation in Figure 8(b) indicates that the deintercalation of lithium from the $Li_{1-\delta}NiO_2$ is seriously retarded as the initial electrode potential is lowered below *ca.* 3.75 V *vs.* Li/Li^+. This implies that some other factor along with the potential step considerably influences lithium deintercalation below 3.75 V *vs.* Li/Li^+.

IV. PHYSICAL ORIGIN OF THE CURRENT TRANSIENTS

1. Linear Relation Between Current and Electrode Potential

The most remarkable point which should be mentioned in relation to the physical aspects of the CTs obtained with transition metal oxides and

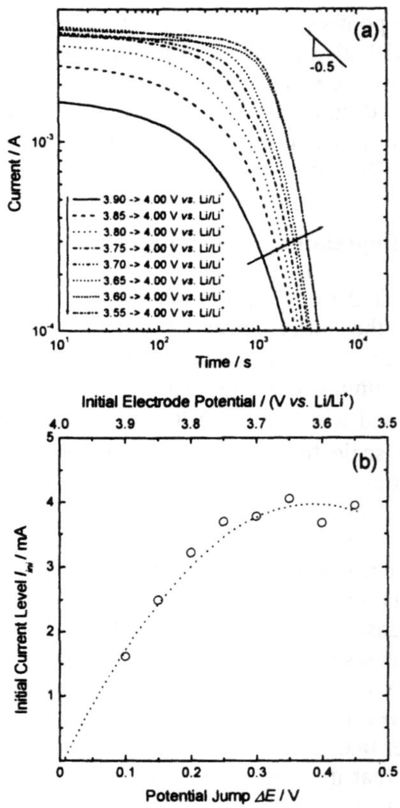

Figure 8. (a) Logarithmic anodic current transients obtained experimentally with a $Li_{1-\delta}NiO_2$ electrode at potential jumps of various initial values of electrode potential to 4.00 V *vs.* Li/Li^+. (b) plot of initial current level I_{ini} against potential jump ΔE obtained from Figure 8(a). Reprinted from [45] with permission from Elsevier Science.

graphite is first the linear relation between the initial current level I_{ini} and the potential step ΔE. In this review, the term *initial current level* I_{ini} is defined as the value of current at a time of 2-10 s during the CT experiments, rather than at the moment of application of the potential step. The reason for this will be discussed in the following Section IV.2.

The solid circle of Figures 9(a)-(e) denotes the initial current levels I_{ini} at various potential steps ΔE, calculated from the corresponding CTs of Figures 2(a), 4, 7(a), 6(a), and 3(a), respectively. Undoubtedly, all the I_{ini} *vs.* ΔE plots shows a linear relation. It should be mentioned that even the *diffusion controlled* CTs can exhibit this type of linear relationship in the case where the electrode potential curves vary linearly with lithium stoichiometry, i.e. $\Delta E \propto (c^1 - c^0)$. However, the linear relation between I_{ini} and ΔE is still valid for the electrodes (*e.g.* $Li_{1+\delta}[Ti_{5/3}Li_{1/3}]O_4$; $Li_{1-\delta}CoO_2$; $Li_\delta V_2O_5$) where the *electrode potential vs. lithium stoichiometry* curves deviate strongly from the linear relationship.

The above argument along with the evidences presented in Section III is indicative of other transport mechanisms than *diffusion controlled* lithium transport dominating during the CT experiments of transition metal oxides and graphite. Furthermore, the Ohmic relation between I_{ini} and ΔE indicates that the internal cell resistance plays a critical role in lithium intercalation/deintercalation. If this is the case, it is reasonable to say that the interfacial flux of lithium is determined by the difference between the applied potential E_{app} and the actual instantaneous electrode potential $E(t)$, divided by the internal cell resistance R_{cell}, and that therefore lithium hardly undergoes the *real potentiostatic* constraint at the electrode/electrolyte interface (see Section II). This condition is referred to as the *cell-impedance controlled* lithium transport.

Next let us examine the current-potential relation with increasing lithium intercalation/deintercalation time. For this purpose, the values of current at the times when various amounts of cathodic or anodic charge have passed (open symbols in Figures 9a-e) were obtained as a function of potential step. For example, in order to determine the open circle A in Figure 9(a), the *cumulative charge vs. time* plot is first obtained from the CT (—— in Figure 2a) for the $Li_{1-\delta}NiO_2$ electrode at the potential drop of 4.0 to 3.9 V *vs.* Li/Li^+ (ΔE = 0.1 V). Then, the time when the cumulative charge reaches 0.5 C is estimated. Finally the value of current on the CT is determined just at the time when the cumulative charge reaches 0.5 C.

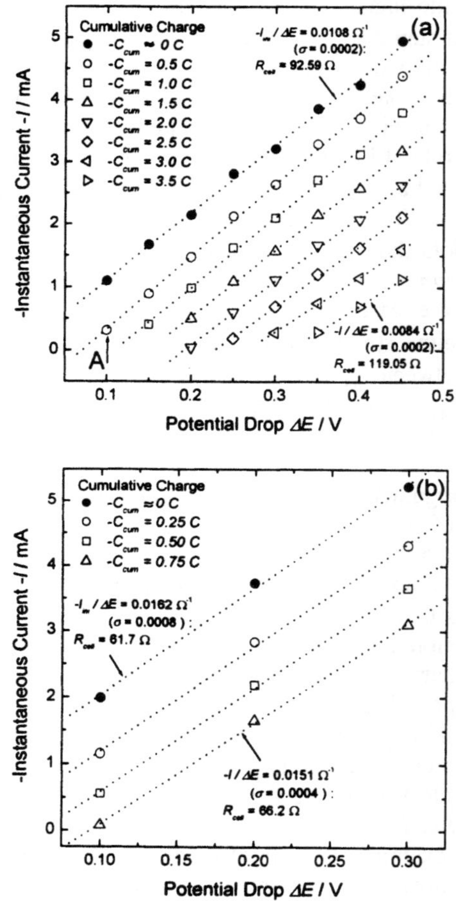

Figure 9. The dependence of instantaneous current level I on potential step ΔE at various values of cumulative charge, reproduced from the current transients of (a) $Li_{1-\delta}NiO_2$ (Figure 2a), (b) $Li_{1-\delta}CoO_2$ (Figure 4), (c) $Li_{\delta}V_2O_5$ (Figure 7a), (d) $Li_{1+\delta}[Ti_{5/3}Li_{1/3}]O_4$ (Figure 6a), and (e) graphite (Figure 3a).

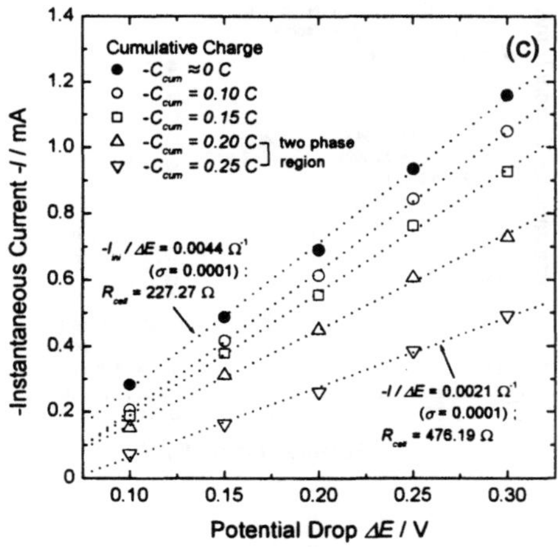

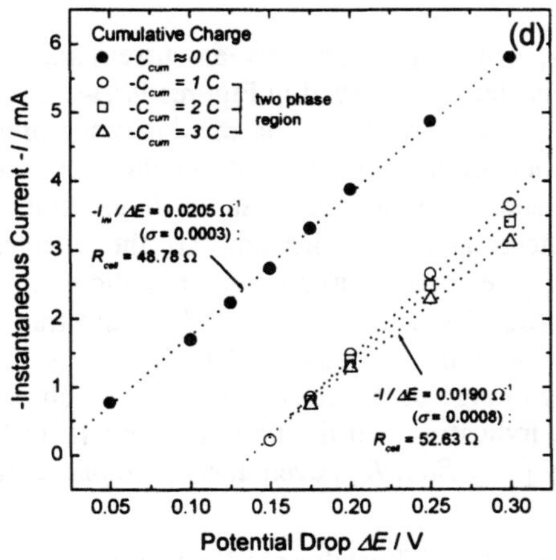

Figure 9. (continued)

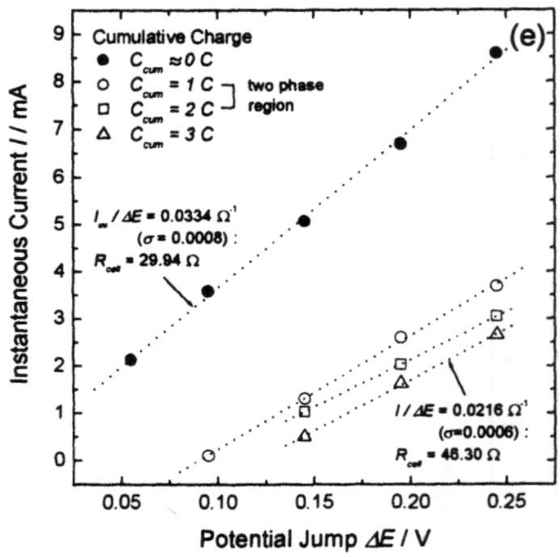

Figure 9. (continued)

Surprisingly, a linear relation between current and potential, i.e. an Ohmic relation, is clearly observed in Figures 9(a)-(e), regardless of the amount of the cumulative charge or the intercalation/deintercalation time elapsed. This implies that the interfacial flux is determined by $[E_{app}-E(t)]/R_{cell}$ and then it takes an infinite time to reach the equilibrium concentration of lithium at the electrode/electrolyte interface during the entire lithium intercalation/deintercalation, in contrast to the case of *diffusion controlled* lithium transport. It should be stressed here that the validity of Ohm's law is extended to the cumulative charge range of the coexistence of two phases, as shown in Figures 9(c)-(e), indicating that the lithium intercalation/deintercalation is also limited by $[E_{app}-E(t)]/R_{cell}$ even during the phase transition.

2. Comparison of Cell Resistances Determined by the Current Transient Technique and by Electrochemical Impedance Spectroscopy

For the sake of clarity of the above argument about the *cell-impedance controlled* lithium transport, it is very useful to determine experimentally the internal cell resistance as a function of the electrode potential using electrochemical impedance spectroscopy (EIS), and to compare it with the cell resistance determined by the CT technique. Figure 10(a) exhibits typical Nyquist plots obtained with $Li_{1-\delta}NiO_2$ at various electrode potentials in a 1M $LiClO_4$-PC solution. The impedance spectra consist of two separate arcs in the high and intermediate frequency ranges.

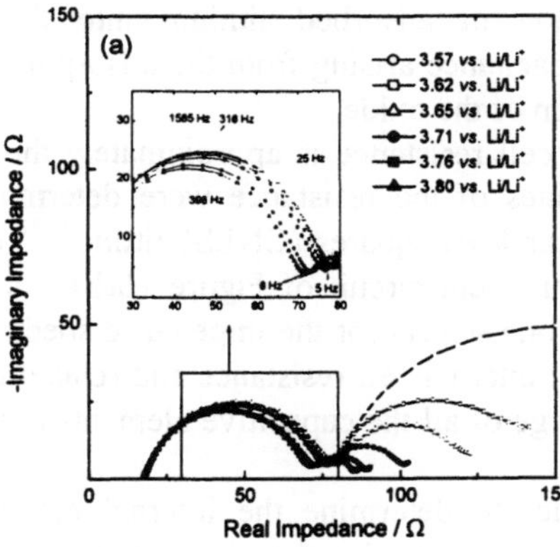

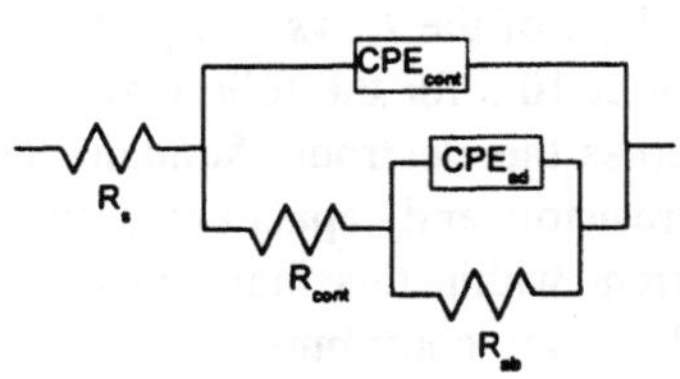

(b)

Figure 10. (a) Nyquist plots obtained with a $Li_{1-\delta}NiO_2$ electrode at various electrode potentials. (b) Equivalent circuit used for analysis of the electrochemical lithium intercalation reaction into $Li_{1-\delta}NiO_2$. $CPE_i = C_i(j\omega)^{\eta_i}$ (i = contact or adsorption), where C_i is a constant with the dimension $F \cdot s^{i-}$ ($0.5 < \eta_i < 1.0$). Reprinted from [45] © (2001), with permission from Elsevier Science.

Among the various models[30,91-94] proposed in order to explain two-arc behavior in the Nyquist plot for the insertion electrode, the simple equivalent circuit[94] depicted in Figure 10(b) was used to analyze the impedance spectra. Here, R_s is the sum of the electrolyte and conducting substrate resistances; R_{cont} is the resistance associated with the particle-to-particle contact among the oxide particles; C_{cont} is the contact capacitance due to the accumulation of charged species at the surface of the oxide particles; R_{ab} is the resistance associated with the absorption reaction of adsorbed lithium into the oxide and C_{ad} represents the capacitance arising from the adsorption of lithium in the near-surface region of the oxide.

The internal cell resistance is approximately the sum of R_s, R_{cont}, and R_{ab}. The values of the resistance were determined by using the complex non-linear least squares (CNLS) fitting[95-97] of the impedance spectra to the equivalent circuit of Figure 10(b). As a matter of fact, whatever models one selects for the impedance spectra in Figure 10(a), the values of total internal cell resistance and relaxation time necessary for charge/discharge of all the capacitive elements remain constant (see below).

Next, in order to determine the internal cell resistance R_{cell} at various electrode potentials by the CT technique, the initial current level I_{ini} is plotted against the potential drop ΔE at various initial electrode potentials E_{ini} (Figure 11a). R_{cell} is evaluated to be the reciprocal of the slope of the I_{ini} *vs*. ΔE plot. Here, the values of initial current were taken at 10's for the following reason: when the potential step is applied across the electrode, faradaic and non-faradaic currents pass through the resistors and capacitors, respectively. This means that the measured current within this time range includes the non-faradaic current, which does not contribute to lithium transport through the electrode.

Only after all the capacitors are fully charged/discharged, does the current pass through the resistors, and thus the measured current can be regarded as the purely diffusional current. In other words, from the moment the capacitors in Figure 10(b) are fully charged/discharged, the CTs can be effectively utilized to follow lithium transport through the electrode.

The charging/discharging time for capacitors in Figure 10(b) can be estimated from the relaxation time, τ, which is independent of the potential step. Figure 11(b) gives the variation in the simulated current

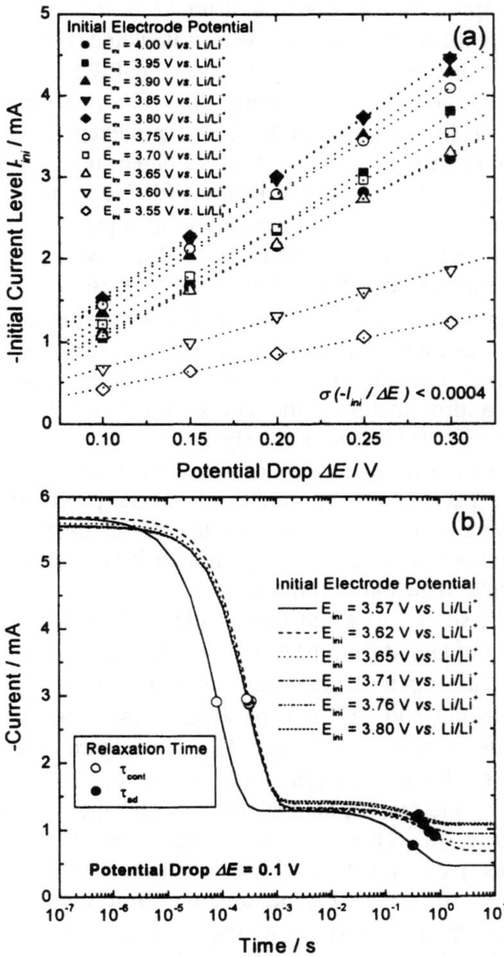

Figure 11. (a) Plots of initial current level I_{ini} against potential drop ΔE at various initial electrode potentials, reproduced from the cathodic current transients obtained from the $Li_{1-\delta}NiO_2$ electrode. (b) variation in current with time up to 10 s by application of a potential drop ΔE of 0.1 V at various initial electrode potentials, calculated using the simulation program with integrated circuit emphasis (SPICE) for the electric circuit of Figure 10(b), by taking the values of resistances and capacitances determined by CNLS fitting of the impedance spectra of Figure 10(a) to the electric circuit of Figure 10(b).

with time by application of potential drop ΔE of 0.1 V at various initial electrode potentials. The curves were calculated using a simulation program with integrated circuit emphasis (SPICE) for the electric circuit of Figure 10(b), by taking the values of resistances and capacitances determined by the CNLS fitting method.

As a matter of fact, it is very hard to determine exactly the values of capacitances in this case. That is, the CPE (constant phase element) exponent η_i in Figure 10(b) (i = contact or adsorption) was estimated to be less than one, indicating that the CPE_i deviate from a purely capacitive behavior and thus should be expressed in the form of $C_i(j\omega)^{\eta_i}$. Here, the CPE coefficient C_i, which has a dimension of $[F{\cdot}s^{\eta_i-1}]$, may no longer be thought as a purely capacitive component. Nevertheless, C_i is regarded as a purely capacitive component in this review. This approximation is thought to be reasonable since the semi-circles in Figure 10(a) are not severely depressed.

The relaxation time τ, especially τ_{ad}, was determined to be shorter than 1s, irrespective of the initial electrode potential. All of the capacitors were fully charged before the initial time of 10 s, which indicates that the time of 10 s is long enough for one to analyze the CTs in terms of bulk diffusion of lithium.

Figure 12 shows, as a function of electrode potential, the internal cell resistance obtained from EIS (solid circles), from the I_{ini} *vs.* ΔE plot of Figure 11(a) (open circles) and from the internal cell resistance obtained from the I *vs.* ΔE plot of Figure 9(a) (open triangles). It is noted that the cumulative charge C_{cum} in Figure 9(a) was simply converted into lithium content $(1 - \delta)$ in $Li_{1-\delta}NiO_2$, which again corresponds to electrode potential based upon the electrode potential curve of Figure 1(a). Thus, the value of the internal cell resistance was calculated easily as a function of electrode potential. It is noted that the *internal cell resistance vs. E* plots obtained from the EIS and from the I_{ini} *vs.* ΔE plot, take the form of a simple parabola in shape and value. The internal cell resistances estimated from the I *vs.* ΔE plot approximate satisfactorily those values obtained from both the EIS and from the I_{ini} *vs.* ΔE plot in the cumulative charge range investigated.

The physical model presented in Sections IV.1 and IV.2 tells us explicitly that the flux of lithium at the electrode/electrolyte interface is mainly limited by the internal cell resistance throughout the entire lithium intercalation/deintercalation.

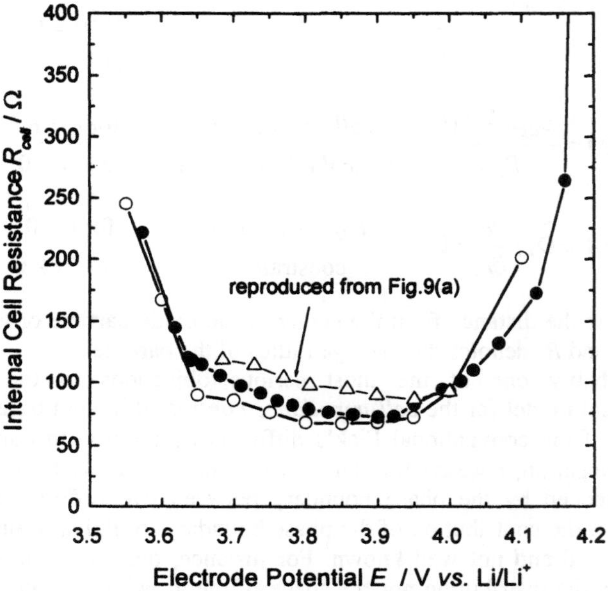

Figure 12. Internal cell resistances as a function of electrode potential, obtained from the I_{ini} *vs.* ΔE plot of Figure 11(a) (open circle) and the EIS (solid circle) along with internal cell resistance obtained from I *vs.* ΔE plot of Figure 9(a) (open triangle). Reprinted from [45] © (2001), with permission from Elsevier Science.

V. THEORETICAL DESCRIPTION OF CELL-IMPEDANCE CONTROLLED LITHIUM TRANSPORT

1. Governing Equation and Boundary Conditions

The governing equation for the *cell-impedance controlled* lithium transport is Fick's diffusion equation. The initial condition (I.C.) and the boundary conditions (B.C.) are given as

$$\text{I.C.:} \qquad c = c_{Li}^{0} \qquad \begin{matrix}\text{for } 0 \le r \le R^* \\ \text{at} \quad t = 0\end{matrix} \tag{7}$$

$$\text{B.C.}: I(t) = \frac{E_{app} - E(t)}{R_{cell}} \qquad \begin{matrix}\textit{cell impedance} \\ \textit{controlled} \text{ constraint}\end{matrix} \qquad \begin{matrix}\text{for } r = R^* \\ \text{at} \quad t > 0\end{matrix} \tag{8}$$

$$-zFA_{ea}\tilde{D}_{Li}\frac{\partial c}{\partial r} = 0 \qquad \begin{matrix}\textit{impermeability} \\ \text{constraint}\end{matrix} \qquad \begin{matrix}\text{for } r = 0 \\ \text{at} \quad t \ge 0\end{matrix} \tag{9}$$

where r is the distance from the center of the oxide particle or graphite particle and R^* denotes the average radius of the particle.

Probably one of the most serious objections to the above theoretical model for the *cell-impedance controlled* lithium transport is the use of the conventional Fick's diffusion equation even during the phase transition, because lithium diffusion inside the electrode should be influenced by the phase boundary between two different phases. However, the contribution of the phase boundary to lithium transport is complicated and not well known. For instance, one can neither know precisely the distribution nor the shape of the growing/shrinking phase during the phase transition.

So far as lithium intercalation/deintercalation into/from transition metal oxides and graphite proceeds under the *cell-impedance controlled* constraint Eq. (8), it is unlikely that the disturbance of lithium diffusion inside the electrode due to the presence of the phase boundary and the phase boundary movement causes any significant change in the CTs. It is likely predicted from Eq. (8) that unlike the case of the *diffusion controlled* phase transformation, the flux of lithium at the electrode/electrolyte interface under the *cell-impedance controlled* constraint is hardly dependent on the location of the phase boundary within the electrode.

For this reason, it is excessive to include the contribution of the phase boundary in the governing equations for the theoretical CT. Nevertheless, the effect of the phase boundary on the lithium diffusion inside the electrode should be still clarified.

2. Calculation Procedure of the Cell-Impedance Controlled Current Transients

The model parameters are determined in the following manner: the functional relations $E = f(1 - \delta)$ and $R_{cell} = f(E)$, which are incorporated into the B.C. of Eq. (8), are obtained by the polynomial regression analysis of the electrode potential curves (Figue 1a-e) and the R_{cell} *vs.* E curves (*e.g.* Figure 12) determined from the I_{ini} *vs.* ΔE plots, respectively. It should be borne in mind that $(1 - \delta)$ does not represent the average lithium content in the electrode, but the lithium content at the surface of the electrode. In other words, the electrode potential $E(t)$ in Eq. (8) is the potential at the electrode surface. Since the relation $E = f(1 - \delta)$ includes the information about the phase transition, the effect of the phase transition on the theoretical CT can be considered with the functional relation $E = f(1 - \delta)$, without using any of the intercalation isotherm.

The chemical diffusivity of lithium, $\tilde{D}_{Li}$, in the transition metal oxides and graphite is estimated by GITT and/or EIS. The average radius of particle R^* of transition metal oxides and graphite can be determined by *e.g.*, microscopic investigation. In this review, the diffusivity $\tilde{D}_{Li}$ and the average radius R^* were taken as 10^{-8}–10^{-9} cm^2 s^{-1} from the earlier investigations[76,94,98-100] and as 1-10 μm through SEM observation, respectively. The electrochemical active area A_{ea} was calculated from the radius R^* and the theoretical density of the particles considered, under the assumption that A_{ea} is identical to the total surface area of the electrode comprised of spherical particles.

As a matter of fact, the A_{ea} value adopted in this review was overestimated: Since the composite electrodes consist of active material, conducting material, and binder material, the area of active material exposed to the electrolyte solution is much lower than the geometrical area assumed as A_{ea} in this review. This is due primarily to intimate contacts among the particles of the active material and the conducting or binder materials, and to close contacts between the particles themselves.[77] Moreover, in the case of electrodes with layered structure, *e.g.* $Li_{1-\delta}NiO_2$, $Li_{1-\delta}CoO_2$, $Li_\delta V_2O_5$ or $Li_\delta C_6$, especially, the highly anisotropic layered property of the electrode material could additionally reduce the effective surface area for lithium intercalation, depending on the distribution of the specific crystallographic planes of the particles exposed to the electrolyte.

The uncertainty regarding the exact A_{ea} value, does not permit exact determination of the value of $\tilde{D}_{Li}$, because in many electrochemical analyses A_{ea} is one of the important parameters determining the $\tilde{D}_{Li}$ value. At the same time, the fact that $\tilde{D}_{Li}$ might be a strong function of electrode potential, makes it much more difficult to analyze CTs quantitatively.

However, in the case of *cell-impedance controlled* lithium transport, it is found that the variation of $\tilde{D}_{Li}$ and A_{ea} (or R^*) in the range of one order of magnitude does not affect markedly the CT in value and shape. The dependence of the shape and magnitude of the simulated CTs on the model parameters $\tilde{D}_{Li}$ and A_{ea} (or R^*) will be discussed in Section V.4.

The calculation procedure of CTs is as follows: at $t = 0$, the values of lithium content $(1 - \delta)$ over the electrode (or the electrode potential E) and of internal cell resistance R_{cell} are first initialized to be those values at initial electrode potential E_{ini}. And then when the infinitesimal time Δt has elapsed, *i.e.*, just after the potential step from E_{ini} to the final electrode potential $E_{fin} = E_{app}$ is applied, the flux at the electrode/electrolyte interface $r = R^*$ (or current I) is calculated by $I = (E - E_{app})/R_{cell}$ of Eq. (8). After that, the lithium concentration at the electrode surface is evaluated. Next, E, R_{cell} and $(1 - \delta)$ inside the electrode are calculated and re-evaluated after Δt. The above procedure is repeated until the desired time elapses. Finally, the theoretical CTs of I *vs.* t plot are obtained.

3. Theoretical Current Transients and their Comparison with Experimental Current Transients

Figures 13(a) and 14(a) depict, on a logarithmic scale, the theoretical CTs of $Li_{1-\delta}NiO_2$ and graphite electrodes, respectively, determined from the numerical solution of the diffusion equation for the conditions of Eqs. (7)-(9), by taking the values described in Section V.2. The theoretical CTs of Figures 13(a) and 14(a) compare fairly well with the corresponding experimental CTs of Figures 2(a) and 3(a), respectively, in value and shape.

First, there is no Cottrell region during the entire lithium intercalation/deintercalation, which is substantiated by no region with constant value of $I(t)\cdot t^{1/2}$ in the $I(t)\cdot t^{1/2}$ *vs.* ln t plots of Figures 13(b) and 14(b). Moreover, Figures 13(b) and 14(b) show shoulders and more than one local maxima at the potential drops where the electrode undergoes a phase transition, in the same manner as the experimental $I(t)\cdot t^{1/2}$ *vs.* ln t plots of Figures 2(b) and 3(b), respectively.

The intersection of cathodic and anodic CTs from $Li_{1-\delta}CoO_2$ (Figure 4) and $Li_{1+\delta}[Ti_{5/3}Li_{1/3}]O_4$ (Figure 5) is also clearly observed in the *cell-impedance controlled* CTs of Figures. 15 and 16, respectively.

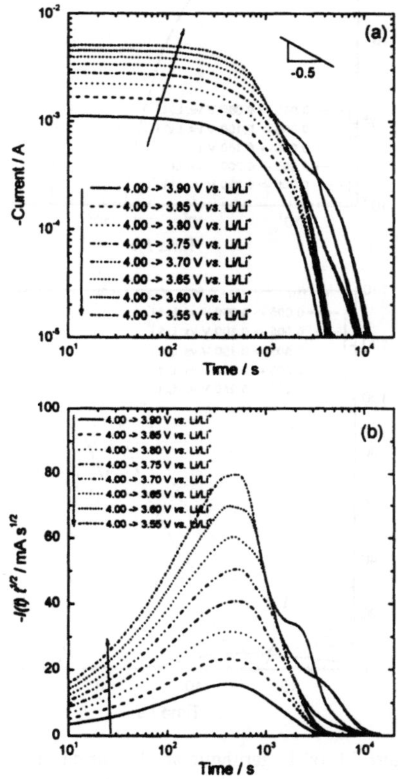

Figure 13. (a) Logarithmic cathodic current transients of a $Li_{1-\delta}NiO_2$ electrode, theoretically determined by means of numerical analysis based upon the *cell-impedance controlled* lithium intercalation and (b) $I(t)\cdot t^{1/2}$ *vs. ln t* plots obtained from Figure 13(a). Reprinted from [45] © (2001), with permission from Elsevier Science.

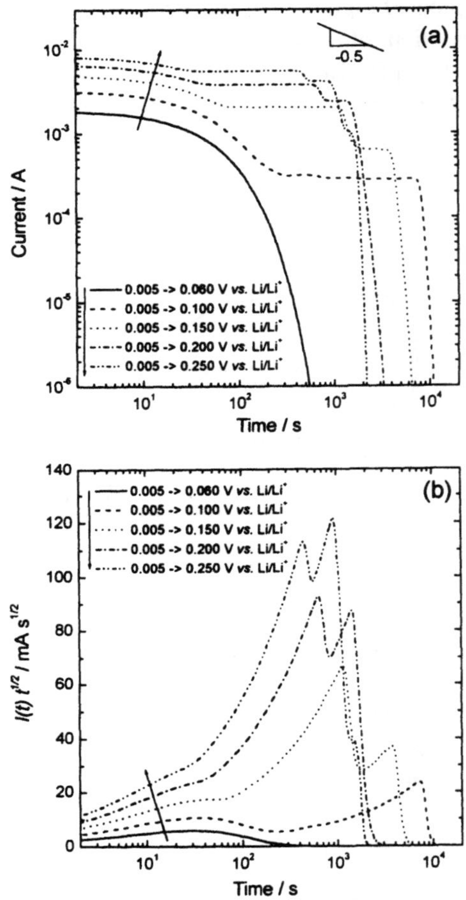

Figure 14. (a) Logarithmic anodic current transients of the graphite electrode, determined theoretically by means of numerical analysis based upon the *cell-impedance controlled* lithium deintercalation. (b) $I(t)\cdot t^{1/2}$ *vs.* ln t plots reproduced from Figure 14(a).

Now the intersection behavior in Figures. 4 and 5 can be interpreted, based upon the *cell-impedance control* concept, in the following way: the electrode potential curves in the potential ranges of 3.9 to 4.2 V *vs.* Li/Li^+ in Figure 1(b) and 1.575 to 1.700 V *vs.* Li/Li^+ in Figure 1(d) are characterized by an upward concave shape. Bearing in mind that both the cathodic and anodic CTs are given in value as B.C. by the quotient of the difference between applied potential and instantaneous electrode potential divided by the internal cell resistance, the rapidly decreasing tendency of the electrode potential in the vicinity of 4.2 V *vs.* Li/Li^+ (Figure 1b) and 1.700 V *vs.* Li/Li^+ (Figure 1d) determines the rapidly decreasing cathodic CTs in value and shape before the time of intersection of both CTs.

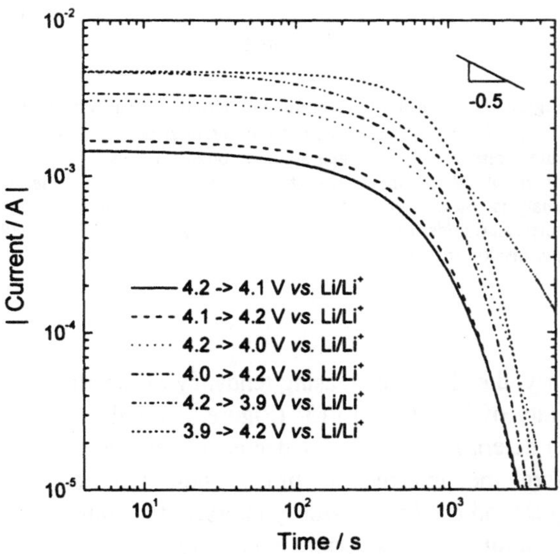

Figure 15. Logarithmic cathodic and anodic current transients of a $Li_{1-\delta}CoO_2$ electrode, determined theoretically by means of numerical analysis based upon the *cell-impedance controlled* lithium intercalation/deintercalation. Reprinted from [40] © (1999) with permission from Elsevier Science.

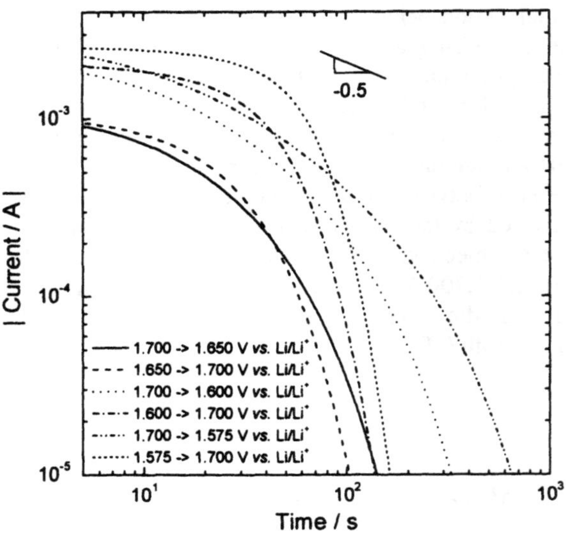

Figure 16. Logarithmic cathodic and anodic current transients of a $Li_{1+\delta}[Ti_{5/3}Li_{1/3}]O_4$ electrode at the potential drop and potential jump, respectively, between applied potentials above the plateau potential, determined theoretically by means of numerical analysis based upon the *cell-impedance controlled* lithium intercalation/deintercalation. Reprinted from [44] © (2001), with permission from Elsevier Science.

Similarly, the slowly increasing tendency of the electrode potential in the vicinity of 3.9 V *vs.* Li/Li$^+$ (Figure 1b) and 1.575 V *vs.* Li/Li$^+$ (Figure 1d) determines the slowly decreasing anodic CTs in value and shape before the intersection time. Thus, the relatively rapidly descending cathodic CTs necessarily intersect the comparatively slowly descending anodic CTs at a certain point of time.

The theoretical CTs from $Li_{1+\delta}[Ti_{5/3}Li_{1/3}]O_4$ (Figure 17) and $Li_\delta V_2O_5$ (Figure 18) feature a clear current plateau during the phase transition when the potential steps encounter the plateau potential, as the corresponding experimental CTs of Figures 6(a) and 7(a) do. The same current plateaus during the phase transition are observed in the *cell-impedance controlled* CTs from $Li_{1-\delta}NiO_2$ (Figure 13a) and

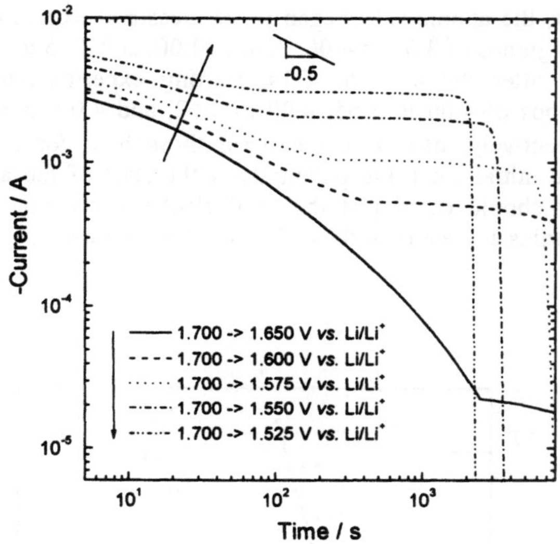

Figure 17. Logarithmic cathodic current transients of the $Li_{1+\delta}[Ti_{5/3}Li_{1/3}]O_4$ electrode at the potential drops from 1.700 V *vs.* Li/Li^+ to various lithium injection potentials below the plateau potential, determined theoretically by means of numerical analysis based upon the *cell-impedance controlled* lithium intercalation. Reprinted from [44] © (2001), with permission from Elsevier Science.

graphite (Figure 14a). The occurrence of the current plateau in the *cell-impedance controlled* CTs is due to the nearly invariant electrode potential E or potential difference (E_{app}-E) during the phase transition, *i.e.*, constant driving force for lithium intercalation/ deintercalation during the phase transition. Nevertheless, it is noted that the current plateaus in the theoretical CTs appear more clearly as compared with those in the experimental CTs. The most propable reason for this slight discrepancy is the size distribution of the oxide and graphite particles.

Finally, Figure 19 presents the theoretical anodic CTs obtained for the $Li_{1-\delta}NiO_2$ electrode. The CTs agree fairly well with the anodic CTs measured experimentally (Figure 8a). Nonetheless, it should be noted first, that the theoretical CTs are much more depressed at the initial potentials below 3.65 V *vs.* Li/Li^+ as compared to the experimental

CTs. In fact, the anodic CTs began to be measured experimentally at the potential jumps of 3.65 to 4.00, 3.60 to 4.00, and 3.55 to 4.00 V *vs.* Li/Li$^+$, just after the cathodic CTs, starting experimentally at the potential drops of 4.00 to 3.65, 4.00 to 3.60, and 4.00 to 3.55 V *vs.* Li/Li$^+$, respectively, had been measured throughout for 2×10^4 s. Namely, the initial electrode potentials at the start of the anodic CT experiments should equal just the final electrode potentials that the electrode attains at the end of the cathodic CT experiments.

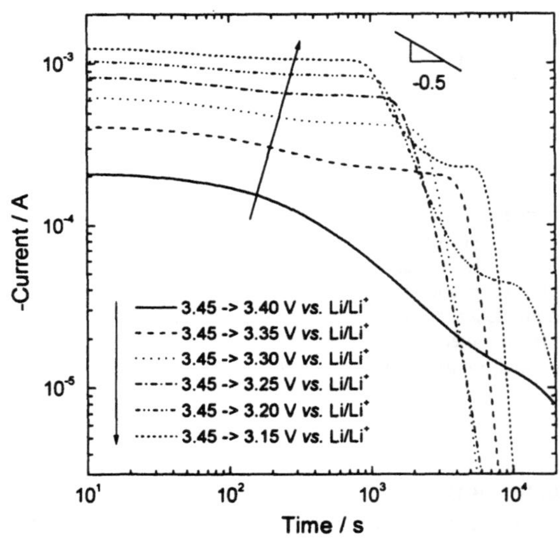

Figure 18. Logarithmic cathodic current transients of the $Li_\delta V_2O_5$ electrode at the potential drops from 3.45 V *vs.* Li/Li$^+$ to various lithium injection potentials below the plateau potential, determined theoretically by means of numerical analysis based upon the *cell-impedance controlled* lithium intercalation. Reprinted from [18] © (2001) and [44] © (2001), with permission from Elsevier Science.

The new equilibrium is so strongly disturbed that the final electrode potentials are never really attained in the ranges below 3.65 V *vs.* Li/Li^+ even if one holds the final electrode potentials throughout for 2×10^4 s, prior to lithium deintercalation. For the theoretical calculation of the anodic CTs, we actually inserted in the numerical value a nominal potential jump (E_{app}-E) slightly higher than what the real value was and at the same time a nominal internal cell resistance R_{cell} much higher than what the real value was. Since such a rise in R_{cell} outweighs the rise in (E_{app}-E), the instantaneous currents are practically calculated to be lower than those currents determined experimentally.

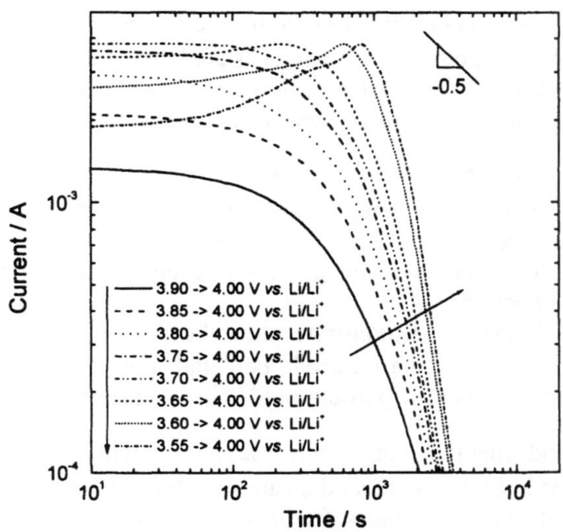

Figure 19. Logarithmic anodic current transients of the $Li_{1-\delta}NiO_2$ electrode at the potential jumps of various initial values of electrode potential to 4.00 V *vs.* Li/Li^+, determined theoretically by means of numerical analysis based upon the *cell-impedance controlled* lithium deintercalation. Reprinted from [45] ©(2001), with permission from Elsevier Science.

Moreover, the calculated current increases rather than decreases with time to *ca.* 200, 600, and 800 s at the potential jumps of 3.65 to 4.00, 3.60 to 4.00, and 3.55 to 4.00 V *vs.* Li/Li$^+$, respectively. This current increase in the *cell-impedance controlled* anodic CTs of $Li_{1-\delta}NiO_2$ is due to the fact that the reduction in the potential difference $(E_{app}-E)$ is much exceeded by the fall in internal cell resistance R_{cell} during lithium deintercalation, to enhance the driving force for lithium transport. This is readily predicted in the R_{cell} *vs.* E plot of Figure 12.

4. Some Model Parameters Affecting the Shape and Magnitude of the Cell-Impedance Controlled Current Transients

As mentioned in Section V.2, in this review the electrochemical active area A_{ea} was overestimated and the chemical diffusivity of lithium $\tilde{D}_{Li}$ was also oversimplified in spite of the dependence of $\tilde{D}_{Li}$ on the electrode potential. It can be stated that the *cell-impedance control* model can not be validated as a tool for analyzing the CTs from transition metal oxides and carbonaceous materials, until the effect of A_{ea} and $\tilde{D}_{Li}$ on the *cell-impedance controlled* CTs has been clearly established. At the same time, it should be verified that the variation of internal cell resistance R_{cell} affects crucially the *cell-impedance controlled* CTs, to complete the discussion about the current depression presented in Section V.3.

Figures 20(a)-(b) demonstrate the theoretical cathodic CTs of $Li_{1-\delta}NiO_2$ by dropping the electrode potential from 4.00 to 3.55 V *vs.* Li/Li$^+$ at various average particle radii R^* (*i.e.*, electrochemical active area A_{ea}) and chemical diffusivities $\tilde{D}_{Li}$, respectively. One can see from Figures 20(a) and (b) that the current values of the theoretical CTs are varied at most by about a factor of two as the parameters R^* and $\tilde{D}_{Li}$ are varied by one order of magnitude.

Now, let us consider the *cell-impedance controlled* CTs under the assumption that the value of R_{cell} is constant, to compare the CTs under the condition that the value of R_{cell} is varied with E. Figure 21(a) shows the theoretical cathodic CTs of $Li_{1-\delta}NiO_2$ at the potential drop from 4.00 to 3.55 V *vs.* Li/Li$^+$, assuming that internal cell resistance R_{cell} remains constant (solid line) at the R_{cell} value at 4.00 V *vs.* Li/Li$^+$ over the lithium intercalation time and that R_{cell} is varied with E (dotted line).

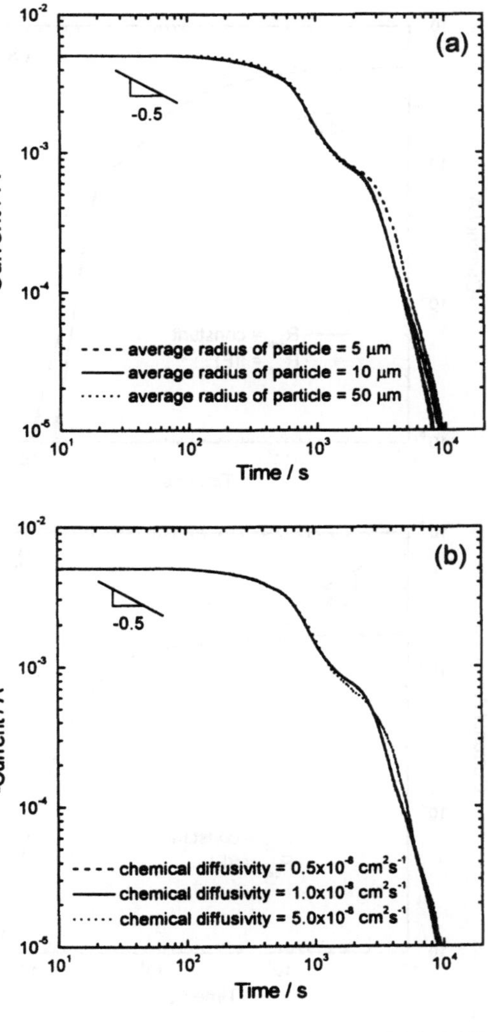

Figure 20. Logarithmic current transients of a $Li_{1-\delta}NiO_2$ electrode at the potential drop of 4.00 to 3.55 V *vs.* Li/Li^+ at (a) various average particle radii R^* and (b) various chemical diffusivities $\tilde{D}_{Li}$ of lithium in the electrode, respectively, determined theoretically by means of numerical analysis based upon the *cell-impedance controlled* lithium intercalation.

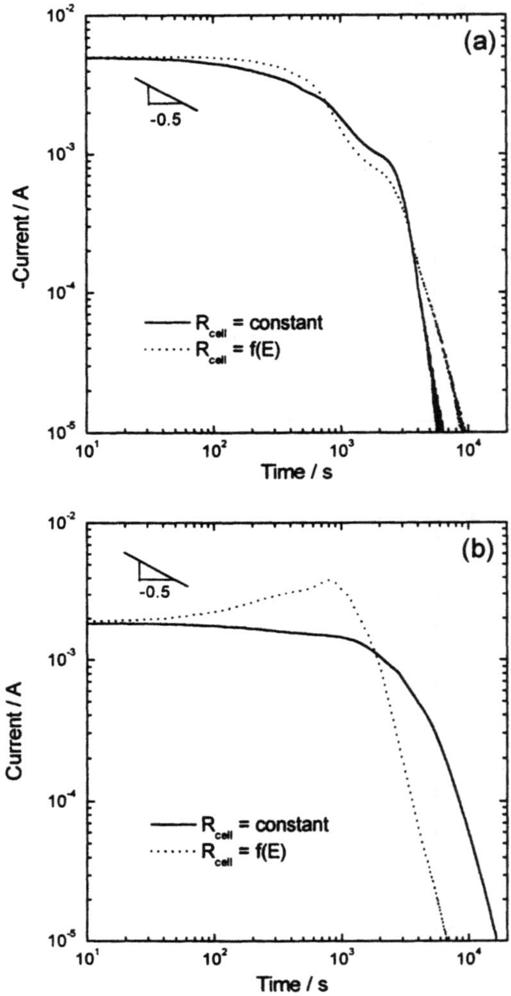

Figure 21. (a) Logarithmic cathodic current transients. (b) Logarithmic anodic current transients of a $Li_{1-\delta}NiO_2$ electrode, determined theoretically by means of numerical analysis based upon the *cell-impedance controlled* lithium intercalation, where the solid line and the dotted line indicate the current transients calculated theoretically under the two conditions that R_{cell} is constant and that R_{cell} is varied with E, respectively.

The cathodic CT calculated theoretically under the condition that R_{cell} is constant, departs strongly from that theoretically calculated under the condition that R_{cell} is varied with E.

Figure 21(b) presents the theoretical anodic CTs of $Li_{1-\delta}NiO_2$ at the potential jump from 3.55 to 4.00 V *vs.* Li/Li^+, assuming that the internal cell resistance R_{cell} remains constant (solid line) at the R_{cell} value at 3.55 V *vs.* Li/Li^+ over the lithium deintercalation time and that R_{cell} is varied with E (dotted line). The anodic CT under constant R_{cell} value deviates strongly from the CT under the condition of $R_{cell} = f(E)$, without any increase in current with time in the short time range.

The above results of Figures 20 and 21 indicate clearly that both the instantaneous electrode potential, E, and the internal cell resistance R_{cell} play major roles in the flux of lithium at the electrode/electrolyte interface in the *cell-impedance controlled* lithium transport model, characterized by the B.C. of Eq. (8).

VI. CONCLUDING REMARKS

The present chapter summarized first briefly the conventional and modified *diffusion controlled* lithium transport models for explaining the CTs of intercalation compounds. Second, this article presented the anomalous features of the experimental CTs for various transitional metal oxides and graphite, and then discussed the physical origin of the experimental CTs in terms of internal cell resistance. Finally, this article gave the theoretical CTs under the assumption that internal cell resistance plays a major role in lithium intercalation/deintercalation, to compare those with experimental CTs. On the basis of the *cell-impedance controlled* lithium transport concept, one can analyze quantitatively the CTs for transition metal oxides and graphite, and readily understand the anomalous features which one can hardly interprete under the *diffusion control* concept.

On the other hand, from the results of CTs, it is likely that the rate of the lithium insertion/desertion should be limited by the internal cell resistance during potential scanning as well, not by lithium diffusion in the electrode. In fact, it has been reported by a few authors[59,101-103] that the linear sweep/cyclic voltammograms from the intercalation compounds were difficult to analyze under the *diffusion control* concept. In addition, the voltammetric results in a number of references[9,35,104-106] seem to show abnormal trends without any explanation offered.

The atypical features in the linear sweep/cyclic voltammograms include the asymmetry in shape between anodic and cathodic current peaks, and the abnormal dependence of peak current on potential scan rate. The *cell-impedance controlled* lithium transport concept is expected to provide a reasonable explanation of these atypical voltammetric responses, and give us a new insight into the lithium transport behavior during potential scanning.[103]

ACKNOWLEDGEMENTS

The receipt of research grant under the internal research program Technological Development of High Performance Lithium Battery from the Korea Advanced Institute of Science and Technology (KAIST) 2000/2002 is gratefully acknowledged. This work was partly supported by the Brain Korea 21 project.

REFERENCES

[1]D. M. MacArther, *J. Electrochem. Soc.* **117** (1970) 729.
[2]C. J. Wen, B. A. Boukamp, R. A. Huggins, and W. Weppner, *J. Electrochem. Soc.* **126** (1979) 2258.
[3]A. J. Bard and L. R. Faulkner, *Electrochemical Methods: Fundamentals and Applications*, Wiley, New York, 1980, p. 142.
[4]J. Barker, D. Baldwin, and D. C. Bott, *Synth. Met.* **28** (1989) D127.
[5]M. Ohashi, W. Gloffke, and M. S. Whittingham, *Solid State Ionics* **57** (1992) 183.
[6]Q. Xu and G. Wan, *J. Power Sources* **41** (1993) 315.
[7]J. Barker, R. Pyneburg, and R. Koksbang, *J. Power Sources* **52** (1994) 185.
[8]H. Lindström, S. Södergren, A. Solbrand, H. Rensmo, J. Hjelm, A. Hagfeldt, and S.-E. Lindquist, *J. Phys. Chem. B* **101** (1997) 7710.
[9]H. Sato, D. Takahashi, T. Nishina, and I. Uchida, *J. Power Sources* **68** (1997) 540.
[10]S. L. de Albuquerque Maranhão, and R. M. Toressi, *Electrochim. Acta* **43** (1998) 257.
[11]J. M. McGraw, C. S. Bahn, P. A. Pariaal, J. D. Perkins, D. W. Reasey, and D.S. Ginley, *Electrochim. Acta* **45** (1999) 187.
[12]M. Nishizawa, H. Koshika, R. Hashitani, T. Itoh, T. Abe, and I. Uchida, *J. Phys. Chem. B* **103** (1999) 4933.
[13]M. Martos, J. Morales, and L. Sánchez, *Electrochim. Acta* **46** (2000) 83.
[14]Z.-W. Fu and Q.-Z. Qin, *J. Electrochem. Soc.* **147** (2000) 4610.
[15]S. Waki, K. Dokko, T. Itoh, M. Nishizawa, T. Abe, and I. Uchida, *J. Solid State Electrochem.* **4** (2000) 205.

[16]J. Isidorsson, M. Strømme, R. Gahlin, G. A. Niklasson, and C. G. Granqvist, *Solid State Commun.* **99** (1996) 109.
[17]M. Strømme Mattsson, G. A. Niklasson, and C. G. Granqvist, *Phys. Rev. B* **54** (1996) 2968.
[18]S.-I. Pyun, M.-H. Lee, and H.-C. Shin, *J. Power Sources* **97-98** (2001) 473.
[19]Y.-M. Choi, S.-I. Pyun, and J. M. Paulsen, *Electrochim. Acta* **44** (1998) 623.
[20]J. S. Bae and S.-I. Pyun, *Solid State Ionics* **90** (1996) 251.
[21]S.-I. Pyun and Y.-M. Choi, *J. Power Sources* **68** (1997) 524.
[22]S.-I. Pyun and Y.-G. Ryu, *J. Power Sources* **70** (1998) 34.
[23]H.-C. Shin and S.-I. Pyun, *Electrochim. Acta* **44** (1999) 2235.
[24]A. Funabiki, M. Inaba, T. Abe, and Z. Ogumi, *Carbon* **37** (1999) 1591.
[25]A. Funabiki, M. Inaba, T. Abe, and Z. Ogumi, *Electrochim. Acta* **45** (1999) 865.
[26]A. Funabiki, M. Inaba, T. Abe, and Z. Ogumi, *J. Electrochem. Soc.* **146** (1999) 2443.
[27]M. D. Levi and D. Aurbach, *J. Electroanal. Chem.* **421** (1997) 79.
[28]M. D. Levi and D. Aurbach, *J. Phys. Chem. B* **101** (1997) 4630.
[29]M. D. Levi and D. Aurbach, *Electrochim. Acta* **45** (1999) 167.
[30]M. D. Levi, G. Salitra, B. Markovsky, H. Teller, D. Aurbach, U. Heider, and L. Heider, *J. Electrochem. Soc.* **146** (1999) 1279.
[31]D. Zhang, B. N. Popov, and R. E. White, *J. Electrochem. Soc.* **147** (2000) 831.
[32]M. Noel and V. Rajendran, *J. Power Sources* **88** (2000) 243.
[33]A.-K. Hjelm, G. Lindbergh, and A. Lundqvist, *J. Electroanal. Chem.* **506** (2001) 82.
[34]A.-K. Hjelm, G. Lindbergh, and A. Lundqvist, *J. Electroanal. Chem.* **509** (2001) 139.
[35]P. Krtil and D. Fattakhova, *J. Electrochem. Soc.* **148** (2001) A1045.
[36]M. S. Mattsson and G. A. Niklasson, *J. Appl. Phys.* **85** (1999) 8199.
[37]M. S. Mattsson, J. Isidorsson, and T.Lindström, *J. Electrochem. Soc.* **146** (1999) 2613.
[38]M. S. Mattsson, *Solid State Ionics* **131** (2000) 261.
[39]Z.-W Fu and Q.-Z. Qin, *J. Phys. Chem. B* **104** (2000) 5505.
[40]H.-C. Shin and S.-I Pyun, *Electrochim. Acta* **45** (1999) 489.
[41]S.-I. Pyun and H.-C. Shin, *Mol. Cryst. & Liq. Cryst.* **341** (2000) 147.
[42]S.-I. Pyun and H.-C. Shin, *J. Power Sources* **97-98** (2001) 277.
[43]J.-Y. Go, S.-I. Pyun, and H.-C. Shin, *J. Electroanal. Chem.* **527** (2002) 93.
[44]H.-C. Shin, S.-I. Pyun, S.-W. Kim, and M.-H. Lee, *Electrochim. Acta* **46** (2001) 897.
[45]M.-H. Lee, S.-I. Pyun, and H.-C. Shin, *Solid State Ionics* **140** (2001) 35.
[46]S.-I. Pyun and S.-W. Kim, *J. Power Sources* **97-98** (2001) 371.
[47]S.-I. Pyun, S.-W. Kim, and J.-M. Ko, *J. New Mater. for Electrochem. Systems* **5** (2002) 135.
[48]S.-W. Kim and S.-I. Pyun, *Electrochim. Acta* 47 (2002) 2843.
[49]S.-I. Pyun, S.-B. Lee, and W.-Y. Chang, *J. New Mater. for Electrochem. Systems* in press.
[50]J. Nagai and T. Kamimori, *Jap. J. Appl. Phys.* **22** (1983) 681.
[51]A. Honders, E. W. A. Young, A. . van Heeren, J. H. W. de Wit, and G. H. J. Broers, *Solid State Ionics* **9&10** (1983) 375.
[52]A. Honders and G. H. J. Broers, *Solid State Ionics* **15** (1985) 173.
[53]A. Honders, J. M. der Kinderen, A. H. van Heeren, J. H. W. de Wit, and G. H. J. Broers, *Solid State Ionics* **15** (1985) 265.
[54]J. Nagai, T. Kamimori, and M. Mizuhashi, *Solar Energy Mater.* **13** (1986) 279.
[55]R. Tossici, R. Marassi, M. Berrettoni, S. Stizza, and G. Pistoia, *Solid State Ionics* **57** (1992) 227.
[56]T. Kudo, Y. Aikawa, Y. M. Li, and A. Kishimoto, *Solid State Ionics* **62** (1993) 99.

[57]Y. M. Li, Y. Aikawa, A. Kishimoto, and T. Kudo, *Electrochim. Acta* **39** (1994) 807.
[58]Y. M. Li and T. Kudo, *Solid State Ionics* **86-88** (1996) 1295.
[59]K. A. Striebel, C. Z. Deng, S. J. Wen, and E. J. Cairns, *J. Electrochem. Soc.* **143** (1996) 1821.
[60]T. Uchida, Y. Morikawa, H. Ikuta, M. Wakihara, and K. Suzuki, *J. Electrochem. Soc.* **143** (1996) 2606.
[61]Z.-I. Takehara, K. Kanamura, S.-I. Yonezawa, and T. Hanawa, *J. Power Sources* **25** (1989) 277.
[62]K. Kanamura, S. Yonezawa, S. Yoshioka, and Z.-I. Takehara, *J. Phys. Chem.* **95** (1991) 7939.
[63]B. B. Mandelbrot, *The Fractal Geometry of Nature*, Freeman, New York, 1980, p. 1.
[64]L. Nyikos and T. Pajkossy, *Electrochim. Acta* **31** (1986) 1347.
[65]L. Nyikos and T. Pajkossy, *Electrochim. Acta* **33** (1989) 171.
[66]A. Imre, T. Pajkossy, and L. Nyikos, *Acta. Metall. Mater.* **40** (1992) 1819.
[67]T. Pajkossy, A. P. Borosy, A. Imre, S. A. Martemyanov, G. Nagy, R. Schiller, and L. Nyikos, *J. Electroanal. Chem.* **366** (1994) 69.
[68]Y. Dassas and P. Duby, *J. Electrochem. Soc.* **142** (1995) 4175.
[69]M. M. Gómez, L. Vázquez, R. C. Salvarezza, J. M. Vara, and A. J. Arvia, *J. Electroanal. Chem.* **317** (1991) 125.
[70]P. Ocon, P. Herrasti, L. Vázquez, R. C. Salvarezza, J. M. Vara, and A. J. Arvia, *J. Electroanal. Chem.* **319** (1991) 101.
[71]W. Jost, *Diffusion in Solids, Liquids, Gases*, Chapter 1, Academic Press, New York, 1960, p. 61.
[72]M. B. Armand, "Intercalation Electrodes" in: *Materials for Advanced Batteries*, D.W. Murphy, J. Broadhead, and B.C.H. Steele, Eds., Plenum Press, New York, 1980, p. 145.
[73]K. P. Rajan and V. D. Neff, *J. Phys. Chem.* **86** (1982) 4361.
[74]S.-I. Pyun, M.-H. Lee, H.-C. Shin, and S.-W. Kim, *Mol. Cryst. & Liq. Cryst.* **341** (2000) 163.
[75]Y.-M. Choi, S.-I. Pyun, J.-S. Bae, and S.-I. Moon, *J. Power Sources* **56** (1995) 25.
[76]Y.-M. Choi, S.-I. Pyun, and S.-I. Moon, *Solid State Ionics* **89** (1996) 43.
[77]S.-I. Pyun and J.-S. Bae, *Electrochim. Acta* **41** (1996) 919.
[78]S.-I. Pyun, S.-W. Kim, and H.-C. Shin, *J. Power Sources* **81-82** (1999) 248.
[79]M. D. Levi, E. A. Levi, and D. Aurbach, *J. Electroanal. Chem.* **421** (1997) 89.
[80]M. D. Levi and D. Aurbach, *J. Phys. Chem. B* **101** (1997) 4641.
[81]B. Markovsky, M. D. Levi, and D. Aurbach, *Electrochim. Acta* **43** (1998) 2287.
[82]D. Aurbach, M. D. Levi, O. Lev, J. Gun, and L. Rabinovich, *J. Appl. Electrochem.* **28** (1998) 1051.
[83]D. Aurbach, M. D. Levi, E. Levi, H. Teller, B. Markovsky, and G. Salitra, *J. Electrochem. Soc.* **145** (1998) 3024.
[84]W. Yang, G. Zhang, S. Lu, J. Xie, and Q. Liu, *Solid State Ionics* **121** (1999) 85.
[85]Z. Lu, M. D. Levi, G. Salitra, Y. Gofer, E. Levi, and D. Aurbach, *J. Electroanal. Chem.* **491** (2000) 211.
[86]F. Varsano, F. Decker, E. Masetti, and F. Croce, *Electrochim. Acta* **46** (2001) 2069.
[87]M. D. Levi, Z. Lu, and D. Aurbach, *Solid State Ionics* **143** (2001) 309.
[88]M. Strømme Mattsson, G. A. Niklasson, and C. G. Granqvist, *Phys. Rev. B* **54** (1996) 17884.
[89]G. A. Niklasson, D. Rönnow, M. Strømme Mattsson, L. Kullman, H. Nilsson, and A. Roos, *Thin Solid Film* **359** (2000) 203.
[90]Y.-M. Choi and S.-I.Pyun, *Solid State Ionics* **109** (1998) 159.

[91]N. Yoskiike, M. Ayusawa, and S. Kondo, *J. Electrochem. Soc.* **131** (1984) 2600.
[92]M. G. S. R. Thomas, P. G. Bruce, and J. B. Goodenough, *J. Electrochem. Soc.* **132** (1985) 1521.
[93]J.-S. Bae and S.-I. Pyun, *J. Alloys and Comps.* **217** (1995) 52.
[94]Y.-M. Choi and S.-I. Pyun, *Solid State Ionics* **99** (1997) 173.
[95]J. R. Macdonald and J. A. Garber, *J. Electrochem. Soc.* **124** (1977) 1022.
[96]J. R. Macdonald, *Impedance Spectroscopy: Emphasizing Solid Materials and Systems*, Wiley, New York, 1987, p. 180.
[97]J.-S. Bae and S.-I. Pyun, *J. Mater. Sci. Lett.* **13** (1994) 573.
[98]P. G. Dickens and G. J. Reynolds, *Solid State Ionics* **5** (1981) 331.
[99]K. Zaghib, M. Simoneau, M. Armand, and M. Gauthier, *J. Power Sources* **81-82** (1999) 300.
[100]N. Takami, A. Satoh, M. Hara, and T. Ohsaka, *J. Electrochem. Soc.* **142** (1995) 371.
[101]R. C. D. Peres and M. de Paoli, *J. Power Sources* **40** (1992) 299.
[102]I. Uchida, H. Fujiyoshi, and S. Waki, *J. Power Sources* **68** (1997) 139.
[103]H.-C. Shin and S.-I. Pyun, *Electrochim. Acta* **46** (2001) 2477.
[104]H. Kanoh, Q. Feng, Y. Miyai, and K. Ooi, *J. Electrochem. Soc.* **142** (1995) 702.
[105]I. Uchida and H. Sato, *J. Electrochem. Soc.* **142** (1995) L139.
[106]M. Morcrette, P. Barboux, J. Perrière, T. Brousse, A. Traverse, and J. P. Boilot, *Solid State Ionics* **138** (2001) 213.

Index

MIX
Papier aus verantwortungsvollen Quellen
Paper from responsible sources
FSC® C105338

If you have any concerns about our products,
you can contact us on
ProductSafety@springernature.com

In case Publisher is established outside the EU,
the EU authorized representative is:
Springer Nature Customer Service Center GmbH
Europaplatz 3, 69115 Heidelberg, Germany

Printed by Libri Plureos GmbH
in Hamburg, Germany